Clostridia

Edited by
H. Bahl and P. Dürre

Clostridia

Biotechnology and Medical Applications

Edited by
H. Bahl and P. Dürre

WILEY-VCH

Weinheim – New York – Chichester – Brisbane – Singapore – Toronto

Editors:

Prof. Dr. Hubert Bahl
Universität Rostock
FB Biowissenschaften
Institut für Molekulare Physiologie und Biotechnologie
Gertrudenstrasse 11a
D-18051 Rostock

Prof. Dr. Peter Dürre
Mikrobiologie und Biotechnologie
Universität Ulm
D-89069 Ulm

Cover Illustration:
Cell-cycle of Clostridia (adapted from K. Christian Schuster, cover illustration of *Journal of Molecular Microbiology and Biotechnology*, Vol. 2, Issue 1, January 2000, with permission).
Electron micrographs by K. Christian Schuster, Technische Universität Wien, Austria, and sample preparation by Doug Hopcroft, Hort Research New Zealand, Palmerston North, New Zealand.

Library of Congress Card No: applied for

British Library Cataloguing-in-Publication Data: A catalogue record for this book is available from the British Library

Die Deutsche Bibliothek – CIP-Cataloguing-in-Publication Data
A catalogue record for this publication is available from Die Deutsche Bibliothek
ISBN 3-527-30269-7

Printed in the Federal Republic of Germany

Printed on acid-free paper.

Composition Hagedorn Kommunikation, Viernheim, Germany
Printing Strauss Offsetdruck GmbH, Mörlenbach, Germany
Bookbinding J. Schäffer, Grünstadt, Germany

ISBN 3-527-30175-5

Preface

"What a bad smell!" – a typical reaction when students in our laboratories open a tube in which a *Clostridium* strain is growing. The odor of butyric acid is not the only bad reputation clostridia have. In the general public, clostridia are synonymous with pathogens, since everybody has to be vaccinated against tetanus, and food is jeopardized to be poisoned by clostridia through one of the most potent natural toxins, botulinum. In science, many researchers do not like working with obligately anaerobic bacteria (special techniques for cultivation have to be applied) and they prefer investigating easy to handle model organisms such as *Escherichia coli* or *Bacillus subtilis*.

In this book the other (good) side of clostridia is highlighted. After a historical perspective and a general overview, the diversity of clostridia is presented to the reader in two chapters. In the following chapter, the industrial acetone–butanol fermentation by *Clostridium acetobutylicum* is described, as performed in South Africa up to the 1970s. Since the possible re-introduction of the applied acetone–butanol fermentation will depend on the ability to engineer the metabolism of *C. acetobutylicum*, genetic tools developed in the last decade for these clostridia are reviewed in another chapter. Molecular analysis of clostridia in general is a prerequisite for the application of their different potentials. The last part of the book deals with clostridial toxins, neurotoxins, and their medical applications. Furthermore, the use of recombinant clostridia in cancer therapy as outlined in the final chapter is another exciting example showing how bacteria, misconceived as being a biological threat to humans, can be very beneficial.

We are indebted to each of the contributors for making this book possible. Finally, we hope that this book will contribute to a wider appreciation of the useful capabilities which are present in members of the genus *Clostridium*.

Rostock and Ulm,
February 2001

Hubert Bahl
Peter Dürre

Contents

List of Authors

Prof. Dr. Josef Anné
University of Leuven
Rega Institute for Medical Research
Minderbroedersstraat 10
B-3000 Leuven
Belgium

Prof. Dr. George N. Bennett
Department of Biochemistry
and Cell Biology
Rice University
M.S. 140, P.O. Box 1892
Houston, TX 77251-1892
USA

Prof. Dr. J. Martin Brown
Division of Radiation Biology
Department of Radiation Oncology
Stanford University
School of Medicine
Stanford, CA 94305-5152
USA

Prof. Dr. Peter Dürre
Mikrobiologie und Biotechnologie
Universität Ulm
D-89069 Ulm
Germany

Dr. Latonia M. Harris
Department of Chemical Engineering
Northwestern University
Evanston, IL 60208
USA

Dr. Hans Hippe
Zur Scharfmühle 46
D-37083 Göttingen
Germany

Prof. Dr. David T. Jones
Department of Microbiology
University of Otago
P.O. Box 913
Dunedin
New Zealand

Prof. Dr. Philippe Lambin
Department of Radiation Oncology RTIL
Academic Hospital
Maastricht Dr. Translaan 12
NL-6221 HX Maastricht
The Netherlands

Prof. Dr. Bruce A. McClane
Department of Molecular Genetics
and Biochemistry
University of Pittsburgh
School of Medicine
E1240 Biomedical Science Tower
Pittsburgh, PA 15261
USA

Dr. Nigel P. Minton
Department of Molecular Microbiology
Centre for Applied Microbiology
and Research
Porton Down,
Salisbury, Wiltshire SP4 0JG
United Kingdom

Dr. Wilfried J. Mitchell
Department of Biological Sciences
Heriot-Watt University
Riccarton
Edinburgh, EH14 4AS
United Kingdom

Prof. Dr. Eleftherios T. Papoutsakis
Department of Chemical Engineering
Northwestern University
Evanston, IL 60208
USA

Dr. Conrad Quinn
Department of Molecular Microbiology
Centre for Applied Microbiology
and Research
Proton Down,
Salisbury, Wiltshire SP4 0JG
United Kingdom

Dr. Julian I. Rood
Department of Microbiology
Monash University
Clayton 3168
Australia

Dr. Frederick B. Rudolph
Department of Biochemistry
and Cell Biology
Rice University
Houston, TX 77251
USA

Prof. Dr. Erko Stackebrandt
Deutsche Sammlung von Mikroorganismen und Zellkulturen GmbH
Mascheroder Weg 1b
D-38124 Braunschweig
Germany

Dr. Christopher Tomas
Department of Chemical Engineering
Northwestern University
Evanston, IL 60208
USA

Dr. Seshu B. Tummala
Department of Chemical Engineering
Northwestern University
Evanston, IL 60208
USA

Dr. Neil E. Welker
Department of Biochemistry,
Molecular Biology and Cell Biology
Northwestern University
Evanston, IL 60208
USA

1
From Pandora's Box to Cornucopia: Clostridia – A Historical Perspective

Peter Dürre

In the public, bacteria of the genus *Clostridium* are often misconceived as being a biological threat and a foe to mankind. It is true that within the more than 150 validly described clostridial species of this heterogeneous genus, there are some that produce the most potent natural toxins known on earth. However, these are only a minority within a group of organisms that offers an enormous potential for biotechnological processes and medical treatments. And even the dangerous toxins became meanwhile valuable tools in the treatment of severe diseases. It is one of the aims of this book to correct the misconception of generally "bad" clostridia.

1.1
Historic reports on effects caused by clostridia and on isolation of various species

The first report on clostridia probably stems from the Greek physician Hippocrates from the island of Kos (about 460-370 BC), when he described in his *Epidemics III* a disease that can be diagnosed as a gas gangrene caused by *Clostridium histolyticum* [1]. The symptoms were a hot, yellowish-red swelling of the skin after accidents or small wounds, followed by a copious and varied flux. In some cases bones were bared of flesh or even entire limbs were lost. These are effects that can be reproduced in animal experiments after injection of *C. histolyticum* [1].

The first description of tetanus or lockjaw, caused by a neurotoxin of *C. tetani*, can be found in Sir Charles Bell's *Essays on the Anatomy and Philosophy of Expression* [2]. There, he perceptively depicted opisthotonus of a soldier, wounded at Corunna (La Coruña). The British were at that time engaged in the Peninsular War against French troops supporting the newly installed Spanish King Joseph, Napoleon's brother. After landing an expeditionary force under the command of Sir Arthur Wellesley (the later Duke of Wellington) in Portugal, the French forces led by Delaborde and Junot were defeated and evacuated Portugal as part of the Convention of Sintra. As Wellesley had to face criticisms of this Convention in Britain, Sir John Moore took over in command and proceeded into Spain. Being outnumbered, Moore had to retreat to Corunna, where he fought a masterly defensive battle against Soult on January 16th, 1809, and, although mortally wounded, succeeded in evacuating the rest of his troops by sea. It was one of the wounded soldiers whom the anatomist Bell treated medically and eventually portrayed after their arrival in Portsmouth at the end of January 1809 [3]. The original of this painting can be seen in the Royal College of Surgeons in Edinburgh [4].

However, recognition of clostridia as bacteria started only with the pioneering work of Louis Pasteur. In 1861, he described a microbe capable of growing without air [5]. This was a mere sensation at that time. Pasteur designated the organism *Vibrion butyrique*, due to the major fermentation product butyrate, and coined the term "anaerobic" to indicate life without free oxygen [6, 7]. He could also show that butanol was formed during fermentation [8], which represents the first description of the biological synthesis of this solvent. Although Pasteur probably did not work with a pure culture, his reports suggest that the dominant species was a butyric *Clostridium* such as *C. acetobutylicum* or *C. butyricum*. The latter bacterium forms butanol according to the same regulatory pattern that is known for *C. acetobutylicum* [9, 10]. The history of solventogenic clostridia has been recently summarized [11, 12] and will thus not be repeated here. Other early reports on presumably similar or identical clostridia include *Bacillus amylobacter* [13] and *Bacillus butylicus* [14, 15]. Van Tieghem soon concluded that his *B. amylobacter* was identical to Pasteur's *Vibrion butyrique* [16]. In the current taxonomic literature, usually Adam Prazmowski is credited with having introduced the name *Clostridium butyricum* for the afore-mentioned organism. Indeed, in his Ph. D. thesis he gave a detailed description of such a bacterium, isolated by himself by making use of its ability of spore formation. Prazmowski proposed that Pasteur's *Vibrion butyrique*, van Tieghem's *Bacillus amylobacter* and a third organism (*Amylobacter*) are all identical and should be named *Clostridium butyricum* [17]. However, Prazmowski was not the one to introduce the designation *Clostridium*, which he correctly made clear in his thesis. There, he referred to an earlier publication by A. Trécul who had published observations on particles, developing from rotting plant material, that he called *Amylobacter* ([18], the same text plus some additional notes was published again by Trécul in 1867 [19]). This name was derived from their staining violet after iodine treatment, a feature that is characteristic for starch (Greek: amylon = starch and bakterion = rod). Trécul distinguished three different morphological forms: the "proper" *Amylobacter* with a cylindrical shape, *Urocephalon* that resembled a tadpole (Greek: ura = tail and kephale = head, i. e. "tail with head"), and *Clostridium*. This latter designation is derived from the Greek word "kloth" (being the root for "to spin" and "spinner"). "kloth-ster" according to Greek language rules became "kloster", which means "spindle". "klostridion" (latinized into "clostridium") finally represents a diminution form (meaning "small spindle"). By the way, Trécul was in favor of spontaneous generation of these particles from the organic material, which was strongly opposed by his colleague W. Nylander in a series of publications [20-22], who pointed out that *Amylobacter* did not show significant differences to *Bacterium*. This caused a sharp response by Trécul who insisted on his original ideas [23]. However, both Trécul and Prazmowski used the term *Clostridium* only to differentiate according to morphology. Consequently, Prazmowski described in his thesis *C. polymyxa* (the later *Bacillus polymyxa*), irrespective of the fact that this is an aerobic bacterium. The metabolic distinction now being used of *Bacillus* (aerobic) and *Clostridium* (anaerobic) was only introduced much later. And the recent findings of anaerobic growth of *B. subtilis* render this definition again obsolete (reviewed in [24]).

The first clostridia being isolated in pure culture after Prazmowski's *C. butyricum* belonged to pathogenic as well as to apathogenic species [25]. Kitasato obtained pure liquid preparations of *C. tetani* in 1889 (although he was unable to grow the organism on solid media [26]) and Welch and Nuttall discovered *C. perfringens* (originally designated *Bacillus aerogenes capsulatus*) in samples taken during autopsy of a died patient from blood and various organs [27]. Beijerinck (1893) described *Granulobacter butylicum* and *G. saccharobutyricum* (thought to be identical to Fitz's *Bacillus butylicus*), anaerobic bacteria forming butanol, hydrogen, carbon dioxide, and butyrate from maltose and glucose. It is worth to note that he still used the term "*Clostridium*" just to indicate the morphology of his isolates. A microbiological landmark represents Winogradsky's report on free-living bacteria able to fix molecular nitrogen, because up to that time this metabolic property was only known from symbiotic bacteria [29]. The detailed description of the respective organism, *C. pasteurianum* (originally designated *C. pastorianum*), followed seven years later [30]. However, Winogradsky admitted that six weeks before his first published report by the Académie des Sciences in Paris in 1893, Berthelot had presented similar results in the same journal [29]. And before the turn of the century, van Ermengem was able to obtain *C. botulinum* in pure culture [31]. The twentieth century then saw the isolation of far more than hundred different clostridial species, some of which proved to be of enormous biotechnological and medical importance.

1.2
Historical use of biotechnological processes involving clostridia

Dyeing with indigo, by firmly attaching the color to clothing, is a very old process. It was obviously practiced in the Kingdom of Judah until about 570 BC and in Northern Europe in the Iron Age. The use of indigo for dyeing (which requires a water-soluble form) was however unknown to the Romans who only knew it as a pigment to be applied by artists or for staining of the skin [32]. One of the first written reports on this blue color stems from Julius Caesar who describes in his commentaries on the Gallic Wars his enemies after the landing in Britain (54 BC): "All the Britons, indeed, stain themselves with woad, which produces a bluish color, and thereby have a more terrible appearance in fight." [33]. Such an appearance was perceptively depicted in Mel Gibson's movie "Braveheart" (1995) portraying the Scotch patriot William Wallace. During the Middle Ages, indigo dyeing became an important business in England, France, Germany, and Italy [34]. The dye was derived from the woad plant (*Isatis tinctoria*) in a relative complicated process. The leaves were harvested in several pickings from about July to October. They were then crushed and comminuted in woad mills, until a pulp developed from which balls with about 5-15 cm in diameter could be formed. These balls were then dried and, thus, could be stored for a long period. The next step involved couching of the balls, sprinkling the powdered mass with water, and allowing it to ferment under aerobic conditions for up to nine weeks. Sometimes, even urine was added, which contributed to the emitted disgusting odors. It was important that the temperature of the process did not exceed about 52 °C [34]. Resulting was a dark clay-like substance, which was formed to apple-sized balls and dried again. The actual dyeing process was performed in the so-called woad vat. The dried balls were broken up and incubated with constant heating in a mixture of water, potash, and even urine [35]. Addition of madder and bran accelerated the fermentation, which due to vessel construction and volume was now largely anaerobic. The aerobic surface layer contained some floating water-insoluble blue pigment, which could be used for staining (e. g., faces and bodies, as recorded by Caesar) [34]. Clothing to be dyed, however, had to be immersed into the solution for up to six hours. When taking it out, it had a yellowish color, which turned blue during drying in air. This process flourished until the turn from the 16th to the 17th century when large amounts of *Indigofera tinctoria* (a plant containing a much higher amount of dye precursors) were imported by Dutch and English merchants from India and Southeast Asia. Woad was no longer cultivated and finally *Indigofera* was replaced by synthetic indigo in the 19th century.

The underlying microbial processes of the woad fermentation have recently been elucidated and are depicted in Figure 1. Woad contains two precursors of indigo, isatan B (indoxyl-5-ketogluconate) and indican (indoxyl-β-D-glucoside) in a ratio of approximately 5:1 [36]. During the first, aerobic degradation step, the 5-ketogluconate and glucose moieties are split off and used for bacterial growth. A number of different organisms was found to be associated with woad, but only few were able to hydrolyze the dye precursors. *Bacillus* and *Arthrobacter* species could cleave indican, whereas a *Cytophaga*-like bacterium revealed low hydrolyzing activity towards isatan B. Only *Enterobacter agglomerans* hydrolyzed both precursors by means of esterase and glucosidase activities and used glucose and 5-ketogluconate for growth [36] The microbial splitting of indican into indoxyl and glucose had already been observed by Beijerinck in his investigations on indigo fermentation from woad and *Indigofera* [37, 38]. The major activity was exerted by a bacterium that he described as *Aerobacter* [38], which resembles closely the recently found *E. agglomerans*. The aerobic transformation left the water-insoluble indoxyl in the solid material, which led to the dark blue appearance of the apple-sized balls due to the oxygen-catalyzed condensation of two such molecules to indigo. For dyeing purposes, this material needs to be reduced in order to become water-soluble and thus being attached to the fibers. This step was achieved during the anaerobic fermentation, creating reducing conditions. The responsible organism could be identified as a *Clostridium* [39]. It turned out to be a moderate thermophile, fermented sugars, produced mainly acetate, formate, and lactate, and was designated *C. isatidis* [40]. The characteristics of this strain nicely explain the cri-

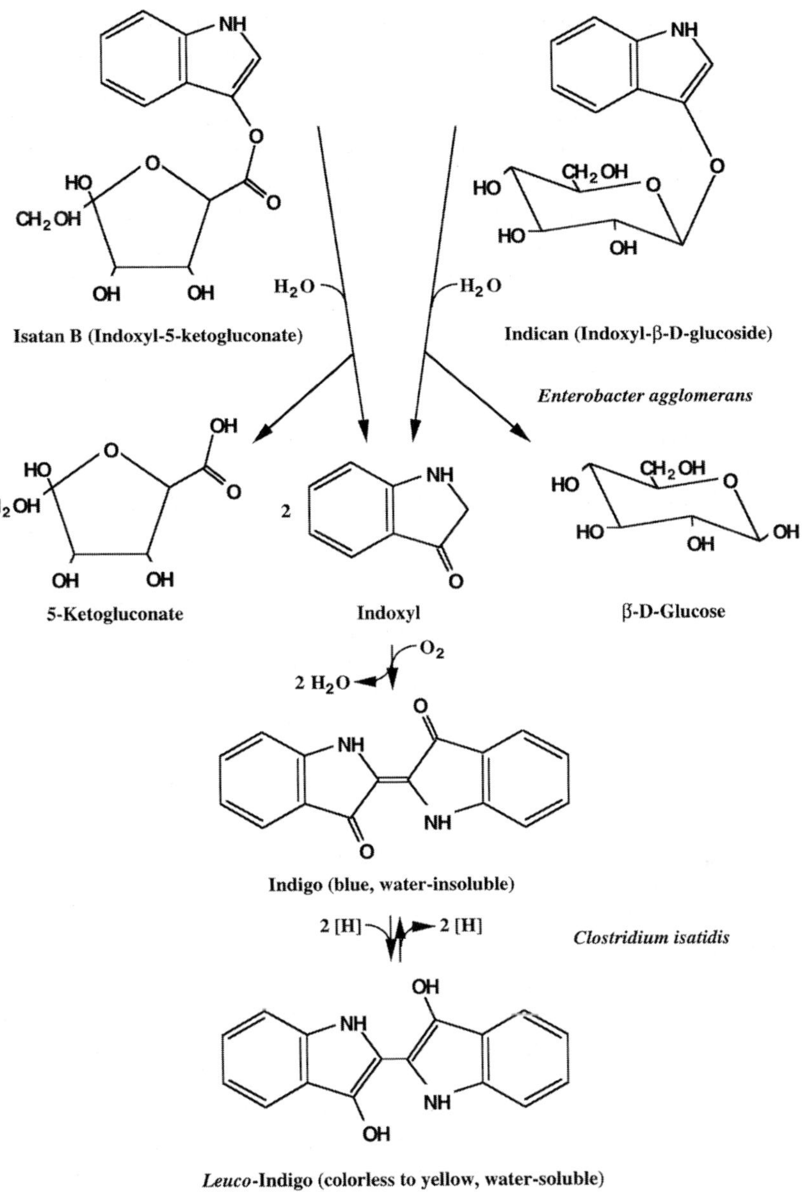

Figure 1. Indigo formation by combined action of *Enterobacter agglomerans* and *Clostridium isatidis*. 2 [H] – reducing equivalents.

tical steps required in the traditional woad dyeing process: anaerobic conditions for allowing the *Clostridium* to grow, keeping the temperature in the optimal range for this moderate thermophile, keeping the conditions alkaline by adding potash (potassium carbonate) (later often lime (calcium oxide) has been used) and urine (adds also to the nitrogen supply) to counteract the toxic effects of the acids formed during fermentation, and adding additional polysaccharide substrates for growth (bran, madder) to maintain the woad vat in operation for months [39].

Another important biotechnological process in the past was flax retting. Archaeological remains provide evidence that flax has been used by mankind since about 6000 BC in the Middle East and Egypt. The plant was processed for production of clothing (linen) and oil (linseed oil). Following the harvesting, seeds (for oil production) were separated from the stems, which were incubated in water for several weeks (retting). When they reached the stage of "ripening", the bundles were pressed and dried. Then, wooden components were fragmented and removed from the fibers. These fibers were further purified and used for clothing fabrication. The Greek historian Herodotus (Greek: Herodotos; about 485-425 BC) from Halikarnassos (now Bodrum in Turkey) reported that Thracians also wore clothing made from hemp in a similar way and that the different materials were hardly discernible. The most important step of the whole process was the ripening, starting with aerobic, but soon reaching anaerobic conditions.

The microbiology of the retting process was already investigated at the turn of the 19th century. The essential step is the separation of phloem-derived fiber bundles from other stem tissue by degradation of the connecting material, pectin. This is due to metabolic activity of anaerobic clostridia. Winogradsky and Fribes described a large species with terminal spores causing swelling of the mother cell, which could degrade a number of sugars in addition to pectin [41]. Nine years later, Beijerinck and van Delden published very similar results [42]. They found two clostridia being involved in flax retting in different proportions. The dominant organism was designated *Granulobacter pectinovorum*, whereas the other was called *G. urocephalum*. Beijerinck and van Delden stated the identity of the former *Clostridium* to Winogradsky's organism. Meanwhile, a large number of clostridial species is known that express pectinolytic activity [43]. A recent investigation on the similar hemp retting process led to the isolation of 121 pectinolytic clostridia that showed high similarity to *C. acetobutylicum*, *C. aurantibutyricum*, and *C. butyricum* [44].

An undesired occurrence in an old biotechnological process is clostridial contamination during cheese ripening. The responsible organism is *C. tyrobutyricum*. Like propionic acid bacteria (which, together with lactic acid bacteria, carry out the essential metabolic reactions for cheese formation and ripening) this organism uses lactic acid for growth. Major fermentation products are butyrate, which ruins the cheese flavor, and carbon dioxide. This prolific gas formation results in the so-called "late blowing", which also renders the cheese unsalable. Contamination might occur by spores associated with grass silage. In the past, growth of vegetative cells was effectively inhibited by addition of nitrate to the milk. Nitrate, after its reduction to nitrite, reacts with iron-sulfur clusters to form iron-nitric oxide complexes, with resultant destruction of the iron-sulfur cluster [45]. Primary targets are clostridial ferredoxin and pyruvate-ferredoxin oxidoreductase [46]. Since new regulations limit the use of nitrate, current approaches aim at a quick detection of contaminating *C. tyrobutyricum* in raw milk.

The most important biotechnological process employing clostridia in recent history is without any doubt the acetone/butanol (AB) fermentation performed with *C. acetobutylicum*. This subject has been extensively reviewed in the past (e. g., [11, 12, 47]) and one chapter of this book is devoted to it, so that only a short mentioning seems to be appropriate in this place. The AB fermentation was performed at a massive industrial scale during the first half of the 20th century and ranked second in size only to ethanol fermentation by yeast. Its introduction had enormous consequences in both industrial production and politics. Unfortunately, this event was caused by martial requirements, somehow confirming the saying of the Greek philosopher Heraclitus (Greek: Herakleitos) from Ephesus: "War is father of all." The biological formation of acetone needed for ammunition production in Great Britain during World War I was the most important source of this solvent. Since the inventor of the process, Charles (later Chaim) Weizmann refused any honors, but made clear that he was in favor of the idea of a Jewish homeland in Palestine, there is no doubt that his merits contributed substantially to the Balfour Declaration in 1917. And Weizmann became the first president of the state of Israel after its foundation in 1948. Due to the lack of the by-product amyl acetate from the ethanol fermentation, which was banned in the US by prohibition since 1920, butanol made its way as the most important feedstock for industrial solvents (e. g., laquers in Henry Ford's assembly-line work automobile production). In 1945, 66 % of the total butanol and 10 % of the total acetone production were obtained by fermentation [48]. Rising substrate costs (molasses) and availability of cheap crude oil made solvent production by the petrochemical industry much more economical and led to the decline of the biotechnological process. However, the oil crisis in 1973, following the Jom Kippur War of Israel against Egypt and Syria, led to intensified research on the fermentation. As a consequence, international workshops on the regulation of metabolism, genetics, and development of solvent-forming clostridia have been established since 1990 and our knowledge in these fields has increased enormously. In addition, a number of technological improvements with respect to solvent recovery and process operation have been introduced. These developments put the AB fermentation now in an economical range [49]. And the year 2000 saw again dramatically rising oil prices ...

1.3
Current uses and future potential of clostridia

Clostridia produce a number of different toxins, including enterotoxins from *C. perfringens*, cytotoxins from *C. difficile*, and neurotoxins from *C. botulinum* and *C. tetani* [50]. The respective genes are located on the chromosomes as well as on plasmids [51]. Due to their importance in medicine, two chapters of this book are devoted to clostridial toxins and only few remarks will be added here. Tetanus and botulinum toxins are the most poisonous natural compounds known. Tetanus toxin binds to peripheral nerve terminals, is then transported across synaptic junctions, and becomes fixed to gangliosides at the presynaptic inhibitory motor nerve endings in the central nervous system. There it blocks the release of inhibitory neurotransmitters across the synaptic cleft, thus producing the generalized muscular spasms characteristic of lockjaw. Botulinum toxins (seven different serotypes are known) are similar in structure, but affect the peripheral nervous system at peripheral nerve endings of stimulatory motor neurons by preventing the release of acetylcholine and causing weakness or flaccid paralysis. Both proteins are zinc-dependent endopeptidases (for recent reviews see [52, 53]). Prophylactic immunization against the former can be achieved by injection (with later boostering) of tetanus toxoid, prepared by chemical modification of the toxin. Passive protection could be administered with tetanus antitoxin. Since there are several different botulinum toxins, neutralization can only be achieved by quick intravenous injection of specific antitoxin. Once the botulinum toxin has bound to nerve endings, its activity is unaffected by the antitoxin. A heptavalent antitoxin for serotypes A-G and a pentavalent toxoid for serotypes A-E have been developed [54, 55]. Tetanus usually results from wounds that disrupture the oxygen supply to the tissue and become contaminated with *C. tetani* spores, germinating into vegetative cells that produce the toxin. Botulism on the other hand is not an infection, but an intoxication, since it results from the ingestion of foods that contain the toxin due to the activity of growing *C. botulinum*. A true infection is infant botulism. Spore-contaminated food can lead to establishment of vegetative *C. botulinum* cells in the bowel of infants. This is the reason why babies should not receive honey during the first year of their life. Contamination of honey with clostridial spores probably results from nectar collection, when the bees come into contact with spores on flowering plants [56]. Significant proportions of honey samples have been found to be contaminated with *C. botulinum* spores [57-60]. At later ages, stomach and gastrointestinal tract are well developed and will not allow development of *C. botulinum* cells.

Since botulinum toxin exerts its lethal effect in a very short time after ingestion, quick detection methods are sought for. These techniques need to aim at the protein, not the producing bacterium. Currently, the mouse lethality test is still the only accepted method for detection of the toxin. The detection limit is about 10 pg/ml or less than 5 mouse MLD_{50}/ml. A disadvantage is that the toxin serotype, which would be required for antiserum treatment, cannot be determined with this test [61]. A number of enzyme-linked immunosorbent assays (ELISAs) have been developed, most of which, however, are insufficiently sensitive [62-64]. Reverse transcription-polymerase chain reaction allows to detect expression of the toxin genes [65]. Finally, endopeptidase and proteolytic activities have been assayed *in vitro* [66, 67]. Recently, an *in vitro* bioassay for botulinum toxin B has been developed that is more sensitive than the mouse bioassay [68]. Accepted use of this test would greatly reduce assay time and number of laboratory animals.

The extreme poisonous activity of botulinum toxin has led to its use as a biological weapon of bioterrorists. Together with smallpox, plague, and anthrax it tops the list of the most dangerous agents [69]. Possibly, tularemia and arenaviruses could be added. This is not a theoretical threat as after the Gulf War in 1990, Iraqi documents were obtained that acknowledged production of 20,000 liters of botulinum toxin. SCUD missiles with a range of 300 to 600 km and carrying 400-lb bombs had been outfitted with warheads containing this compound (and anthrax as well). Also, the Japanese religious sect Aum Shinrikyo, known from the 1995 sarin gas attack on the Tokyo subway, had actually sought to aerolize botulinum toxin throughout metropolitan Tokyo [69]. Fortunately, aerosol dissemination of large areas by botulinum toxin is a less likely prospect. Production and dispensation of substantial quantities pose virtually insurmountable problems. Nevertheless, the scientific community should inform people and policymakers everywhere about the serious threats by biological weapons, to which botulinum toxin belongs, too.

However, despite its potentially lethal effects, botulinum toxin A has meanwhile become an extremely valuable therapeutic and cosmetic agent. Highly diluted suspensions, injected into spastic muscles, generally relieve muscle spasm within three days. Such toxin preparations have been found very useful in treating muscle spasms affecting vision (blepharospasm and strabismus). Other disorders to be treated in such a way include cerebral palsy, other dystonias, and pain syndromes (for recent reviews see [4, 70]). Cosmetic applications are removal of frown lines and "crow's feet" wrinkles on the face. As the muscle weakens, the skin overlying the muscle relaxes and the wrinkles in the skin gradually soften and often disappear. The effect of each injection lasts approxomately for three months. The botulinum toxin preparations are produced with extreme care and are sold under the trademarks "Botox" in the US and "Dysport" in Europe. Another recent therapy is treatment of hyperhidrosis (localized excessive sweating) with botulinum toxin by blocking the respective perspiratory glands [71].

Other clostridial toxins may become valuable drugs as well. Tetanus toxin enters the central nervous system from the circulatory system faster than any other known protein. This property is currently exploited to enable targeted delivery of other proteins to the central nervous system [70]. The *C. botulinum* ADP-ribosylating C_2 toxin has been demonstrated to induce degranulation of attached mast cells and might thus be used to modulate allergic responses. *C. perfringens* enterotoxin is able to destroy certain tumor cell lines, as does *C. difficile* toxin A. The latter has also been suggested for possible use in the treatment of inflammatory disorders [70].

Not only clostridial proteins are useful in medical therapies, but also whole cells. Since 1968, the Miyairi 588 strain of *C. butyricum* is commercially available in Japan. In clinical studies it has been approved as a probiotic agent against infections with enterohemorrhagic *Escherichia coli* (EHEC O157:H7). It also inhibited growth of *C. difficile, Salmonella typhimurium,* and *Vibrio cholerae.* Experiments with gnotobiotic mice showed a therapeutic as well as a preventive effect [72].

The most fascinating therapeutic approach is the recent proposal to use recombinant clostridial spores for cancer treatment. Apathogenic clostridia such as *C. acetobutylicum* are used as tumor-specific vectors to selectively deliver therapeutic proteins to solid tumors [73]. The specificity for solid tumor cells is based on the hypoxic and necrotic nature of such tumors. Recombinant spores, carrying genetic information for therapeutic proteins, are injected into tumor-bearing organisms. Healthy tissues are well oxygenated and do not allow germination and growth of the strictly anaerobic bacteria. However, the vicinity of solid tumors is hypoxic, allowing spores to germinate and vegetative cells to proliferate at the expense of necrotic tissue only. Therapeutic proteins will then be expressed and kill the tumor cells. Currently, two different approaches are under investigation: Enzymes converting innocuous prodrugs into highly cytotoxic drugs [74, 75] and cytokines such as tumor necrosis factor [76]. This topic is detailed in another chapter of this book.

Clostridia produce a large battery of industrially useful enzymes. Among these are polymer-degrading proteins such amylases, cellulases, pectinases, pullulanases, and proteinases, just to name a few. An enzyme important for special medical treatments is already commercially marketed. This is a preparation containing collagenase proteins from *C. histolyticum* (Santyl, Knoll Pharmaceutical Co.). Since collagen accounts for approximately 75 % of the dry weight of skin tissue, use of this enzyme is particularly effective in the removal of detritus. Collagenase thus contributes towards the formation of granulation tissue and subsequent epithelization of dermal ulcers and severely burned areas. Collagen in healthy tissue or in newly formed granulation tissue is not attacked.

Indigo dyeing is of high economic importance due to the everlasting success of blue denim materials (e. g., in blue jeans). This has also resulted in the construction of recombinant bacteria (*Escherichia coli*) producing indoxyl from tryptophane [77]. Modern dyeing uses indigo and sodium dithionite as a reducing agent. During the process, high concentrations of oxidized and non-oxidized sulfur components are released into the wastewater, creating enormous ecological and technical (sewer corrosion) problems [78]. A revived use of *C. isatidis* might help to solve these problems by avoiding the use of synthetic reducing materials [39].

Whether flax will become again a widely used crop is both a political as well as a market consideration. Almost the complete plant could be used in various applications. Long fibers (14-19 % of the plant material) are the starting point for fabrication of clothing, yarn, ropes, and textile wallpaper. Short fibers (3-13 %) can be used in combination with cotton or synthetic fibers, improve paper quality, serve as biological degradable insulation material in e. g., automobile production, and are used in filter compositions. Seeds (10-18 %) yield foodgrade oil. The removed wooden particles (35-50 %) can be used in furniture production or for heating. New technologies significantly reduced the former time-consuming harvest and processing. If these cost reductions in combination with EU agricultural guidelines will result in increased flax cultivation, clostridial strains with high pectinolytic activities or even purified enzymes will be of enormous importance for the retting process.

Clostridial solvent formation might be ready to enter commercial production again. However, the process based on newly developed technologies would probably serve only a niche market, using cheap, low-grade agricultural substrates. Economic considerations are in favor of the reintroduction of the process [49]. On the other hand, in the US considerable amounts of ethanol, derived from corn fermentation, are used for gasoline blending. Mixtures of butanol, isopropanol, and ethanol can be obtained from fermentation with *C. beijerinckii* at competitive prices and result in greatly reduced NO_x emissions when used as engine fuel [79]. Butanol alone has also better octane numbers, lower vapor pressure, and a lower heat of vaporization than ethanol. In addition, it mixes better with both gasoline and diesel as a fuel additive [80]. This could eventually result in converting existing ethanol fermentation plants into butanol ones. In fact, a recent newspaper article already addressed the possibility of constructing a new butanol plant in Kahoka, Mo., USA [81]. Another bulk fermentation employing homoacetogenic clostridia could be the production of calcium magnesium acetate from hydrolyzed cornstarch [82]. *C. thermoaceticum* and *C. thermoautotrophicum* produce up to 45 g acetate per liter. Calcium magnesium acetate has ideal properties as a de-icing agent (e.g., for roads, highways) and would avoid the ecological problems caused by the use of sodium chloride.

An extremely versatile organism seems to be *C. acetobutylicum*. Besides its prominent abilities in cancer therapy and solvent production, this bacterium has also been suggested for application in microbial-enhanced raw oil reçovery [83, 84]. The principle is as follows: First, bacteria or spores are injected along with nutrient into the candidate reservoir, usually one that has been water-flooded. Second, all access ports are sealed off for a few months. This time period allows the anaerobic bacteria to grow and to migrate to even technically unattainable regions of the oil reservoir. Wherever they grow at the expense of the provided nutrients, they produce gases and mobility-modifying metabolites such as acids, solvents, biopolymers, and possibly even biosurfactants. On opening the well, the produced gas will push out some of the residual oil. Then, water-flooding can be used to retrieve the remaining oil. The experiments in a Turkish oil well resulted in a 12% increase in oil recovery and are thus very promising [84]. Another interesting feature of *C. acetobutylicum* is its ability to degrade trinitrotoluene (TNT) and related explosives [85, 86]. TNT can also be reduced by carbon monoxide dehydrogenase from *C. thermoaceticum* [87]. In addition to explosives, nitroaromatics are widely used in the synthesis of dyes, pesticides, and pharmaceuticals. Unfortunately, they are very recalcitrant and thus represent an environmental problem. Since anaerobic technology is technically cheaper to handle, clostridia might become the premier choice for clean-up processes of production and dumping sites.

Like other anaerobic bacteria, clostridia are able to perform specific reduction reactions. However, the spectrum of reactions found in this genus is extremely large. Stereospecific enoate reductases allow the production of chiral compounds in high yield and at low cost. Bile acids and steroids can be selectively epimerized. This ability might prove useful for the development of new oral contraceptives or hormones. Dehalogenation of insecticides such as lindane (γ-hexachlorocyclohexane) has also been reported. A detailed summary of these reactions is provided in a former review [43]. A recent project by the Brookhaven National Laboratory investigates the ability of *Clostridium* sp. BC1 to reduce oxidized uranium compounds.

The rationale of these experiments is that the oxidized uranyl ion [U(VI)] is much more water-soluble than the reduced form, U(IV). Besides its radioactivity, U(VI) also is highly toxic. Due to its good solubility, it could easily contaminate ground water supplies. Reduction to the less soluble form would greatly facilitate its concentration and decontamination.

Finally, the scientific knowledge stemming from investigations with clostridia is immense. In a review from 1993, Morris lists a number of important microbiological and biochemical principles elucidated by experiments with clostridial species [25], such as the afore-mentioned nitrogen fixation by free-living bacteria, the Stickland reaction (fermentation of amino acid pairs by different oxidative and reductive pathways), the finding that fermentation is not necessarily degradative in character as exemplified by *C. kluyveri*, the discovery of the coenzyme forms of vitamin B_{12}, the C_1-carrier role of tetrahydrofolates, the discovery of iron-sulfur proteins such as ferredoxin and rubredoxin, the demonstration of selenoenzymes and of NADH-ferredoxin reductase. Certainly, one could add further key discoveries from the past, but I would like to focus on some recent findings.

An extremely important step forward is the sequencing of several clostridial genomes. *C. acetobutylicum* has been completely sequenced by Genome Therapeutics Corp. as a component of the US Department of Energy Microbial Genome Project and the sequence is currently being annotated, *C. difficile* is being sequenced in the Sanger Centre in Great Britain, *C. perfringens* in Japan, and *C. tetani* in the Göttingen Genomics Laboratory in Germany. Publication of these data will provide a wealth of information for comparative and functional genomic studies. However, even the already available sequences allow significant conclusions. The year 1999 saw the scientific answer to a question that remained open for decades: How do anaerobic microbes detoxify oxygen when shortly exposed to it? Jenney et al. provided the answer by experiments with *Pyrococcus furiosus* and comparisons with available genome sequences [88], e. g., from *C. acetobutylicum*. A superoxide reductase converts superoxide to hydrogen peroxide which most probably is further metabolized to water by action of a peroxidase. Reducing equivalents are carried to the superoxide reductase by reduced rubredoxin, also defining (for the first time) a definite function for this iron-sulfur protein in anaerobes. Genes encoding superoxide reductase have been detected in all available genomes of anaerobic prokaryotes, with *Methanobacterium thermoautotrophicum* being the only exception. Aerobes and facultative anaerobes are devoid of such a gene and employ different detoxification mechanisms.

The second finding of my subjective list refers to one of World's largest bacteria, *Epulopiscium fishelsonii*, which belongs phylogenetically to the clostridia [89]. Although still not cultivatable in the laboratory, morphological investigations revealed an unusual mode of cellular reproduction. *Epulopiscium fishelsonii* produces internally multiple offspring, designated as daughter cells. Formation is initiated in the tips of the mother cell and offspring grow until they completely fill the mother cell. A closely related organism, *Metabacterium polyspora*, phylogenetically also belonging to the *Clostridium* clusters, shows a very similar behavior, except that it forms multiple spores instead of daughter cells [90]. This opens the way to study the relationship and evolution of viviparous offspring production and sporulation. Unfortunately, *Metabacterium polyspora* is also uncultivable, but *Anaerobacter polyendosporus* shows a very similar morphology (up to 5 spores per mother cell). This bacterium is another member of the *Clostridium* cluster and can be grown in axenic culture [91]. Thus, *Anaerobacter polyendosporus* might become the model organism for such studies.

The third recent example that I would like to mention is the proposal that the ancestor of animal mitochondria and many primitive amitochondrial eukaryotes was a fusion microbe composed of a *Clostridium*-like eubacterium and a *Sulfolobus*-like archaebacterium [92]. This suggestion was based on comparison of genome signatures, expanded energy metabolism of the fusion organism, and ability of endospore formation by *Clostridium* (one of the few cell differentiation processes among prokaryotes) as a starting point for nucleus development. Future experiments will have to prove whether this proposal can compete with other evolution hypotheses.

I hope, all these examples have demonstrated that there is an enormous biotechnological and medical potential of clostridia, which partially awaits application, and an even much larger potential, which still awaits elucidation and exploitation. Thus, clostridia will continue to be of utmost scientific, industrial, and medical interest. They really emerged from foes to friends and will certainly play an important role in future.

Acknowledgements

I express my sincere thanks to Bodo Gatz and Max Sussman for their help and valuable information on Greek nomenclature and ancient reports on clostridial activities.

Experimental work carried out in the author's laboratory was supported by the Deutsche Forschungsgemeinschaft and the Commission of the European Communities.

References

[1] Sussman, M., A description of *Clostridium histolyticum* gas-gangrene in The Epidemics of Hippocrates, *Med. Hist.* **1958**, 2, 226.

[2] Bell, C., Sir, *Essays on the Anatomy and Philosophy of Expression*, London, Murray, **1824**.

[3] Gordon-Taylor, G., Sir, and Walls, E. W., *Sir Charles Bell. His Life and Times*, E. & S. Livingstone, Edinburgh, **1958**.

[4] Arnon, S. S., Human tetanus and human botulism, in: *The Clostridia: Molecular Biology and Pathogenesis* (J. I. Rood, B. A. McClane, J. G. Songer, R. W. Titball, Eds.), Academic Press, San Diego, CA, **1997**, pp. 95-115.

[5] Pasteur, L., Animacules infusoires vivant sans gaz oxygène libre et déterminant des fermentations, *C. R. Hedb. Séances Acad. Sci.* **1861**, 52, 344-347.

[6] Pasteur, L., Expériences et vues nouvelles sur la nature des fermentations, *C. R. Hebd. Séances Acad. Sci.* **1861**, 52, 1260-1264.

[7] Pasteur, L., Recherches sur la putréfaction, *C. R. Hebd. Séances Acad. Sci.* **1863**, 56, 1189-1194.

[8] Pasteur, L., Quelques résultats nouveaux relatifs aux fermentations acétique et butyrique, *Bull. Soc. Chim. Paris*, **1862**, May 1862, 52-53.

[9] Crabbendam, P. M., Neijssel, O. M., and Tempest, D. W., Metabolic and energetic aspects of the growth of *Clostridium butyricum* on glucose in chemostat culture, *Arch. Microbiol.* **1985**, 142, 375-382.

[10] Sauer, U., Fischer, R., and Dürre, P., Solvent formation and its regulation in strictly anaerobic bacteria, *Curr. Topics Mol. Genet.* (Life Sci. Adv.) **1993**, 1, 337-351.

[11] Dürre, P., and Bahl, H., Microbial production of acetone/butanol/isopropanol, in: *Biotechnology* 2nd Edn. Vol. 6 (M. Roehr, Ed.), VCH, Weinheim, **1996**, pp. 229-268.

[12] Dürre, P., New insights and novel developments in clostridial acetone/butanol/isopropanol fermentation, *Appl. Microbiol. Biotechnol.* **1998**, 49, 639-648.

[13] van Tieghem, P., Sur le *Bacillus amylobacter* et son role dans la putréfaction des tissus végétaux, *Bull. Soc. Bot. France* **1877**, 24, 128-135.

[14] Fitz, A.,Ueber Schizomyceten-Gährungen III, *Ber. Dtsch. Chem. Ges.* **1878**, 11, 42-55.

[15] Fitz, A., Ueber Spaltpilzgährungen. VII. Mittheilung, *Ber. Dtsch. Chem. Ges.* **1882**, 15, 867-880.

[16] van Tieghem, P., Identité du *Bacillus amylobacter* et du *Vibrion butyrique* de M. Pasteur, *C. R. Hebd. Séances Acad. Sci.* **1879**, 89, 5-8.

[17] Prazmowski, A., Untersuchungen über die Entwickelungsgeschichte und Fermentwirkung einiger Bacterien-Arten, *Thesis*, University of Leipzig, Germany, **1880**.

[18] Trécul, A., Production de plantules amylifères dans les cellules végétales pendant la putréfaction. Chlorophylle cristallisée, *C. R. Hebd. Séances Acad. Sci.* **1865**, 61, 432-436.

[19] Trécul, A., Matière amylacée et cryptogames amylifères dans les vaisseaux du latex de plusieurs apocynées (Sur les *Amylobacter*), *Ann. Sci. Nat. Bot. 5ième Série* **1867**, 7, 204-214.

[20] Nylander, W., Circa Amylobacteria Tréc. Notula, *Flora* **1865**, 33, pp. 521 (reprinted in *Ann. Sci. Nat. Bot.* 5ième Série **1867**, 7, 214-218).

[21] Nylander, W., Adhuc Circa Amylobacteria Adnotatio. *Flora* **1865**, 37, pp. 579 (reprinted in *Ann. Sci. Nat. Bot.* 5ième Série **1867**, 7, 218-219).

[22] Nylander, W., Sur les *Amylobacter, Bull. Soc. Bot. France* **1865**, 12 (reprinted in Ann. *Sci. Nat. Bot.* 5ième Série **1867**, 7, 219-220).

[23] Trécul, A., Résponse à trois notes de M. Nylander concernant la nature des *Amylobacter, C. R. Hebd. Séances Acad. Sci.* **1867**, 65, 513-521.

[24] Nakano, M. M., and Zuber, P., Anaerobic growth of a "strict aerobe" (*Bacillus subtilis*). *Annu. Rev. Microbiol.* **1998**, 52, 165-190.

[25] Morris, J. G., History and future potential of the clostridia in biotechnology, in: *The Clostridia and Biotechnology* (D. R. Woods, Ed.), pp. 1-23, **1993**. Stoneham, MA, Butterworth-Heinemann.

[26] Kitasato, S., Ueber den Rauschbrandbacillus und sein Culturverfahren, *Z. Hyg.* **1889**, 6, 105-116.

[27] Welch, W. H., and Nuttall, G. H. F., A gas-producing *Bacillus* (*Bacillus aerogenes capsu-*

latus, nov. spec.) capable of rapid development in the blood-vessels after death, *Bull. Johns Hopkins Hosp.* **1892**, 3, 81-91.

[28] Beijerinck, M. W., Ueber die Butylalkoholgährung und das Butylferment, *Verhand. Kon. Akad. Wetenschappen* Amsterdam **1893**, 2. Sect., Part 1, No. 10, 1-51.

[29] Winogradsky, S., Recherches sur l'assimilation de l'azote libre de l'atmosphère par les microbes, *Arch. Sci. Biol.* **1895**, T. III, Lf. 4 (reprinted in *Microbiologie du Sol* (S. Winogradsky, Ed.) (1949), pp. 367-402. Paris, Masson.

[30] Winogradsky, S., *Clostridium pastorianum*, seine Morphologie und seine Eigenschaften als Buttersäureferment, *Cbl. Bakteriol. Parasitenkd, Infektionskrankh.* **1902**, II. Abt. 9, 43-54, 107-112.

[31] van Ermengem, E., Untersuchungen über Fälle von Fleischvergiftung mit Symptomen von Botulismus, *Cbl. Bakteriol. Parasitenkd. Infektionskrankh.* **1896**, I. Abt. 19, 442-444.

[32] Clark, R. J. H., Cooksey, C. J., Daniels, M. A. M., and Withnall, R., Indigo, woad, and Tyrian Purple: important vat dyes from antiquity to the present, *Endeavour* **1993**, 17, 191-199.

[33] Caesar, G. J. (54 BC). *De Bello Gallico.* German translation by C. Woyte, Philipp Reclam jun, Stuttgart, **1951**.

[34] Hurry, J. B.,*The Woad Plant and its Dye.* Oxford University Press, London, Humphrey Milford, **1930**.

[35] Döscher, I., Schuster, C., and Meyer, O., Indigo modern, *Chem. i. u. Zeit* **1992**, 26, 54-55.

[36] Ewerdwalbesloh, I., and Meyer, O., Bacteriology and enzymology of the woad fermentation, *Proc. 2nd Int. Symp. on Woad, Indigo and other Natural Dyes: Past, Present and Future,* June 8-12. Toulouse, France, **1995**.

[37] Beijerinck, M. W., On the formation of indigo from the woad (*Isatis tinctoria*), *Proc. Sect. Sci.* Kon. Akad. Wetenschappen Amsterdam, **1899**, 2, 120-129.

[38] Beijerinck, M. W., On indigo-fermentation, *Proc. Sect. Sci.* Kon. Akad. Wetenschappen Amsterdam **1900**, 2, 495-512.

[39] Padden, A. N., Dillon, V. M., John, P., Edmonds, J., Collins, M. D., and Alvarez, N., *Clostridium* used in medieval dyeing, *Nature* **1998**, 396, 225.

[40] Padden, A. N., Dillon, V. M., Edmonds, J., Collins, M. D., Alvarez, N., and John, P., An indigo-reducing moderate thermophile from a woad vat, *Clostridium isatidis* sp. nov, *Int. J. Syst. Bacteriol.* **1999**, 49, 1025-1031.

[41] Winogradsky, S., and Fribes, V.,Sur le rouissage du lin et son agent microbien, *C. R. Hebd. Séances Acad. Sci.* **1895**, 121, 742-744 (reprinted in *Microbiologie du Sol* (S. Winogradsky, Ed.) (1949), pp. 407-409. Paris, Masson.

[42] Beijerinck, M. W., and van Delden, A., Sur les bactéries actives dans le rouissage du lin, *Arch. Néerl. Sci. Exactes et Naturelles,* Haarlem, **1904**, Série II, 9, 418-441.

[43] Bahl, H., and Dürre, P., Clostridia, in: *Biotechnology* 2nd Edn., Vol. 1 (H. Sahm, Ed.), VCH, Weinheim, **1993**, pp. 285-323.

[44] Mastromei, G., Daly, S., Perito, B., Vandini, C., Ranalli, P., and Polsinelli, M., *Clostridium* strains involved in the hemp retting process. Poster presentation, IUMS (Int. Union Microbiol. Soc.) Congress, August 9-20. Sydney, Australia, **1999**.

[45] Reddy, D., Lancaster, L. R., Jr., and Cornforth, D. P., Nitrite inhibition of *Clostridium botulinum*: electron spin resonance detection of iron-nitric oxide complexes, *Science* **1983**, 221, 769-770.

[46] Carpenter, C. E., Reddy, D. S. A., and Cornforth, D. P., Inactivation of clostridial ferredoxin and pyruvate-ferredoxin oxidoreductase by sodium nitrite, *Appl. Environ. Microbiol.* **1987**, 53, 549-552.

[47] Jones, D. T., and Woods, D. R., Acetone-butanol fermentation revisited, *Microbiol. Rev.* **1986**, 50, 484-524.

[48] Rose, A. H., *Industrial Microbiology,* London, Butterworths, **1961**.

[49] Gapes, J. R., The economics of acetone-butanol fermentation: theoretical and market considerations, *J. Mol. Microbiol. Biotechnol.* **2000**, 2, 27-32.

[50] Rood, J. I., McClane, B. A., Songer, J. G., and Titball, R. W., *The Clostridia: Molecular Biology and Pathogenesis,* Academic Press, San Diego, CA, **1997**.

[51] Braun, V., and von Eichel-Streiber, C., Virulence-associated mobile elements, in: bacilli and clostridia, in: *Pathogenicity Islands and Other Mobile Virulence Elements* (J. B. Kaper, J. Hacker, Eds.), American Society for Microbiology, Washington, DC, **1999**, pp. 233-264.

[52] Wictome, M., and Shone, C. C., Botulinum neurotoxins: mode of action and detection, *J. Appl. Microbiol. Symp. Suppl.* **1998**, 84, 87S-97S.

[53] Pellizzari, R., Rossetto, O., Schiavo, G., and Montecucco, C., Tetanus and botulinum neurotoxins: mechanism of action and therapeutic uses, *Phil. Trans. R. Soc. London B* **1999**, 354, 259-268.

[54] Brown, J. E., and Williamson, E. D., Molecular approaches to novel vaccines for the control of clostridial toxemias and infections, in: *The Clostridia: Molecular Biology and Pathogenesis* (J. I. Rood, B. A. McClane, J. G. Songer, R. W. Titball, Eds.), Academic Press, San Diego, CA, **1997**, pp. 505-524.

[55] Atlas, R. M., The medical threat of biological weapons, *Crit. Rev. Microbiol.* **1998**, 24, 157-168.

[56] Centorbi, H. J., Aliendro, O. E., Demo, N. O., Dutto, R., Fernandez, R., and Puig de Centorbi, O. N., First case of infant botulism associated with honey feeding in Argentina, *Anaerobe* **1999**, 5, 181-183.

[57] Sugiyama, H., Mills, D. C., and Kuo, L. J. C., Number of *Clostridium botulinum* spores in honey, *J. Food Prot.* **1978**, 41, 848-850.

[58] Huhtanen, C. N., Knox, D., and Shimanuki, H., Incidence and origin of *Clostridium botulinum* spores in honey, *J. Food Prot.* **1981**, 44, 812-814.

[59] Monetto, A. M., Francavilla, A., Rondini, A., Manca, L., Siravegna, M., and Fernandez, R., A study of botulinum spores in honey, *Anaerobe* **1999**, 5, 185-186.

[60] Schocken-Iturrino, R. P., Carneiro, M. C., Kato, E., Sorbara, J. O. B., Rossi, O. D., and Gerbasi, L. E. R., Study of the presence of the spores of *Clostridium botulinum* in honey in Brazil, *FEMS Immunol. Med. Microbiol.* **1999**, 24, 379-382.

[61] Fernández, R. A., and Ciccarelli, A. S., Botulism: laboratory methods and epidemiology, *Anaerobe* **1999**, 5, 165-168.

[62] Shone, C. C., Wilton-Smith, P., Appleton, N., Hambleton, P., Modi, N., Gattey, S., and Melling, J., Monoclonal antibody-based immunoassay for type A *Clostridium botulinum* toxin is comparable to the mouse bioassay, *Appl. Environ. Microbiol.* **1985**, 50, 63-67.

[63] Doellgast, G. J., Triscott, M. X., Beard, G. A., Bottoms, J. D., Cheng, T., Roh, B. H., Roman, M. G., Hall, P. A., and Brown, J. E., Sensitive enzyme-linked immunosorbent assay for detection of *Clostridium botulinum* neurotoxins A, B, and E using signal amplification via enzyme-linked coagulation assay, *J. Clin. Microbiol.* **1993**, 31, 2402-2409.

[64] Szílagyi, M., Rivera, V. R., Neal, D., Merrill, G. A., and Poli, M. A., Development of sensitive colorimetric capture elisas for *Clostridium botulinum* neurotoxin serotypes A and B, *Toxicon* **2000**, 38, 381-389.

[65] McGrath, S., Dooley, J. S. G., and Haylock, R. W., Quantification of *Clostridium botulinum* toxin gene expression by competitive reverse transcription-PCR, *Appl. Environ. Microbiol.* **2000**, 66, 1423-1428.

[66] Hallis, B., James, B. A. F., and Shone, C. C., Development of novel assays for botulinum type A and B neurotoxins based on their endopeptidase activities, *J. Clin. Microbiol.* **1996**, 32, 1911-1917.

[67] Pellizzari, R., Rossetto, O., Washbourne, P., Tonello, F., Nicotera, P. L., and Montecucco, C., In vitro biological activity and toxicity of tetanus and botulinum neurotoxins, *Toxicol. Lett.* **1998**, 102/103, 191-197.

[68] Wictome, M., Newton, K., Jameson, K., Hallis, B., Dunnigan, P., Mackay, E., Clarke, S., Taylor, R., Gaze, J., Foster, K., and Shone, C., Development of an *in vitro* bioassay for *Clostridium botulinum* type B neurotoxin in foods that is more sensitive than the mouse bioassay, *Appl. Environ. Microbiol.* **1999**, 65, 3787-3792.

[69] Henderson, D. A., The looming threat of bioterrorism, *Science* **1999**, 283, 1279-1282.

[70] Johnson, E. A., Clostridial toxins as therapeutic agents: benefits of nature's most toxic proteins, *Annu. Rev. Microbiol.* **1999**, 53, 551-575.

[71] Reye, B., Parfum aus der Achselhöhle, *Der Spiegel* **2000**, 35, 236-238.

[72] Kamiya, S., Takahashi, M., Yamaguchi, H., Taguchi, H., and Nakamura, S., Preventive effect of *Clostridium butyricum* on enterohemorrhagic *Escherichia coli* O157:H7 infection, *Proc. 3. Int. Meeting Mol. Genet. Pathogenesis Clostridia*, June 8-11. Kazusa, Japan, **2000**, p. 52.

[73] Minton, N. P., Mauchline, M. L., Lemmon, M. J., Brehm, J. K., Fox, M., Michael, N. P., Giaccia, A., and Brown, J. M., Chemotherapeutic tumor targeting using clostridial spores, *FEMS Microbiol. Rev.* **1995**, 17, 357-364.

[74] Fox, M. E., Lemmon, M. J., Mauchline, M. L., Davis, T. O., Giaccia, A. J., Minton, N. P., and Brown, J. M., Anaerobic bacteria as a delivery system for cancer gene therapy: *in vitro* activation of 5-fluorocytosine by genetically engineered clostridia, *Gene Ther.* **1996**, 3, 173-178.

[75] Lemmon, M. J., van Zijl, P., Fox, M. E., Mauchline M. L., Giaccia, A. J., Minton, N. P., and Brown, J. M., Anaerobic bacteria as a gene delivery system that is controlled by the tumor microenvironment, *Gene Ther.* **1997**, 4, 791-796.

[76] Theys, J., Nuyts, S., Landuyt, W., van Mellaert, L., Dillen, C., Böhringer, M., Dürre, P., Lambin, P., and Anné, J., Stable *Escherichia coli-Clostridium acetobutylicum* shuttle vector for secretion of murine tumor necrosis factor alpha, *Appl. Environ. Microbiol.* **1999**, 65, 4295-4300.

[77] Ensley, B. D., Ratzkin, B. J., Osslund, T. D., Simon, M. J., Wackett, L. P., and Gibson, D. T., Expression of naphthalene oxidation genes in *Escherichia coli* results in the biosynthesis of indigo, *Science* **1983**, 222, 167-169.

[78] Kabdasli, I., Tünay, O., and Orhon, D., Sulfate removal from indigo dyeing textile wastewaters, *Water Sci. Technol.* **1995**, 32, 21-27.

[79] Schoutens, G. H., Groot, J. W., and Hoebeek, J. B. W., Application of iso-propanol-butanol-ethanol mixtures as an engine fuel, *Process Biochem.* **1985**, February 1986, 30.

[80] Ladisch, M. R., Fermentation-derived butanol and scenarios for its uses in energy-related applications, *Enzyme Microb. Technol.* **1991**, 13, 280-283.

[81] Husar, E., Butanol plant may come to Kahoka, *The Quincy Herald-Whig*, **1999**, November 28.

[82] Ljungdahl, L. G., Hugenholtz, J., and Wiegel, J., Acetogenic and acid-producing clostridia, in: *Biotechnology Handbooks*, Vol. 3, *Clostridia* (N. P. Minton, D. J. Clarke, Eds.), Plenum Press, New York, **1989**, pp. 145-191.

[83] Jang, L.-K., Chang, P. W., Findley, J. E., and Yen, T. F., Selection of bacteria with favorable transport properties through porous rock for the application of microbial-enhanced oil recovery, *Appl. Environ. Microbiol.* **1983**, 46, 1066-1072.

[84] Behlulgil, K., Mehmetoglu, T., and Donmez, S., Application of microbial enhanced oil recovery technique to a Turkish heavy oil, *Appl. Microbiol. Biotechnol.* **1992**, 36, 833-835.

[85] Hughes, J. B., Wang, C., Bhadra, R., Richardson, A., Bennett, G. N., and Rudolph, F. B., Reduction of 2,4,6-trinitrotoluene by *Clostridium acetobutylicum* through hydroxylamino-nitrotoluene intermediates, *Environ. Tech. Chem.* **1998**, 17, 343-348.

[86] Hughes, J. B., Wang, C., Yesland, K., Richardson, A., Badra, R., Bennett, G. N., and Rudolph, F. B., Bamberger rearrangement during TNT-metabolism by *Clostridium acetobutylicum*, *Environ. Sci. Tech.* **1998**, 32, 494-500.

[87] Huang, S., Lindahl, P. A., Wang, C., Bennett, G. N., Rudolph, F. B., and Hughes, J. B., 2,4,6-Trinitrotoluene reduction by carbon monoxide dehydrogenase from *Clostridium thermoaceticum*, *Appl. Environ. Microbiol.* **2000**, 66, 1474-1478.

[88] Jenney, F. E., Jr., Verhagen, M. F. J. M., Cui, X., and Adams, M. W. W., Anaerobic microbes: oxygen detoxification without superoxide dismutase, *Science* **1999**, 286, 306-309.

[89] Angert, E. R., Clements, K. D., and Pace, N. R., The largest bacterium, *Nature* **1993**, 362, 239-241.

[90] Angert, E. R., Brooks, A. E., and Pace, N. R., Phylogenetic analysis of *Metabacterium polyspora*: clues to the evolutionary origin of daughter cell production in *Epulopiscium* species, the largest bacteria, *J. Bacteriol.* **1996**, 178, 1451-1456.

[91] Siunov, A. V., Nikitin, D. V., Suzina, N. E., Dmitriev, V. V., Kuzmin, N. P., and Duda, V. I., Phylogenetic status of *Anaerobacter polyendosporus*, an anaerobic, polysporic bacterium, *Int. J. Syst. Bacteriol.* **1999**, 49, 1119-1124.

[92] Karlin, S., Brocchieri, L., Mrázek, J., Campbell, A. M., and Spormann, A. M., A chimeric prokaryotic ancestry of mitochondria and primitive eukaryotes, *Proc. Natl. Acad. Sci. USA* **1999**, 96, 9190-9195.

2
Taxonomy and Systematics

Erko Stackebrandt and Hans Hippe

"Systematics ... includes not only the service functions of identification and classifying but the comparative study of all aspects of organisms as well as interpretation of the role of lower and higher taxa in the economy of nature and in evolutionary history. It is a synthesis of many kinds of knowledge, theory, and method, applied to all aspects of classification. The ultimate task of the systematist is not only to describe the diversity of the living world but also to contribute to its understanding" [1]. In bacteriology, the truth of this citation has been shown valid not earlier than at the beginning of the second century of bacterial classification. To understand this revolution one has to consider the situation that existed before the evolutionary history of the prokaryotes became determinable, exemplified by the taxonomy of the clostridia.

For more than one century of identification of prokaryotes properties used to characterize members of *Clostridium* have proven their taxonomic value by allowing rapid affiliation of isolates to the genus and species. Previously organisms with non-*Clostridium* phenotypes were not considered to be closely related to *Clostridium* species, such as the anaerobic species *Desulfotomaculum, Peptococcus, Peptostreptoccus,* and *Ruminococcus* (they are related to *Clostridium* species), or the aerobic species *Planococcus, Caryophanon, Filibacter,* and *Staphylococcus* (they are related to members of *Bacillus*). Emphasis on a few morphological, physiological, and ultrastructural traits as the basis for classification at the species level was the accepted strategy – as long as data on the intra- and intergeneric relationships were unavailable.

Phylogenetic evidence, mainly based upon RNA analysis of the small subunit of ribosomes (named 16S rDNA in the following) indeed seem to indicate that, except for morphology, the taxonomic criteria used to affiliate bacteria to the genus *Clostridium* are ancient phylogenetic traits. Moreover, the formation of heat-stable endospores appears to be a monophyletic trait that is expressed in a few genera of the *Clostridium/Bacillus* subphylum of Gram-positive bacteria, i. e., *Clostridium, Bacillus, Sarcina, Desulfotomaculum,* and a few new genera recently described for organisms which were phylogenetically distinct from the genera *Clostridium* and *Bacillus*. In contrast to the traditional taxonomic treatment of *Clostridium*, however, molecular analysis clearly indicated the phylogenetic incoherence of the genus. Not only were endospores found in species of different lines of descent of the subphylum. Also the loss of spore formation and changes in morphology observed in phylogenetic neighbors of several *Clostridium* species of different lineages led to phenotypes that were clearly recognizable as being different from that of spore-forming species. This finding on the evolution of phenetic properties does not influence taxonomic decisions *per se*, except when systematists are trying to bring taxonomy in congruence with phylogeny. A recent trend in systematics, starting some 20 years ago shortly after the introduction of 16S rDNA sequencing into bacteriology, is indeed using the emerging phylogenetic branching pattern(s) as a skeleton for deriving taxonomic conclusions. The success story of the polyphasic approach to bacterial systematics, especially the definition of taxa above the genus level, is to a great extent due to the availability of 16S rDNA sequences, generated for almost each novel described species. In many cases this situation is true where the separate phylogenetic position of a species cannot be immediately used to draw taxonomic conclusions, i.e., by describing a new genus as the novel position is not immediately accompanied by the availability of exclusive phenotypic properties of this species which would allow its reclassification. Actually, the elucidation of genomic relationships and the determination of the approximate phylogenetic position of a strain often is easier than the determination of its phenotypic properties. It is the latter task that classification is not faster and less demanding than during the preceding decades, despite the recognition of a much wider spectrum of molecular and non-molecular data than previously available.

2.1
Information of some phenotypic characteristics of members of *Clostridium*

The hallmark of members of the genus *Clostridium* is a combination of phenotypic characteristics [2], i.e., rod-shaped morphology, endospore formation, Gram-positive staining behavior and anaerobic, fermentative metabolism in which sulfate is not reduced dissimilatorily. This combination has been used for more than hundreds years to affiliate new species to the taxon. Nevertheless, some exceptions were permitted by taxonomists when a single of the five main properties did not match. Examples are:

- The formation of coccoid cells by *C. coccoides*.
- Endospores may be difficult to detect or not formed under laboratory conditions, e.g., in some strains of *C. perfringens, C. barati, C. leptum, C. nexile, C. ramosum,* and *C. spiroforme*.
- A few species are able to grow under air at atmospheric pressure, e.g., *C. tertium, C. carnis, C. histolyticum,* and *C. intestinalis*. Catalase is absent except for trace activity, occasionally observed in a few species/strains.

- Several species are Gram-positive by staining in young or very young cultures only. In the stationary phase of growth these cells may stain Gram-negative. Some species always stain Gram-negative, e. g., *C. symbiosum.* The cell wall is of the Gram-positive type. Exceptions were determined in members of RNA cluster IX; but after the reclassification of *C. quercicolum* as *Dendrosporobacter quercicola* [3] this cluster does not contain a *Clostridium* species anymore.

"Minimal Standards" published as a guide for describing several genera and their species are not available for clostridia. However, a large number of tests has been introduced mainly by Holdeman et al. [2, 4] which are recommended by *Clostridium* taxonomists for the characterization of new isolates and description of species of anaerobic bacteria including clostridia. For the identification and taxonomic affiliation the tests may include:

- Anaerobic growth.
- Absence of dissimilatory sulfate-reduction.
- Morphology of cells and motility.
- Formation of endospores. This property should be investigated carefully using different methods; there is no single medium which is optimal for sporulation of all *Clostridium* species. Tests should therefore include poor and rich media, presence and absence of carbohydrates, poorly and well-utilized carbohydrates, liquid and solid media, optimal and suboptimal growth temperatures, and extended incubation time. In case that spores cannot be observed due to a low sporulation rate, survival of a culture after pasteurization (75 °C, 10 min) or ethanol treatment should be tested.
- Determination of physiological and biochemical properties should include determination of temperature and pH optimum and range for growth; lipase, lecithinase, indole, acetylmethylcarbinol and urease production, esculin and gelatin hydrolysis, meat digestion and action on milk; utilization of various sugars and polymeric carbohydrates, especially soluble starch; determination of volatile and nonvolatile fermentation products following growth in carbohydrate-containing and carbohydrate-free media.

Information on chemotaxonomic properties of clostridia are fragmentary as they have just been studied randomly. Information is available on:

1. *Cell wall peptidoglycan:* The peptidoglycan of the cell wall of most clostridia studied so far belongs to the meso-diaminopimelic acid (A_2pm) containing directly linked type peptidoglycan [5-12, Kandler, pers.comm.]. About 17 species contain either LL- A_2pm, lysin or ornithine instead of meso- A_2pm at position 3 of the peptide subunit; alanine, glycine, serine, or aspartate may form an interpeptide bridge. These additional types of peptidoglycan were hitherto found only in *Clostridium* species of the RNA clusters I, IV, XI, XVI, and in *C. hydroxybenzoicum.*

2. *Respiratory quinones of the menaquinone type:* These have only been found in the mesophilic species *C. formicoaceticum* [13] and the thermophilic species *C. thermocellum, C. thermobutyricum, C. stercorarium* and *C. thermolacticum* [14], as well as in reclassified former clostridia, such as *C. thermoaceticum, C. thermoautotrophicum,*

C. thermosaccharolyticum, C. thermohydrosulfuricum, C. thermosulfurigenes, C. thermocopriae, and *C. fervidus* [13-15].

3. Cellular fatty acids and lipids: Studies were applied to a limited number of clostridial species, but the results have demonstrated a great diversity in the composition of cellular fatty acids and lipids of the clostridia [16, 17]. Extension to a significantly higher number of species and careful standardization of the methods is needed before such data can be used as chemotaxonomic traits, serving both characterization of individual species and support of phylogenetic clustering of groups of species (and vice versa).

A large number of clostridia has been isolated in the past which were mainly used as patent strains for the production of solvents like acetone, butanol, and isobutanol on an industrial scale. Various names had been given to these strains which often do not correspond to current standards and rules of nomenclature. Some strains had been clearly misidentified. The history of these strains, some of them are still in use in some solvent-producing application, as well as their taxonomic and phylogenetic affiliation, has been considered in recent thorough studies [18-22]. As a result of these studies, most solvent-producing clostridial strains, irrespective of their historical taxonomic affiliation, could be assigned to two established species, *C. beijerinckii* and *C. acetobutylicum*, and two separate genospecies which may represent new species for which the names "*C. saccharoperbutylacetonicum*" and "*C. saccharobutylicum*" have been proposed. These names, however, have not yet been validated (see below).

The invalidly named "*C. madisonii*" found in the older patent literature should be considered as strain of *C. beijerinckii* [19]. The species "*C. oncolyticum*" (also named Möse strain M55; *C. butyricum* M55; "*C. oncolyticum* subsp. *butyricum*") which produces kininases that selectively destroy tumor tissue in mice [23-25] shows metabolic reactions and electrophoretic cell protein pattern of *C. sporogenes* and may therefore be regarded as a strain of this species ([2] and DSMZ tests).

2.2
Unraveling the phylogenetic position of *Clostridium* species

The most useful molecular markers for taxa between the rank of species meet the following characteristics of phylogenetic markers:

1. Ubiquitous distribution because only they have the potential to cover the complete range of taxa.
2. Homologous origin, i. e., deriving from a common ancestor. This is deduced from the constraint in function and from sequence similarity. A comparable mode of evolution can only be expected for functionally equivalent molecules, such as the products of housekeeping genes of which are genetically stable. Analysis of molecules involved in lateral gene transfer would disturb any phylogenetic conclusions, though these molecules can be used for resolving phylogenetic traits in those lineages which emerged after the panmixis event (horizontal gene exchange).

3. Appropriate size; the finding that each individual sequence position can only carry the information on a rather narrow range of evolutionary time; it can thus be concluded that the larger the number of independently evolving positions the larger the detectable phylogenetic levels.

Presently, the most extensively used phylogenetic marker molecule is the 16S ribosomal RNA and the genes coding for it (16S rDNA). Automated DNA sequence analysis is generally carried out as linear PCR cycle sequencing reaction. Public databases contain more than about 20,000 sequences of cultured and up to now uncultured prokaryotic strains, including the strains of all validly described *Clostridium* species.

2.3
Sequence alignment and treeing algorithms

The quality of phylogenetic trees derived from sequence data firstly depends on the quality of the sequence alignment and secondly on the algorithms used to determine the topology of phylogenetic patterns. A correct alignment arranges homologous nucleotides in columns, i. e., residues which are derived from a common position within the ancestral sequence. Only these positions are recognized as being identical or different. Alignment is difficult or even impossible within variable and highly variable regions which also show the highest degree of length variation. In this case, only highly similar organisms can be compared to each other while these regions are excluded from the analyses of less closely related sequences. Three major types of tree inferring approaches are commonly used: distance in pairs, maximum likelihood and maximum parsimony methods.

In the most widely used approach, the distance method, a matrix of dissimilarity values in pairs is calculated from the sequence alignment. Usually these dissimilarity values are transformed into phylogenetic distances, taking into account that each nucleotide position is permutated, the degree of which depends upon the relative evolutionary distance between a pair of sequences (Jukes and Cantor, Kimura-2 transformations). Based on distance matrices, phylogenetic trees are preferentially reconstructed applying additive tree methods. These methods seek the tree for which expected distances are most similar to those calculated from present-day sequences according to their topology and branch lengths. The tree topology is regarded to reflect the best topology that requires minimal errors with respect to the values of the underlying matrix. A disadvantage of the distance method is that only overall dissimilarity values are used and all information about individual sequence positions is disregarded. Most dendrograms displaying the topology of members of the *Clostridium-Bacillus* subline of descent have been produced by distance matrix algorithms, using either the method of DeSoete [26], Fitch and Margoliash [27], or the neighbor-joining algorithm of Saitou and Nei [28], compiled by Felsenstein [29]. The latter two procedures have been used by Collins et al. to generate the most detailed phylogenetic analysis of *Clostridium* species [30]. This analysis leads to about twenty individual lineages (clusters), almost each of which contained species of the genus *Clostridium* and formed the starting point for the reclassification of the genus. Several of the species which branched distantly from *Clostridium sensu stricto* (cluster I, containing the type species of the genus) and from other species-rich clusters were reclassified (see below). Due to constraint in space these clusters are shown schematically in Figure 1.

In order to demonstrate differences in the branching patterns generated with different algorithms, representatives were selected from each of the ten lineages which contain authentic *Clostridium* species. Though the number of sequences were significantly reduced when compared to those selected by Collins et al [30], not only most of the clusters but also their specific relationship emerge in the neighbor-joining tree, based upon distance matrix analyses (Figure 2a). Clusters XVI, XVIII, and XIX branch deeply, followed by the split lineage of clusters

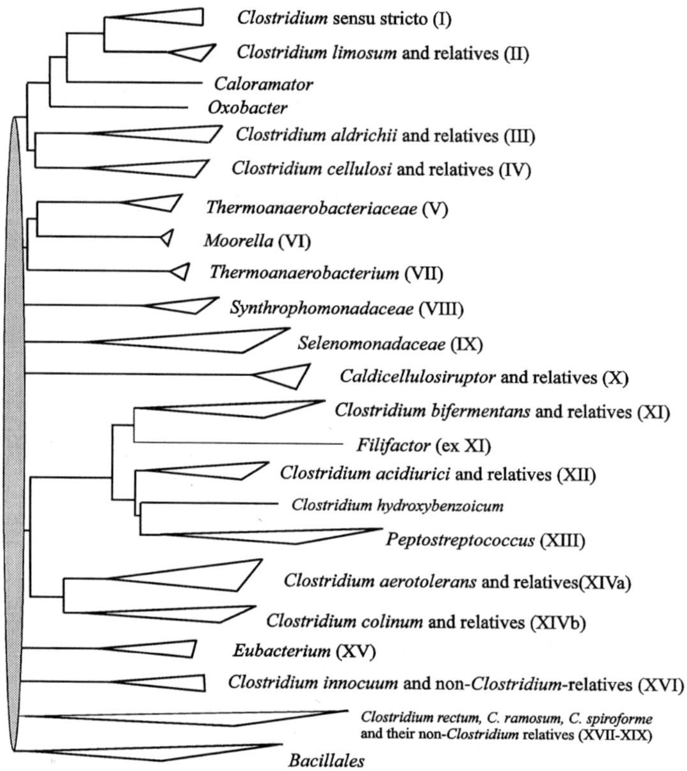

Figure 1. Schematic outline of the approximate phylogenetic diversity of the 16S rDNA clusters (roman numbers, according to ref. [30]) of anaerobic, rod-shaped, mainly Gram-positive bacteria, classified as species of *Clostridium*, some genera that resulted from the beginning of the reclassification process of *Clostridium*, and taxa exhibiting non-clostridial phenotype. The depth of triangles is an approximation of the phylogenetic diversity of members of this cluster. The width of triangles is no indication for the number of species enclosed therein. The ellipsoid area indicates the uncertainty of the order, at which the lineages diverge from each other.

XIVa and b. Exceptions from the tree shown in ref. [30] refer to cluster IV, which does not form a monophyletic cluster, and to the position of the branching points of clusters XI and XII.

Maximum likelihood methods analyze the sequences on a site-by-site basis and incorporate an explicit model of sequence evolution, such as the tree, an assumed mechanism of nucleotide change (for example, changes are independent), and/or the probabilities of observing the nature of its change (for example, transition versus transversion). A tree is considered the best tree which maximizes the congruence of model and data. The advantages of the maximum likelihood approach are that frequency and nature of changes at the individual sites can be taken into consideration and that it can compensate for superimposed changes. Figure 2b is a maximum likelihood dendrogram using the DNAML (transition/transversion rate 2,000) [29]. As judged from the name of organisms listed in the middle of the two trees the DNAML tree is similar to the neighbor-joining tree, though some differences are obvious. These refer to the split of members of cluster II and the order, at which clusters III, IV, XI, and XII branch from each other. While this effect does not put into question the validity of the *Clostridium* species it may indeed influence the decision about the reclassification of the respective species into novel genera.

The third approach are maximum parsimony methods that also use information on the individual positions of aligned sequences. The underlying model of evolution assumes that sequences derived from their ancestors by acquiring a minimal number of changes. The methods seek for the most parsimonious trees among all possible tree topologies by determining the sum of changes which must have occurred to give the sequences in the alignment. Here, the parsimony method is used to generate a consensus tree, in which 100 bootstrapped trees are evaluated to give the most plausible tree. The bootstrap approach randomly re-samples alignment positions in that some are included more often than others, while yet others are not taken into account at all. The procedure is repeated with alternatively truncated or rearranged data sets. The consensus tree (not shown) recovers the main rRNA cluster as depicted in Figure 2, though the order at which these groups are branching differs from that shown in Figures 2, 3, and 4. In addition to the algorithm-dependent changes, deviations refer to

(i) *C. cellulosi*: constituting a separate lineage within cluster IV in the neighbor-joining tree, clusters with members of cluster III; this new position is in accord with the cellulolytic activity in members of this cluster, including *C. cellulosi.*

(ii) C. *sticklandii* and *C. halophilum*: both representing individual lineages in cluster XI in the neighbor-joining tree, they emerge as separate lineages branching at adjacent to cluster XII.

(iii), *C. felsineum, C. formicoaceticum, C. aceticum* and *C. litorale* are disconnected from the core of cluster XI forming a sister cluster to members of group XII, *C. sticklandii* and *C. halophilum.*

These examples show that a taxonomic rearrangement of species-rich clusters is not indicated. Before this can be approached, correlation of phylogenetic patterns with phenotypic and other genomic data are necessary to allow insights into the homogeneity at the genetic- and epigenetic level.

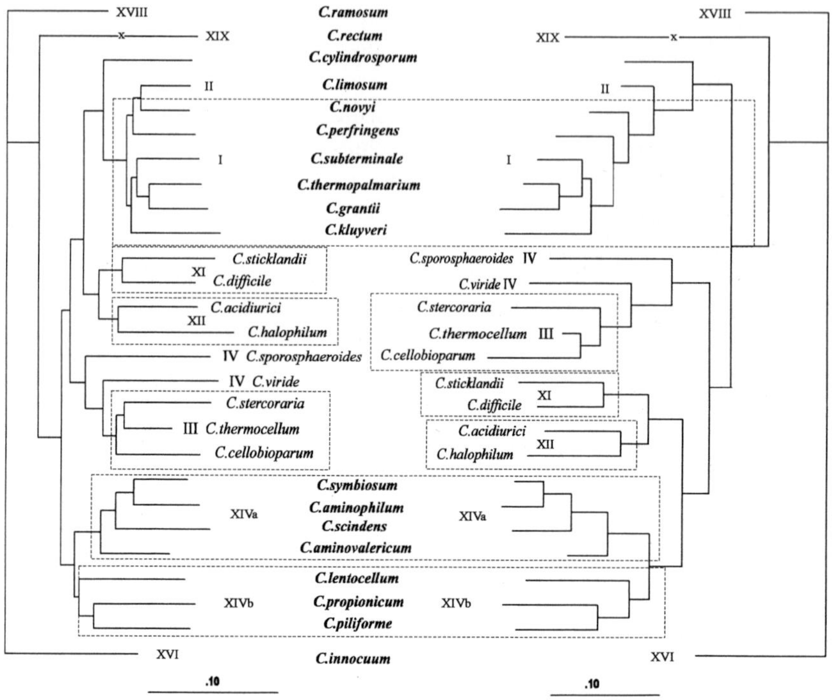

Figure 2. Comparative phylogenetic analysis of representatives of 16S rDNA clusters using distance matrix and consensus parsimony analyses [29]. The boxed area indicates those organisms of which the branching order is effected most obviously from the selection of treeing algorithms. The bar indicates 10% nucleotide substitutions.

a. Neighbor-joining dendrogram based upon distance matrix analyses.
b. Maximum likelihood analysis, using the program DNAML (transition-transversion rate 2,000).

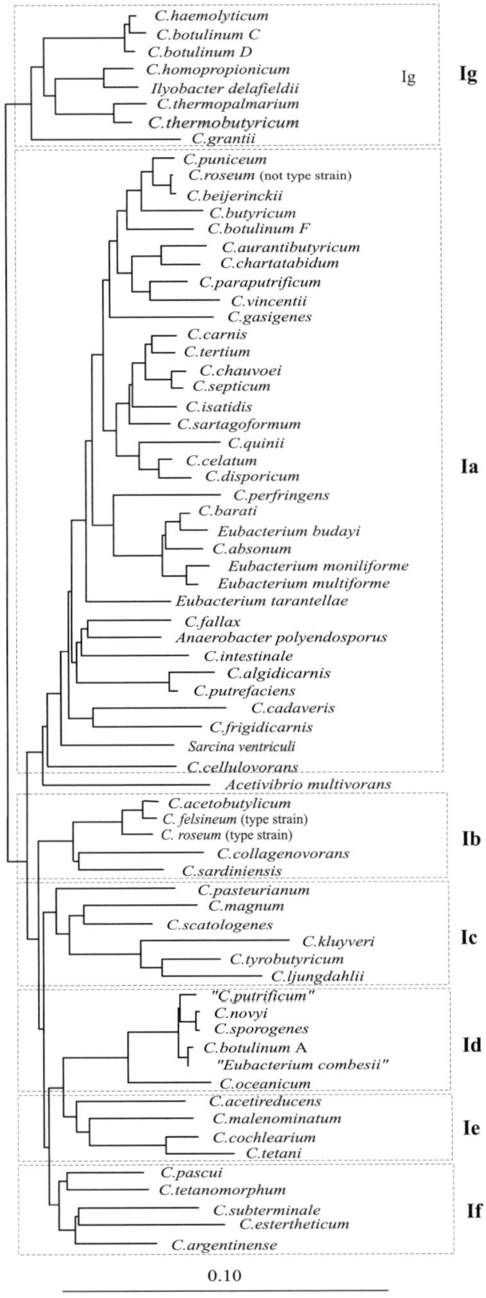

Figure 3. Comparative analysis of 16S rDNA of *Clostridium* species of cluster I and some non-*Clostridium* reference species. Distance matrix analysis using the neighbor-joining method [29] and Jukes & Cantor correction to compensate for different evolutionary rates. Framed areas separate subclusters. The bar indicates 10% nucleotide substitutions.

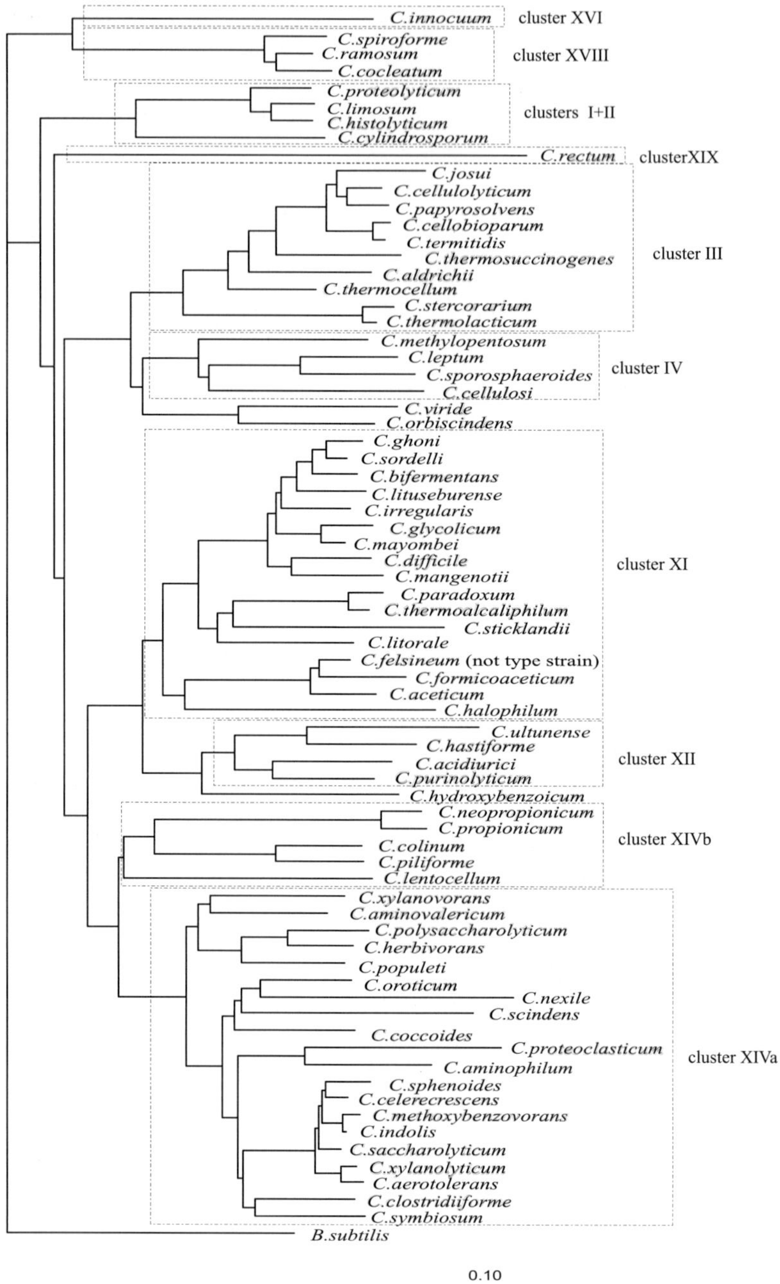

Figure 4. Comparative analysis of 16S rDNA of *Clostridium* species of clusters II-XIX and some non-*Clostridium* reference species. Distance matrix analysis using the neighbor-joining method [29] and Jukes & Cantor correction to compensate for different evolutionary rates. Framed areas separate clusters. The bar indicates 10% nucleotide substitutions.

Phylogenetic markers other than 16S rDNA

Results from phylogenetic studies using genes coding for ribosomal RNA [31, 32], the b-subunit of the ATP synthase and the protein chain elongation factor EF-Tu [33], have demonstrated a high degree of correlation in the branching patterns obtained. This finding indicates not only the genetic stability of these genes, but also the ability of these markers to reflect the phylogeny of the majority of housekeeping genes, hence the phylogeny of the organisms themselves. Phylogenetic analyses on genes other than ribosomal RNA genes were mainly performed on a few *Clostridium* species. Analyses of these genomic properties do therefore not contribute significantly to the phylogeny of the genus *Clostridium*. One exception is the clostridia group with Gram-positive reference organisms exhibiting a low G+C content of DNA, e. g., in studies on RNases H [34], family C DNA polymerases [35, 36], GroEL (chaperon) [37, 38], and σ^{70}-type sigma factors [39]. In contrast, analysis of *nifH* genes [40] showed members of *Clostridium* to cluster separately from *Paenibacillus* species, which themselves were placed adjacent to cyanobacteria.

Some evolutionary aspects

16S rRNA/DNA sequence analysis has revealed that members of the genus *Clostridium* branch deeply within the subphylum of Gram-positive bacteria, the members of which are characterized by a DNA G+C content of less than 55 mol%. Their appearance in the phylogenetic tree predates the evolution of the facultative and aerobic descendents of the clostridial genotype and phenotype such as the bacilli, lactobacilli, and staphylococci. This order of evolution of the respiration chain parallels the situation in certain other major sublines of descent in which the anaerobic phenotypes appear to be more ancient than the aerobic metabolism. Examples are found in the actinomycete subphylum (e. g., *Bifidobacterium* versus mycelium-forming actinomycetes), the *Bacteroides* (anaerobic) and *Cytophaga* (aerobic) phylum, or within the subclasses of *Proteobacteria* where the anaerobic purple sulfur- and non-sulfur photosynthetic organisms represent deeply rooting lineages.

The availability of a phylogenetic framework allows the ranking of recently evolved organisms (today's species) according to their evolutionary history (at present preferably by 16S rDNA sequence analysis [41]) and physiological, biochemical and morphological characters can be superimposed onto the phylogenetic tree. It can then be decided whether a given trait evolved only once in evolution (monophyletic) or whether it evolved independently in several different lineages (polyphyletic). This has not yet been done for metabolic properties but, for example, endospore formation appears to be a monophyletic property (though lost in several lineages during evolution).

A multi-layered peptidoglycan has evolved at least two or three times (deinococci and Gram-positive bacteria of the *Clostridium* and actinomycete subphyla) in the evolution of the bacteria. The presence of Gram-positive peptidoglycan structure in members of the phylogenetically neighboring *Clostridium/Bacillus* lineage and actinomycetes lineage has been interpreted as the divergence of a Gram-positive ancestral organism into two phylogenetic lineages [32]. Support for this theory came from the phylogenetic placement of Gram-positive staining cyanobacteria close to the root of the Gram-positive organisms, in contrast to the deinococci which are related to the genus *Thermus* [42]. This hypothesis of common origin has been challenged by two findings. Firstly, the common ancestry of the *Clostridium/Bacillus* lineage and the actinomycete lineages cannot be recovered from the analysis of all phylogenetic markers, and the unity of these two subphyla is not significant. For example, analyses of 16S rDNA have revealed that the order of divergence of the main lines of descent is not constant, but changes as a function of the particular set of sequences included in the analyses [32]. Similarly, analyses of 23S rDNA and the elongation factor Tu (EF-Tu) show the two subphyla of Gram-positive bacteria to be separated [33]. Secondly, the most deeply branching species in the *Clostridium/Bacillus* lineage are not Gram-positive, but possess a Gram-negative cell wall. This finding was originally observed in members of *Sporomusa, Selenomonas,* and *Megasphaera* [43] and in *Heliobacterium* [44]. However, today this group contains a significantly larger number of genera (>20), including, to name a few, *Zymophilus, Pectinatus, and Dialister.* Of special interest in this context is the report of an outer membrane in members of the endospore-forming genus *Sporomusa* but this property is also present in other members of cluster IX (Hippe, unpublished). Thus, the topology of the *Sporomusa* cell wall does not appear to be a reduction of the Gram-positive wall but rather an ancestral feature of a Gram-negative (pre-proteobacterial) ancestor.

However, even the elucidation of phenotypic properties of the hypothetical ancestor is complicated by the fact that the branching order of deeply rooting lineages is not significant and may change with the inclusion of more sequences. Furthermore, the deeply branching cluster of Gram-negative species of *Acidaminococcus* and *Phascolarctobacterium* are intermixed with Gram-positive species of the genera *Desulfotomaculum, Desulfitobacterium,* and *Peptococcus niger.* The close relatedness between the Gram-negative staining *Tissierella* and *Clostridium hastiforme* [45], and between *Roseburia, Catonella,* and *Johnsonella* [46] species and clostridia of cluster XIVa indicates that a significant reduction of the Gram-positive cell wall may actually occur at any stage of the evolution of the peptidoglycan.

The phylogenetic heterogeneity of the genus *Clostridium* and related genera

The phylogeny of the genus *Clostridium* has recently been described in detail in articles, such as the comprehensive study of Collins et al. [30] and a more recent update [47], description of physiological similar species, such as thermophilic [48] and alkalithermophilic species [49]. The phylogeny has also been covered in overview articles [50, 51] and a dendrogram of relationships can be retrieved from ARB ("A software environment for sequence data") program (retrievable from arb@mikro.biologie.tu-muenchen.de). The recognition of the phylogenetic heterogeneity and the beginning of the dissection of the genus *Clostridium* parallels the history of the use of ribosomal RNA and their genes in phylogenetic studies. The early studies by Johnson and Francis [52] clearly demonstrated that the genus *Clostridium* was composed of a wide range of phylogenetically remotely related species. However, the intermixing of phenotypes could not be recognized because

Table 1. Reclassified clostridial species.

Previous name	***New name***	***Reference***
C. barkeri	*Eubacterium barkeri*	[30]
C. botulinum type G	*C. argentinense*	[55]
C. bryantii	*Syntrophospora bryantii*	[56]
C. durum	*Paenibacillus durus*	[30]
C. fervidus	*Coloramator fervidus*	[30]
C. lentoputrescens	*C. cochlearium*	[57]
C. lortetii	*Sporohalobacter lortetii*	[58]
C. oxalicum	*Oxalophagus oxalicus*	[30]
C. paraperfringens	*C. barati*	[59]
C. perenne	*C. barati*	[59]
C. pfennigii	*Oxobacter pfennigii*	[30]
C. putrificum	*C. sporogenes*	[60]
C. quercicolum	*Dendrosporobacter quercicola (corrig.)*	[3]
C. thermoaceticum	*Moorella thermoacetica*	[30]
C. thermoautotrophicum	*Moorella thermoautotrophica*	[30]
C. thermocopriae	*Thermoanaerobacter thermocopriae*	[30]
C. thermohydrosulfuricum	*Thermoanaerobacter thermohydrosulfuricus*	[61]
C. thermosaccharolyticum	*Thermoanaerobacterium thermosaccharolyticum*	[30]
C. thermosulfurogenes	*Thermoanaerobacterium thermosulfurigenes*	[61]
C. villosum	*Filifactor villosus*	[30]

only members of the genus *Clostridium* were included in the reassociation experiments. The extent to which this occurred was first recognized by the analysis of the cumulative database of partial 16S rRNA sequences [42, 53, 54]. To the surprise of taxonomists, many species of *Clostridium* were found to be more closely related to species of different genera which contained a mixture of clostridial and non-clostridial phenotypes, e. g., spherical form and non-spore-formation (e. g., *Ruminococcus* and *Peptococcus*), or rod-shaped form and non-spore formation (e. g., *Eubacterium*).

By now, nearly all validly described species have been subjected to 16S rDNA analysis (*C. arcticum* and *C. cellulofermentans* have not been deposited in public databases), and the reclassification process has been initiated starting with those species that do not fall into the radiation of the main clusters of *Clostridium* species (Table 1). The table also lists some reclassified species which were united with other *Clostridium* species.

Without any doubt, the emergence of more than twenty distinct groups of mainly anaerobic bacteria within the lineage of Gram-positive bacteria exhibiting

Table 2. Genera of obligately anaerobic sporeformers other than *Clostridium*, including reclassified former species of the genus *Clostridium* and other genera.

Mesophiles	*Sporobacterium*
Acetonema	*Sporohalobacter*
Anaerobacter	*Sporomusa*
Anoxybacillus	*Sporotomaculum*
Bacillus (infernus)	*Syntrophospora* (C.)
Filifactor (C.)	*Syntrophobotulus*
Dendrosporobacter (C.)	
Halobacteroides halobius	**Thermophiles**
Heliobacillus	*Caloramator* (C.)
Heliobacterium	*Moorella* (C.)
Propionispora	*Thermoanaerobacter* (C., Ac.)
Orenia (S.)	*Thermoanaerobacterium* (C.)
Oxalophagus (C.)	
Oxobacter (C.)	**Sulfate-/sulfite-reducers**
Pelospora	*Desulfitobacterium*
Sarcina	*Desulfosporosinus* (D.)
Sporobacter	*Desulfotomaculum*

In parenthesis, the name of the genus is given to which all or some species belonged prior to their reclassification.
Ac. = *Acetogenium*, C. = *Clostridium*; S. = *Sporohalobacter*, D. = *Desulfotomaculum*.

a low base ratio of DNA, 16 of which contained *Clostridium* species, was a major scientific breakthrough. The largest group of these, named cluster I, contained the type species of the genus, while the others received labels between clusters II and XIX. Many of these latter clusters not only contained clostridial species but Gram-positively and Gram-negatively staining species of many other genera, previously not considered to be closely related to *Clostridium* species. Table 2 is a compilation of genera of the *Clostridium-Bacillus* subphylum containing anaerobic spore formers. It also displays the new home of former *Clostridium* species.

In addition to the above-mentioned heterogeneity of the genus *Clostridium*, most of the intermixed genera which contain more than a single species are also heterogeneous. One of the two species of *Butyrivibrio*, *B. fibrisolvens*, is remotely related to *C. aminophilum*, while the other, *B. crossotus* is distantly related to *Eubacterium saburreum* (cluster XIVa). Last but not least, the phenotypically similar *Pseudobutyrivibrio ruminis* is closely related to *B. fibrisolvens*. The phylogenetic heterogeneity of the genus *Ruminococcus*, grouping within clusters IV (including the type species *R. flavefaciens*) and XIVa, has recently been demonstrated independently by Rainey and Janssen [62] and Willems and Collins [63]. Within the latter group, only two of the *Ruminococcus* species, i. e., *R. hansenii* and *R. obeum*, cluster together, while the other species constitute distinct lines of decent. The phylogenetic heterogeneity of the genus *Eubacterium* is not surprising as this genus has been defined as a dumping ground for anaerobic, asporogenous, rod-shaped Gram-positive bacteria lacking additional diagnostic features. While some of the *Eubacterium* species, such as *E. combesii, E. fosser,* (reclassified as *Atopobium fossor* [64]) and *E. lentum* (reclassified as *Eggerthella lenta* [65]), have been found to group with the actinomycetes and their relatives, other species are members of *Clostridium* groups I, IV, IX, XI XII, and XIVa. Of the 12 *Eubacterium* species falling within group XIV, two species are closely related to *C. oroticum*, one species is moderately related to *C. nexile*, while the other species are related to a varying degree either among themselves or to non-*Clostridium* species. In no other group of *Clostridium* species is the lack of significance of endospore formation in classification more impressively demonstrated than in group XIVa. Loss of spore formation, reduction of peptidoglycan layers and changes in morphology have apparently occurred during the evolution of *Clostridium* species immediately following the first occurrence of the clostridial phenotype. The following description of the individual clusters can be kept short because little information can be added to that given in the original description of Collins et al. [30].

The phylogenetic structure of the core of the genus *Clostridium*

Cluster I, equivalent to group I of Johnson and Francis [52], constitutes the largest of the *Clostridium* rRNA clusters and it contains the type species, *C. butyricum*. Except for a few highly related species, such as *C. haemolyticum* and *C. botulinum* producing neurotoxins of types C and D (99.3 % sequence similarity), *C. roseum* and *C. acetobutylicum* (99.8 % similarity), *C. septicum, C. tertium, C. carnis,* and *C. chauvoei, C. disporicum* and *C. celatum* and *C. putrefaciens* with *C. algidicarnis* (all above 98,5 % sequence similarity), the majority of species is phylogenetically well separated. The intracluster divergence is high, leading to the recognition of several subclusters, named Ia–Ig (Figure 3). We refrain from showing bootstrap values as means of statistical support for the branching points as they are no indication of phylogenetic significance and will most likely change with new sequence entries. Generally, bootstrap values are high when the horizontal path leading to a clade (one or several species) is well separated from its sister clade (see [30]). Also, it cannot be excluded that the addition of new deeply rooting species may change the relative branching point of these groups. The delineation of most of these sub clusters is somewhat arbitrary and is not supported by phenotypic traits which cover a wide range of metabolic diversity. Cluster I contains psychrophilic, mesophilic, and thermophilic species, cellulolytic, saccharolytic, and proteolytic representatives and specialists, as well as pathogenic and non-pathogenic species. Members of each subcluster are listed alphabetically in Table 3 to facilitate their affiliation. In addition to the species listed in Figure 3, new species have been described recently, i. e., the acid-tolerant *C. akagii* and *C. acidisoli* [66] which are closely related to *C. pasteurianum*.

Several invalid species have been included in the publication of Collins et al. [30], only some of which have been validly published until today. The invalid species (not shown in Figure 3) are the vesicular cap-forming soil organism "*C. caliptrosporum*" which is highly related to *C. beijerinckii*; the virtually identical pair "*C. favososporum*" and "*C. corinoforum*", being closely related to *C. puniceum*;

Table 3. Species composition of subclusters a through h of 16S rDNA cluster I[a].

Cluster Ia	**Cluster Ia cont'd**	**Cluster Ie**
C. absonum	*C. tertium*	*C. acetireducens*
C. algidicarnis	*C. vincentii*	*C. cochlearium*
C. aurantibutyricum	*Sarcina ventriculi*	*C. malenominatum*
C. barati		*C. tetani*
C. beijerinckii	**Cluster Ib**	
C. botulinum types B, E, F	*C. acetobutylicum*	**Cluster If**
(non-proteolytic)	*C. collagenovorans*	*C. argentinense*
C. butyricum	*C. felsineum*	*C. estertheticum*
C. cadaveris	*C. roseum*	*C. pascui*
C. carnis	*C. sardiniensis*	*C. subterminale*
C. celatum		*C. tetanomorphum*
C. cellulovorans	**Cluster Ic**	
C. chartatabidum	*C. akagii*	**Cluster Ig**
C. chauvoei	*C. acidisoli*	*C. botulinum* types C, D
C. disporicum	*C. kluyveri*	(proteolytic)
C. fallax	*C. ljungdahlii*	*C. grantii*
C. frigidicarnis	*C. magnum*	*C. haemolyticum*
C. gasigenes	*C. pasteurianum*	*C. homopropionicum*
C. intestinale	*C. scatologenes*	*C. thermopalmarium*
C. isatidis	*C. tyrobutyricum*	*C. thermobutyricum*
C. paraputrificum		
C. perfringens	**Cluster Id**	**Tentatively**
C. puniceum	*C. botulinum* types A, F	*C. cylindrosporum*
C. putrefaciens	(proteolytic)	
C. quinii	*C. novyi*	
C. sartagoformum	*C. oceanicum*	
C. septicum	*C. putrificum* (invalid)	
	C. sporogenes	

[a] as defined by Collins et al. (1994). Within each cluster the species are listed alphabetically. For a detailed intracluster relatedness, including non-*Clostridium* species, see Figure 3.

"*C. longisporum*" isolated from the rumen of bisons is highly related to *C. chartatabidum*, while "*C. kainantoi*" is highly similar to *C. butyricum*. "*C. autoethanogenum*" DSM 10061 [47] is phylogenetically indistinguishable from *C. ljungdahlii*. "*C. quinii*" occupies an isolated phylogenetic position which indicates that this organism does merit species status.

C. cylindrosporum branches at the periphery of cluster I showing a distant relationship to members of *Caloramator* and *Thermobrachium celere*. Subcluster Ia contains some non-*Clostridium* (non-spore-forming) species, such as misclassified species of the genus *Eubacterium* and *Acetivibrio*, as well as *Anaerobacter polyendosporus* [67] which, together with the pathogenic species *C. fallax* and *C. intestinalis* forms a phylogenetic group (subcluster Ia). Though *A. polyendosporus* shows some ultrastructural features not reported so far for *Clostridium* species this species should be reclassified as a member of *Clostridium*.

An interesting taxonomic problem arose with the finding that the type species of *Sarcina, S. ventriculi,* is a phylogenetic member of cluster I, also containing the type species of *Clostridium, C. butyricum.* As pointed out by Willems and Collins [68], subsequent application of the now widely accepted use of 16S rDNA similarities for merging and splitting higher taxa would lead to the reclassification of *Clostridium* species as species of the genus *Sarcina,* since the latter genus [69] has priority over the genus *Clostridium* [70]. While such a move would be in agreement with the Bacteriological Code [71] it would probably not be accepted by the majority of microbiologists, especially from the medical field. For such far-reaching and dramatic changes, with their striking impact on nomenclature, the Code provides the possibility to conserve the name *Clostridium* and the type species *C. butyricum* (*nomen periculosum* [Rule 56a]).

Other phylogenetic lineages containing *Clostridium* and *Clostridium*-related species
The composition of *Clostridium* species, arranged alphabetically within clusters, is shown in Table 4.

- *Cluster II:* This cluster contains the highly proteolytic and acetate producing species *C. histolyticum, C. limosum,* and *C. proteolyticum.* As indicated by Collins et al [30] these organisms may constitute a separate genus, but its delimitation from the core cluster I is not yet settled. Members of *Caloramator* [30] containing the former *Clostridium fervidum, Thermobrachium celere* [72], as well as *Clostridium cylindrosporum* are found to be distantly related to members of cluster II. *Oxobacter pfennigii* [30] represents an individual deeply branching lineage.
- *Cluster III:* This cluster comprises 10 cellulolytic mesophilic and thermophilic species, two of which, *C. thermosuccinogenes* and *C. josui,* were added after the recognition of its separate position by Rainey and Stackebrandt [73] and Collins et al. [30]. This cluster may also contain *C. cellulosi* (see above). *Acetivibrio cellulolyticus* branches within the radiation of the cellulolytic clostridia and represents a non-sporulating variant of these bacteria [Hippe, unpublished]. The primary structure of 16S rDNA of *C. thermosuccinogenes* DSM 5807^T significantly differs from those of other *Clostridium* species in the presence of a large insertion located between position 70 and 100. This insertion, comprising about 200 bases, is also larger than those reported for members of *Desulfotomaculum* [74] and *Thermoanaerobacter* [48]. The fact that a readable sequence was obtained from PCR amplified 16S rDNA indicates the lack of sequence heterogeneity in different *rrn* operons as reported for *Desulfotomaculum* strains and for *Clostridium paradoxum* [75].
- *Cluster IV:* The four phenetically very different species, displaying a broad range of DNA G+C content of more than 25 % mol% and representing thermophilic and mesophilic strains are remotely related to *C. methylpentosum* and *C. orbiscindens.* The depth of the branching point of these two subclusters makes their membership to the same cluster tentative. The sequence of *C. orbiscindens* is

Table 4. Species composition of 16S rDNA clusters II through XIX[a].

Cluster II	**Cluster Xl cont'd**	**Cluster XIVa cont'd**
C. histolyticum	*C. glycolicum*	*C. nexile*
C. limosum	*C. halophilum*	*C. oroticum*
C. proteolyticum	*C. irregularis*	*C. polysaccharolyticum*
	C. litorale	*C. populeti*
Cluster III	*C. lituseburense*	*C. proteoclasticum*
C. aldrichii	*C. mangenotii*	*C. saccharolyticum*
C. cellobioparum	*C. mayombei*	*C. scindens*
C. cellulolyticum	*C. paradoxum*	*C. sphenoides*
C. josui	*C. sordelli*	*C. symbiosum*
C. papyrosolvens	*C. sticklandii*	*C. xylanolyticum*
C. stercorarium	*C. thermoalcaliphilum*	*C. xylanovorans*
C. termitidis		
C. thermocellum	**Cluster XII**	**Cluster XIVb**
C. thermolacticum	*C. acidiurici*	*C. colinum*
C. thermosuccinogenes	*C. hastiforme*	*C. lentocellum*
	C. hydroxybenzoicum	*C. neopropionicum*
Cluster IV	*C. purinolyticum*	*C. piliforme*
C. cellulosi	*C. ultunense*	*C. propionicum*
C. leptum		
C. methylpentosum	**Cluster XIVa**	**Cluster XVI**
C. orbiscindens	*C. algidixylanolyticum*	*C. innocuum*
C. sporosphaeroides	*C. aminophilum*	
C. viride	*C. aerotolerans*	**Cluster XVIII**
	C. aminovalericum	*C. cocleatum*
Cluster XI	*C. celerecrescens*	*C. ramosum*
C. aceticum	*C. clostridiiforme*	*C. spiriforme*
C. bifermentans	*C. coccoides*	
C. difficile	*C. fimetarium*	**Cluster XIX**
C. formicoaceticum	*C. herbivorans*	*C. rectum*
C. ghoni	*C. indolis*	
	C. methoxybenzovorans	

[a] as defined by Collins et al. [30]. Within each cluster the species are listed alphabetically. For a detailed intra-cluster relatedness, including non-*Clostridium* species, see Figure 4.

99.7% similar to that of the misclassified species *Eubacterium plautii*. As mentioned above, several *Eubacterium* species are members of this cluster, as are *Fusobacterium prausnitzii*, the two species of *Anaerofilum* [76], and *Sporobacter thermitidis* [77].

- *Clusters V-X:* The former *Clostridium* species of cluster V ("*C. uzoni*", is a member of *Thermoanaerobacter*), VI, VII, VIII (*C. bryantii* has been reclassified as *Syntrophospora bryantii* before the assignment of RNA clusters), and IX have been reclassified (see Figure 1 and Table 1). Cluster VI embraces the members of *Moorella*, established for the former *C. thermoaceticum* [30]. Several lineages branch adjacent to cluster VI, such as the ones containing members of cluster V (*Thermoanaerobacter*), cluster VII (*Thermoanaerobacterium*), cluster VIII (*Syntro-*

phospora, Syntrophomonas, Thermosynthropha [78] and *Anaerobranca* [79]) cluster X (*Caldocellulosiruptor* [80]), as well as the unassigned one, harboring *Coprothermobacter* [81], *Dictyoglomus* [82], and *Dethiosulfovibrio* [83]. Especially cluster IX contains a rich spectrum of different genera. It contains members which stain Gram-negatively, such as *Sporomusa, Acidaminococcus, Pectinatus, Propionispira, Selenomonas, Megasphaera, Phascolarctobacterium,* and *Zymophilus.* This cluster also harbors *Anaerovibrio lipolyticus* [84], *Mitsuokella multacida, Centipedia periodontii* [85], *Anaeromusa acidaminophila* [86], *Succinispira mobilis* [87], *Veillonella* species, *Dendrosporobacter quercicola* [3], *Quinella ovalis* [88], *Dialister pneumosintes* [89], *Anaerosinus glyzcerini* [84], *Anaeroarcus burkinensis* [84], *Acetonema longum* [90], *Succiniclasticum ruminis* [91], and *Schwartzia succinivorans* [92]. Adjacent to the members of cluster IX, members of *Desulfotomaculum, Desulfitobacterium* [93], *Heliobacterium,* and *Syntrophobotulus glycolicus* [94] yet form another separate cluster.

- *Cluster XI: Clostridium aceticum* has been added to the list of species [47] of this cluster established in 1994, in which it groups with *C. felsineum* (non-type strain, see below) and *C. formicoaceticum* (95-97 % similarity). Two thermoalkaliphilic organisms, the non-spore forming *C. thermoalcaliphilum* and the spore former *C. paradoxum* are highly related while the other species are well separated from each other. The sequence of *C. aceticum* contains an insert of about 120 nucleotides between positions 70 and 100. The nucleotide composition cannot be determined from PCR-generated 16S rDNA which indicates that the individual rrn operons differ in the primary structure of the insert. As pointed out by Collins et al. [30] this group is phylogenetically and phenotypically heterogeneous, containing in addition to *Clostridium* species misclassified members of *Peptostreptococcus* and *Eubacterium,* as well as *Fusibacter paucivorans, Acidaminobacter hydrogenoformans* [95], and *Filifactor villosus* [30].
- *Cluster XII:* Four species are members of this cluster which in addition to the three species *C. acidiurici, C. hastiforme,* and *C. purinolyticum,* added by Collins et al. [30], contains *C. ultunense* [47]. This cluster also embraces the invalid species *"Clostridium filamentosum" strain* DSM 6645, reported to contain a ornithineine-containing peptidoglycan. Non-*Clostridium* species belong to the genera *Eubacterium* and *Tissierella.* More distantly related are members of *Peptostreptococcus, Helcococcus* [96], *Eubacterium, Acetobacterium,* as well as *Haloanaerobium praevalens* and *Sporohalobacter lortetii.*
- *Cluster XIV:* This is the second largest *Clostridium* species cluster, which however is significantly intermixed with representatives of other genera, such as *Acetitomaculum* [97], *Eubacterium, Roseburia, Ruminococcus, Johnsonella* and *Catonella* [98], *Syntrophococcus sucromutans* [99], *Pseudobutyrivibrio ruminis* [100], *Coprococcus eutactus* and the giant bacteria of the invalid genus *"Epulopiscium".* Because of its deep bifurcation this cluster was divided into XIVa and XIVb, but these two subclusters may actually deserve the rank of separate higher taxa in a future reclassification scheme. A few new species have been added to subcluster XIVa since its establishment by Collins et al. [30]. This is the pair *C. indolis* and *C. saccharolyticum* (98.7 % similarity) which are members of a morphologically coher-

ent group comprising *C. sphenoides, C. celerecrescens* and the misclassified species *Desulfotomaculum guttoideum* (97.1 %). *C. scindens* and *Syntrophococcus sucromutans* (not shown in Figure 4) stands phylogenetically isolated and has species of *Eubacterium* as remotely related neighbors. Recently, the psychrotolarant species *C. algidixylanolyticum*, a relative of *C. xylanolyticum* and *C. aerotolerans*, has been described (not indicated in Figure 4) [101].

- Clusters XVI, XVIII and XIX: *C. innocuum* clusters with *Eubacterium biforme* and *Streptococcus pleomorphus*; even more distantly related are members of *Erysipelothrix rhusiopathiae* and *Holdemania filiformis* (XVI). *C. cocleatum* is related to *C. spiroforme* and *C. ramosum* (XVIII). Cluster XIX comprises *C. rectum*, together with members of *Fusobacterium, Ilyobacter, Leptotrichia buccalis* and *Propionigenium modestum*. These three separate lineages, together with *Erysipelothrix* and misclassified *Lactobacillus* strains appear to branch at the root of a supercluster containing the orders *Bacillales, Lactobacillales* and *Mycoplasmatales*.

The phylogenetic position of clostridial species of industrial importance

Many clostridial species are well-known for their potential to be used in industrial application. Psychrophilic, alkaliphilic, and thermophilic strains are isolated and searched for proteases for use in washing detergents, while xylanases are screened for to be used in pulp and paper industry. These enzymes and their producers will be covered in a separate chapter. This paragraph is included to demonstrate the power and applicability of molecular methods for solving taxonomic problems in applied microbiology.

The genus *Clostridium* contains a significant number of species used in large-scale acetone-butanol fermentation (*C. acetobutylicum*) as well as in the production of isopropanol and butanol (*C. beijerinckii*), acetone and isopropoanol together with *n*-butanol (*C. aurantibutyricum*) and mainly *n*-butanol (*C. puniceum*). Table 5 is a list of invalid species names used in the fermentation industry since its beginning in the late 1920's. The vast majority of phylogenetic studies centered around the aceton-butanol producing organism, *C. acetobutylicum*, used for starch-based fermentation. With the ability to apply molecular techniques, such as 16S rDNA

Table 5. Names of patented solvent-producing strains of species with no standing in nomenclature, used in different commercial companies from 1929 on [18].

C. inverto-acetobutylicum	*C. amylo-saccharo butyl-propylicum*
C. celerifactor	*C. granulobacter acetobutylicum*
C. madisonii	*C. propyl-butylicum-alpha*
C. saccharo-acetobutylicum	*C. saccharo acetoperbutylicum*
C. saccharo-acetobutylicum-alpha	*C. saccharo-acetobutylicum-beta*
C. saccharo-acetobutylicum-gamma	*C. saccharobutyl-acetonicum*
C. saccharo-butyl-acetonicum-liquefaciens	*C. saccharobutylicum-gamma*
C. saccharobutyl-isopropyl-acetonicum	*C. saccharobutyl-isopropyl-acetonicum-beta*
C. saccharoperbutylacetonicum	*C. viscifaciens*
Bacillus (Clostridium) butacone	*Bacillus (Clostridium) terylium*
Bacillus (Clostridium) saccharobutylicum-beta	

sequence analysis [18-20,], DNA-DNA reassociation ([21, 22], genome size determination and DNA probing [20] the genomic diversity of this species was unraveled. Not unexpected for workers in the field of fermentation, a variety of different organisms carried the label *C. acetobutylicum*. Different genome sizes followed by probing of DNA fragments gave the first indication that this species was comprised of a heterogeneous collection of strains. Wilkinson and Young found three groups [20], one contained strains ATCC 824^{T} and DSM 1731, a second pair contained strains NCIMB 8052 and NI-4081, while strain NCP 262 constituted the third group. Subsequent DNA-DNA reassociation studies on almost 40 strains of *C. acetobutylicum* [22] revealed the presence of 4 groups sharing intergroup relatedness of mostly less than 30 %. While a substantial number indeed clustered with the type strain of the species ATCC 824^{T}, a second group was highly related to the type strain of *C. beijerinckii* ATCC 25752^{T}. The other 2 groups each contained few strains only, NRRL B643 and NCP 262 and "*Clostridium saccharoperbutylacetonicum*" (strains N1-4 and N1-4081). In contrast to the former two groups which were used in starch fermentation, the latter two strains were used in sugar (molasse)-based processes.

Keis et al. applyed a combination of 16S rDNA sequence analysis and DNA fingerprinting to more than 50 strains of solvent-producing clostridia and found 9 groups [19]. The main outcome of this study was the finding that the type strain of *C. acetobutylicum* NCIMB 8052^{T} was different from the type strain ATCC 824^{T}. This finding was also reported by Johnson and Chen [21]. As a consequence, Wilkinson et al. [102] suggested that the name *C. acetobutylicum* should be restricted to those strains which are genetically closely related to strain ATCC 824^{T} (DSM 792 and DSM 1731) while strain NCIMB 8052 is not the type strain of *C. acetobutylicum* but a strain of *C. beijerinckii*. Based upon 16S rDNA analysis and DNA-DNA reassociation the two strains of *C. felsineum*, DSM 794^{T} and DSM 10690, are closely related to the type strain of *C. acetobutylicum* [103]. Hence, the phylogenetic placement of *C. felsineum* in rDNA cluster XI [30] has not been based upon the type strain of this species. Also, the type strain of *C. roseum* is a close relative of *C. acetobutylicum*.

The phylogenetic position of clostridial species of medical importance

The genus *Clostridium* contains 35 species which are considered to have pathogenic potential. The degree of pathogenicity varies and many of the species, listed in risk groups 2 according to the German legislative are not considered pathogenic according to the US legislative. *Clostridium* cluster I, containing the type species of the genus *C. butyricum*, contains half of the pathogenic species, including those considered the major pathogenic agents, i. e., *C. barati, C. botulinum, C. haemolyticum, C. novyi, C. perfringens, C. tetani, C. tertium, C. septicum*, and *C. chauvoei*. It has previously been shown that the dendrograms of relatedness obtained with 16S rDNA [30, 47, 104] and *C. botulinum* neurotoxin genes [105] do only match for very closely related *C. butulinum* strains. While the former genes are most likely not included in lateral gene transfer, genes coding for neurotoxins are discussed to have spread on mobile genetic elements such as transposons, plasmids and bacteriophage genomes [106]. Thus, molecular detection of the host by probing 16S rDNA will not necessarily reveal that the target organism is pathogenic. On the other hand, molecular detection of the neurotoxin may not give evidence about the taxonomic status of the host organism. If the genus *Clostridium* was restricted to clusters I and II, the generic assignment for these pathogens would not be affected. Unless considered *nomen periculosa*, the names of other major and minor pathogens should be changed when the clusters to which they belong phylogenetically are reclassified. These include *C. difficile* and *C. sordellii* and some minor pathogens in cluster XI, and several pathogens of clusters XII, XIV, and XVIII.

C. botulinum constitutes a taxonomic problem as it constitutes a phylogenetically very heterogeneous species. If this species is restricted to the type strain ATCC 25763 (toxin types A, B, and F, proteolytic) [107], the phylogenetically unrelated strains of this species need to be reclassified. As these include strains responsible for botulism in humans (type E, proteolytic) and animals (types C and D) this scientifically logical step would certainly cause significant problems in communication and diagnosis in the field of medical microbiology, and should not be done without prior consultation with medical microbiologists.

For the purpose of diagnosis and identification of bacterial species, including members of *Clostridium*, the ribosomal RNA is of proven value. As part of the ribosome, rRNA occurs as naturally amplified molecules, and hence provides an ideal target for labeled oligonucleotide probes. Analysis of variable stretches of the 23S rDNA of *C. botulinum* strains that produce different types of neurotoxins has confirmed the genomic heterogeneity of this species, evident in the significant differences in metabolic properties [108], rRNA cistron similarity studies [52], DNA reassociation experiments [52], and 16S rDNA sequence analysis [103]. Probes developed against the variable 23S rDNA regions of *C. botulinum* were found to reliably identify

(i) proteolytic strains with toxins of types A, B, and F as well as related species,
(ii) non-proteolytic strains with toxins B and E,
(iii) strains with toxin D, and *C. argentinense* (formerly *C. botulinum*) with toxin of type G [108].

Using a similar approach, the phylogenetic coherence of the species *C. perfringens* has been confirmed [109]. For identification of pathogenic strains, 16S rDNA based determination of phylogenetic identity is not able to identify the toxin type or type group, as the genealogies of ribosomal RNA genes and toxin genes are probably not coupled. Certain toxin genes (C and D) of *C. botulinum,* are contained in the genome of temperate bacteriophages [105,110] and certain non-pathogenic *Clostridium* strains can be transformed to express a toxin gene [111]. A non-virulent strain that has lost its toxin gene(s) may still be identified as a non-pathogenic strain of *C. botulinum* by analysis of 16S rDNA, while the sole application of electroimmunodiffusion and enzyme-linked immunosorbent assays, able to identify the toxin type, will most likely mis-determine the phylogenetic identity of the organism. In the design of probes based on 16S rDNA sequence data for identification of *Clostridium* species or strains, two points need to be considered. Firstly, the quality of the sequence data currently available in the public databases needs careful assessment; in many cases the 100-150 nucleotides of the 5′ end is missing or of low quality, yet this region contains variable regions useful as probe target sites. The second point relates to the presence of multiple rRNA operons, which can differ in sequence composition within one strain. This has been demonstrated in *C. paradoxum* [61] and in members of the genus *Desulfotomaculum* (Spröer et al., unpublished data). The variability of operon sequence composition between strains of the same species has yet to be investigated. Differences in the nucleotide composition of genes and the intergeneric regions, however, can be taxonomically exploited by typing methods, such as ARDRA (amplified random DNA restriction analysis) or riboprinting. These techniques generate patterns which can not only be used to verify the genomic authenticity of a strain but also, provided the availability of an extensive database, for the identification and subsequent classification of isolates.

The classification of uncultured clostridia

Recent advances in molecular methodologies enable microbiologists to search for the presence of organisms that have not yet been cultivated. Applying a series of techniques which include isolation of DNA from an environmental sample, PCR amplification and either cloning of 16S rDNA or separation of PCR products by gradient gel electrophoreses, sequence analysis of cloned inserts or separated fragment, respectively, and subsequent search for nucleotide similarities against thousands of homologous sequences have changed our view of bacterial diversity. Hardly any sequence retrieved from the environment is highly similar to that of cultured strains and a tremendous number of novel sequences and potential new taxa have been described from such molecular ecological studies.

The literature containing information on the presence of novel *Clostridium* sequences is sparse. This is mainly due to the choice of the environmental sample which concentrates mainly on those habitats from which DNA is easily recovered, such as marine sites or soil. As these habitats are mainly aerobic the chance of retrieving DNA from anaerobic organisms is reduced. Also, as environmental clostridial organisms are mostly present as spores, conventional DNA isolation procedures may fail to retrieve DNA from these strains. A recent study in the human gastrointestinal tract [112] revealed the presence of *Clostridium* and *Clostridium*-related strains which were similar, though not identical, to strains derived from the human intestine, e. g., *C. celerecrescens, Ruminococcus obeum* and *R. productus, Eubacterium renale, E. plauti* and *E. formicigenerans, Coprocoocus eutactus,* and *Fusobacterium prausnitzii.*

Analysis of an undisturbed rainforest soil from Hawaii [113] demonstrated the presence of a large number of *Clostridium*-type sequences, which were distantly to closely related to different *Clostridium* species, such as to members of the *C. novyi* subgroup, *C. fallax, C. puniceum, C. butyricum,* and *C. beijerinckii.*

By far the richest diversity of as yet uncultured *Clostridium* species and related taxa have been found so far in anoxic sediments of several marine-types salinity lakes and in a mat sample of a moderately salinity lake which belongs to the Vestfold Hills lake system, Antarctica. Analyses of clone libraries generated from DNA of the sediment samples revealed that about 36 % of all clones belong to the *Clos-*

tridium-Bacillus subphylum, the majority of which clustered outside the rDNA clusters containing *Clostridium* species [114]. Analysis of the extensive mat 16S rDNA clone library from Lake Fryxell [115] revealed that about 43 % and 15 % of the sequences belong to members of the *Clostridium-Bacillus* subphylum and *Clostridium* species, respectively. Among the sequences determined, members of the genera *Clostridium, Acetonema, Eubacterium,* and *Sporomusa* dominated, while those of *Acetivibrio, Acetobacterium, Caloramator, Dendrosporobacter, Desulfosporosinus* and *Rumicoccus* occurred in lower numbers. Of the sequences related to *Clostridium* species the highest degree of diversity was found among sequences which could be affiliated to *C. esterthetium.* Sequences of this organism were also detected in the study of Bowman et al. [114]. All other sequences seem to indicate the presence of novel species which are likely to be related to *C. aminovalericum, C. fallax, C. limosum, C. subterminale, C. termitidis,* and *C. viride.* The only clostridial members cultivated so far from this environment are strains belonging phylogenetically to *C. estertheticum* [116].

Acknowledgement

We wish to thank Jolantha Swiderski for her assistance in generating phylogenetic dendrograms.

References

[1] Mayr, E., Principles of systematic zoology, McGraw-Hill, New York, 1969.

[2] Cato, E. P., George, W. L., Finegold, S. M., The genus *Clostridium,* in: *Bergey's Manual of Systematic Bacteriology,* Vol. 2, (Sneath, H. A., Mair, N. S., Sharpe, M. E., Holt, J. G., Eds.), Williams & Wilkins Co., Baltimore, **1986**.

[3] Strömpl, C., Tindall, B. J., Lünsdorf, H., Wong, T.-Y., Moore, E. R. B., Hippe, H., Reclassification of *Clostridium quercicolum* as *Dendrosporobacter quercicolus* gen. nov., sp. nov., *Int. J. Syst. Evol. Microbiol.* **2000**, 50, 101-106.

[4] Holdeman, L. V., Cato, E. P., Moore, W. E. C., *Anaerobe Laboratory Manual,* Anaerobe Laboratory, Virginia Polytechnic Institute and State University, Blacksburg, Va., **1977**.

[5] Cummins, C. S., Johnson, J. L., Taxonomy of the clostridia: wall composition and DNA homologies in *Clostridium butyricum* and other butyric acid-producing bacteria, *J. Gen. Microbiol.* **1971**, 67, 33-46.

[6] Schleifer, K. H., Kandler, O., Peptidoglycan types of bacterial cell walls and their taxonomic implications, *Bacteriol. Rev.* **1972**, 36, 407-477.

[7] Weiss, N., Schleifer, K. H., Kandler, O., The peptidoglycan types of gram-positive anaerobic bacteria and their taxonomic implications, *Rev. Inst. Pasteur* (Lyon) **1981**, 14, 3-12.

[8] Li, Y., Mandelco, L., Wiegel, J., Isolation and characterization of a moderately thermophilic anaerobic alkaliphile, *Clostridium paradoxum* sp. nov., *Int. J. Syst. Bacteriol.* **1993**, 43, 450-460.

[9] Li, Y., Engle, M., Weiss, N., Mandelco, L., Wiegel, J., *Clostridium thermoalcaliphilum* sp. nov., an anaerobic and thermotolerant facultative alkaliphile, *Int. J. Syst. Bacteriol.* **1994**, 44, 111-118.

[10] Fendrich, C., Hippe, H., Gottschalk, G., *Clostridium halophilum* sp. nov. and *Clostridium litorale* sp. nov., an obligate halophilic and a marine species degrading betaine in the Stickland reaction, *Arch. Microbiol.* **1990**, 154, 127-132.

[11] Zhang, X., Mandelco, L., Wiegel, J., *Clostridium hydrobenzoicum* sp. nov., an amino acid-utilizing, hydroxybenzoate-decarboxylating bacterium isolated from methanogenic freshwater pond sediment, *Int. J. Syst. Bacteriol.* **1994**, 44, 214-222.

[12] Wilde, E., Collins, M. D., Hippe, H., *Clostridium pascui*, a new glutamate-fermenting sporeformer from a pasture in Pakistan, *Int. J. Syst. Bacteriol.* **1997**, 47, 164-170.

[13] Gottwald, M., Andreesen, J. R., LeGall, J., Ljungdahl, L. G., Presence of cytochrome and menaquinone in *Clostridium formicoaceticum* and *Clostridium thermoaceticum, J. Bacteriol.* **1975**, 122, 325-328.

[14] Yamamoto, K., Murakami, R., Takamura, Y., Isoprenoid quinone, cellular fatty acid composition and diaminopimelic acid isomers of newly classified thermophilic anaerobic Gram-positive bacteria, *FEMS Microbiol. Lett.* **1998**, 161, 351-358.

[15] Das, A., Hugenholtz, J., van Halbeek, H., Ljungdahl, L. G., Structure and function of a menaquinone involved in electron transport in membranes of *Clostridium thermoautotrophicum* and *Clostridium thermoaceticum, J. Bacteriol.* **1989**, 171, 5823-5829.

[16] O'Leary, W. M., Wilkinson, S. G., Part 2: Distribution of lipids. 5. Gram-positive bacteria, in: *Microbial Lipids*, Vol. 1 (Radledge, C., Wilkinson, S. G., Eds.), Academic Press, London, **1988**.

[17] Johnston, N. C., Goldfine, H., Isolation and characterization of a novel four-chain ether lipid from *Clostridium butyricum*: the phosphatidylglycerol acetal of plasmenylethanolamine, *Biochim. Biophys. Acta* **1988**, 961, 1-12.

[18] Jones, D. T., Keis, S., Origins and relationships of industria solvent-producing clostridial strains, *FEMS Microbiol.Rev.* **1995**, 17, 223-232.

[19] Keis, S., Bennett, C. F., Ward, V. K., Jones, D. T., Taxonomy and phylogeny of industrial solvent-producing clostridia, *Int. J. Syst. Bacteriol.* **1995**, 45, 693-705.

[20] Wilkinson, S. R., Young, M., Wide diversity of genome size among different strains of *Clostridium acetobutylicum, J. Gen. Microbiol.* **1993**, 139, 1069-1076.

[21] Johnson, J. L., Chen, J.-S., Taxonomic relationships among strains of *Clostridium acetobutylicum* and other phenotypically similar organisms, *FEMS Microbiol. Rev.* **1995**, 17, 233-240.

[22] Johnson, J. L., Toth, J., Santiwatanakul, S., Chen, J.-S., Cultures of "*Clostridium acetobutylicum*" from various collections comprise *Clostridium acetobutylicum, Clostridium beijerinkii,* and two other distinct types based on DNA-DNA reassociation, *Int. J. Syst. Bacteriol.* **1997**, 47, 420-424.

[23] Möse, J. R., Fischer, G., Briefs, C., Die Wirkung von *Clostridium butyricum* (Stamm M55) auf menschliches Kininogen und ihre Bedeutung für den Onkolyseprozess, *Zentralbl. Bakteriol. Parasitenkd.* Abt.I, Orig. A **1972**, 221, 474-491.

[24] Fischer, G., Brantner, H., Platzer, P., The kininase activity of Ehrlichs' ascites solid tumor after treatment with oncolytic clostridia, *Z. Krebsforsch.* **1975**, 84, 203-206.

[25] Brantner, H., Schwager, J., Enzymatische Mechanismen der Onkolyse durch *Clostridium oncolyticum* M55 ATCC 13732, Enzymatische Mechanismen der Onkolyse durch *Clostridium oncolyticum* M55 ATCC 13732 *Zentralbl. Bacteriol. Hyg.* Abt.I, Orig. A **1979**, 243, 113-118.

[26] DeSoete, G., A least squares algorithm for fitting additive trees to proximity data, *Psychometrika* **1983**, 48, 621-626.

[27] Fitch, W. M., Margoliash, Construction of phylogenetic trees: a method based on mutation distances as estimated from cytochrome c sequences is of general applicability, *Science* **1967**, 155, 279-284.

[28] Saitou, N., and Nei, M., The neighbor-joining method: a new method for reconstructing phylogenetic trees, *Mol. Biol. Evol.* **1987**, 4, 406-425.

[29] Felsenstein, J., *Phylip* (Phylogeny Inference Package), version 3.5c., distributed by the author. Department of Genetics, University of Washington, Seattle, **1993**.

[30] Collins, M. D., Lawson, P. A., Willems, A., Cordoba, J. J., Fernandez-Garayzabal, J., Garcia, P., Cai, J., Hippe, H., Farrow, J. A. E., Phylogenetic analysis of some *Aerococcus*-like organisms from clinical sources: description of *Helcococcus kunzii* gen. nov., sp. nov., *Int. J. Syst. Bacteriol.* **1994**, 812-826

[31] Olsen, G. J., Woese, C. R., Overbeek, R., The wind of (evolutionary) change: breathing new life into microbiology *J. Bacteriol.* **1994**, 176, 1-6.

[32] Van de Peer, Y., Neefs, J.-M., de Rijk, P., De Vos, P., de Wachter, R., About the origin of divergence of the major bacterial taxa during evolution, *Syst. Appl. Microbiol.* **1994**, 17, 32-38.

[33] Schleifer, K. H., Ludwig, W., Molecular taxonomy: classification and identification, in: *Bacterial diversity and systematics* (Priest, F. G., and Goodfellow, M., Eds), Plenum Press, New York, **1994**, pp 1-15.

[34] Ohtani, N., Haruki, M., Morikawa, M., Crouch, R. J., Itaya, M., Kanaya, S., Identification of the genes encoding Mn^{2+}-dependent RNase HII and Mg^{2+}-dependent RNase HIII from *Bacillus subtilis*: classification of RNases H into three families, *Biochemistry* **1999**, 38, 605-618.

[35] Huang, Y.-P., Ito, J., The hyperthermophilic bacterium *Thermotoga maritima* has two different classes of family C DNA polymerases: evolutionary implications, *Nucl. Acids Res.* **1998**, 26, 5300-5309.

[36] Gupta, R. S., Bustard, K., Falah, M., Singh, D. J., Sequencing of heat shock protein 70 (DnaK) homologs from *Deinococcus proteolyticus* and *Thermomicrobium roseum* and their integration into a protein-based phylogeny of prokaryotes, *J. Bacteriol.* **1997**, 179, 345-357.

[37] Dale, C. J. H., Moses, E. K., Ong, C.-C., Morrow, C. J., Reed, M. B., Hasse, D., Strugnell, R. A., Identification and sequencing of the *gro*E operon and flanking genes of *Lawsonia intracellularis*: use in phylogeny, *Microbiology* **1998**, 144, 2073-2084.

[38] Viale, A. M., Arakaki, A. K., Soncini, F. C., Ferreyra, R. G., Evolutionary relationships among eubacterial groups as inferred from GroEL (Chaperonin) sequence comparisons, *Int. J. Syst. Bacteriol.* **1994**, 44, 527-533.

[39] Gruber, T. M., Bryant, D. A., Molecular systematic studies of Eubacteria, using σ^{70}-type sigma factors of group 1 and group 2., *J. Bacteriol.* **1997**, 179, 1734-1747.

[40] Achouak, W., Normand, P., Heulin, T., Comparative phylogeny of *rrs* and *nif*H genes in the Bacillaceae, *Int. J. Syst. Bacteriol.* **1999**, 49, 961-967.

[41] Woese, C. R., Bacterial evolution, *Microbiol. Rev.* **1987**, 51, 221-271.

[42] Fox, G. E., Stackebrandt, E., Hespell, R. B., Gibson, J., Maniloff, J., Dyer, T., Wolfe, R. S., Balch, W., Tanner, R., Magrum, L., Zablen, L. B., Blakemore, R., Gupta, R., Luehrsen, K. R., Bonen, L., Lewis, B. J., Chen, K. N., Woese, C. R., The phylogeny of prokaryotes, *Science* **1980**, 209, 457-463.

[43] Stackebrandt, E., Pohla, H., Kroppenstedt, R. M., Hippe, H., Woese, C. R., 16S rRNA analysis of *Sporomusa, Selenomonas,* and *Megasphaera*: on the phylogenetic origin of Gram-positive Eubacteria, *Arch. Microbiol.* **1985**, 143, 270-276.

[44] Woese, C. R., Debrunner-Vossbrinck, A., Oyaizu, H., Stackebrandt, E., Ludwig, W., Gram-positive bacteria: possible photosynthetic ancestry, *Science* **1985**, 229, 762-765.

[45] Farrow, J. A. E., Lawson, P. A., Hippe, H., Gauglitz, U., Collins, M. D., Phylogenetic evidence that the Gram-negative nonsporulating bacterium *Tisseriella* (*Bacteroides*) *praecuta* is a member of the *Clostridium* subphylum of the Gram-positive bacteria and description of *Tisseriella creatinini* sp. nov., *Int. J. Syst. Bacteriol.* **1995**, 45, 436-440.

[46] Willems, A., Collins, M. D., Phylogenetic placement of *Dialister pneumosintes* (formerly *Bacteroides pseumosintes*) within the *Sporomusa* subbranch of the *Clostridium* subphylum of the Gram-positive bacteria, *Int. J. Syst. Bacteriol.* **1995**, 45, 855-857.

[47] Stackebrandt, E., Kramer, I., Swiderski, J., Hippe, H., Phylogenetic basis for a taxonomic dissection of the genus *Clostridium, FEMS Immunol. Med. Microbiol.* **1999**, 24, 253-258.

[48] Rainey, F. A., Ward, N. L., Morgan, H. W., Toalster, R., Stackebrandt, E., Phylogenetic analysis of anaerobic thermophilic bacteria: aid for their reclassification, *J. Bacteriol.* **1993**, 175, 4772-4779.

[49] Wiegel, J., Anaerobic alkalithermophiles, a novel group of extremophiles, *Extremophiles* **1998**, 2, 257-267

[50] Cato, E. P., Stackebrandt, E., Taxonomy and Phylogeny, in: *Clostridia* (Minton, N. P., Clarke, D. J., Eds.), Plenum Press, New York, **1989**, pp. 1-26.

[51] Stackebrandt, E., Rainey, F. A., Phylogenetic relationships, in: *The Clostridia: molecular biology and pathogenesis* (Rood, J. I., McClane, B. A., Songer, J. G., Titball, R. W., Eds.), Academic Press New York, **1997**, pp. 3-13.

[52] Johnson, J. L., Francis, B. S., Taxonomy of the clostridia: ribosomal ribonucleic acid homologies among the species, *J. Gen. Microbiol.* **1975**, 88, 229-244.

[53] Tanner, R. S., Stackebrandt, E., Fox, G. E., Woese, C. R., A phylogenetic analysis of *Acetobacterium woodii, Clostridium barkeri, Clostridium butyricum, Clostridium lituseburense, Eubacterium limosum* and *Eubacterium tenue, Curr. Microbiol.* **1981**, 5, 35-38.

[54] Tanner, R. S., Stackebrandt, E., Fox, G. E., Woese, C. R., A phylogenetic analysis of anaerobic eubacteria capable of synthesizing acetate from CO_2, *Curr. Microbiol.* **1982**, 7, 127-132.

[55] Suen, J. C., Hatheway, C. L., Steigerwaldt, A. G., Brenner, D. J., *Clostridium argentinense* sp. nov.: a genetically homogeneous group composed of all strains of *Clostridium botulinum* toxin type G and some non-toxic strains previously identified as *Clostridium subterminale* and *Clostridium hastiforme, Int. J. Syst. Bacteriol.* **1988**, 38, 375-381.

[56] Zhao, H., Yang, D., Woese, C. R., Bryant, M. P., Assignment of *Clostridium bryantii* to *Syntrophospora bryantii* gen. nov., comb. nov. on the basis of a 16S rRNA sequences analysis of its crotonate-grown pure culture, *Int. J. Syst. Bacteriol.* **1990**, 40, 40-44.

[57] Nakamura, S., Okado, I., Abe, T., Nishida, S., Taxonomy of *Clostridium tetani* and related species, *J. Gen. Microbiol.* **1979**, 113, 29-35.

[58] Oren, A., Pohla, H., Stackebrandt, E., Transfer of *Clostridium lortetii* to a new genus *Sporohalobacter* gen. nov. as *Sporohalobacter lortetti* comb. nov., and description of *Sporohalobacter marismortui* sp. nov, *Syst. Appl. Microbiol.* **1987**, 9, 239-246.

[59] Cato, E. P., Holdeman, L. V., Moore, W. E. C., *Clostridium perenne* and *Clostridium paraperfringens*: later subjective synonyms of *Clostridium barati, Int. J. Syst. Bacteriol.* **1982**, 32, 77-81.

[60] Judicial Commission of the International Committee on Systematic Bacteriology, *Int. J. Syst. Bacteriol.* **1999**, 49, 339.

[61] Lee, Y.-E., Jain, M. K., Lee, C., Lowe, S. E., Zeikus, J. G., Taxonomic distinction of saccharolytic thermophilic anaerobes: description of *Thermoanaerobacterium xylanolyticum* gen. nov., sp. nov, and *Thermoanaerobacterium saccharolyticum* gen. nov., sp. nov; reclassification of *Thermoanaerobium brockii, Clostridium thermosulfurogenes,* and *Clostridium thermohydrosulfuricum* E100-69 as *Thermoanaerobacter brockii* com. nov., *Thermoanaerobacterium thermosulfurigenes* comb. nov., and *Thermoanaerobacter thermohydrosulfuricus* comb. nov., respectively, *Int. J. Syst. Bacteriol.* **1993**, 43, 41-51.

[62] Rainey, F. A., Janssen, P. H., Phylogenetic analysis by 16S ribosomal DNA sequence comparison reveals two unrelated groups of species within the genus *Ruminococcus, FEMS Microbiol. Lett.* **1995**, 129, 69-74.

[63] Willems, A., Collins, M. D., Phylogenetic placement of *Dialister pneumosintes* (formerly *Bacteroides pseumosintes*) within the *Sporomusa* subbranch of the *Clostridium* subphylum of the Gram-positive bacteria, *Int. J. Syst. Bacteriol.* **1995**, 45, 572-575.

[64] Kageyama, A., Benno, Y., Nakase, T., Phylogenetic and phenotypic evidence for the transfer of *Eubacterium fossor* to the genus *Atopobium* as *Atopobium fossor* comb. nov, *Microbiol. Immunol.* **1999**, 43, 389-395.

[65] Wade, W. G., Downes, J., Dymock, D., Hiom, S. J., Weightman, A. J., Dewhirst, F. E., Paster, B. J., Tzellas, N., Coleman, B., The family *Coriobacteriaceae*: reclassification of *Eubacterium exiguum* (Poco et al. 1996) and *Peptostreptococcus heliotrinireducens* (Lanigan 1976) as *Slackia exigua* gen. nov., comb. nov., and *Eubacterium lentum* (Pevot 1938) as *Eggerthella lenta* gen. nov. comb. nov., *Int. J. Syst. Bacteriol.* **1999**, 49, 595-600.

[66] Kuhner, C. H., Matthies, C., Acker, G., Schmittroth, M., Gößner, A. S., Drake, H. L., *Clostridium akagii* sp. nov. and *Clostridium acidisoli* sp. nov.: acid-tolerant, N_2-fixing clostridia isolated from acidic forest soil, *Int. J. Syst. Evol. Microbiol.* **2000**, 50, 873-881.

[67] Siunov, A. V., Nikitin, D. V., Suzina, N. E., Dmitriev, V. V., Kuzmin, N. P., Duda, V. I., Phylogenentic status of *Anaerobacter polyendosporus*, an anaerobic, polysporogenic bacterium, *Int. J. Syst. Bacteriol.* **1999**, 49, 1119-1124.

[68] Willems, A., Collins, M. D., Phylogenetic placement *of Sarcina ventriculi* and *Sarcina maxima* within Group I *Clostridium*, a possible problem for future revision of the genus *Clostridium Int. J. Syst. Bacteriol.* **1994**, 44, 591-593.

[69] Goodsir, J., History of a case in which a fluid periodically ejected from the stomach contained vegetable organisms of an undescribed form. With chemical analysis of the fluid, by George Wilson *Edinb. Med. Surg. J.* **1842**, 57, 430-443.

[70] Prazmowski, A., *Untersuchungen über die Entwicklungsgeschichte und Ferment-*

wirkung einiger Bacterien-Arten, Inaugural Dissertation. Hugo Voigt, Leipzig, Germany, **1880**.

[71] LaPage, S. P., Sneath, P. H. A., Lessel, E. F., Skerman, V. B. D., Seeliger, H. P. R., Clark, W. A., *International Code of Nomenclature of Bacteria*, American Society for Microbiology, Washington, D. C., **1992**.

[72] Engle, M., Li, Y., Rainey, F. A., DeBlois, S., Mai, V., Reichert, A., Mayer, F., Messner, P., Wiegel, J., Isolation and characterization of a novel alkalitolerant thermophile, *Anaerobranca horikoshii* gen. nov., sp. nov., *Int. J. Syst. Bacteriol.* **1996**, 46, 1025-1033.

[73] Rainey, F. A., Stackebrandt, E., 16S rDNA analysis reveals phylogenetic diversity of polysaccharolytic clostridia, *FEMS Microbiol. Lett.* **1993**, 113, 125-128.

[74] Stackebrandt, E., Spröer, C., Rainey, F. A., Burghardt, J., Päuker, O., Hippe, H., Phylogenetic analysis of the genus *Desulfotomaculum*: Evidence for the misclassification *of Desulfotomaculum guttoideum* and description of *Desulfotomaculum orientis* as *Desulfosporosinus orientis* gen. nov. comb. nov., *Int. J. Syst. Bacteriol.* **1997**, 47, 1134-1139.

[75] Rainey, F. A., Ward-Rainey, N., Janssen, P. H., Hippe, H., Stackebrandt, E., *Clostridium paradoxum* DSM 7308^T contains multiple 16S rRNA genes with heterogeneous intervening sequences, *Microbiology* **1996**, 142, 2087-2095.

[76] Zellner, G., Stackebrandt, E., Nagel, D., Messner, P., Weiss, N., Winter, J., *Anaerofilum pentosovorans* gen. nov., spec. nov., and *Anaerofilum agile* sp. nov., two new, strictly anaerobic, mesophilic, acidogenic bacteria from anaerobic bioreactors, *Int. J. Syst. Bacteriol.* **1996**, 46, 871-875.

[77] Grech-Mora, Fardeau, M.-L., Patel, B. K. C., Ollivier, B., Rimbault, A., Prensier, G., Garcia, J.-L., Garnier-Sillam, E., Isolation and characterization of *Sporobacter termitidis* gen. nov., sp. nov., from the digestive tract of the wood-feeding termite *Nasutitermes luja*, *Int. J. Syst. Bacteriol.* **1996**, 46, 512-518.

[78] Svetlitshnyi, V., Rainey, F. A., Wiegel, J., *Thermosyntropha lipolytica* gen. nov., sp. nov., a lipolytic, anaerobic, organoheterotrophic, alkalitolerant thermophile utilizing short- and long chain fatty acids in syntrophic coculture with a methonogen, *Int. J. Syst. Bacteriol.* **1996**, 46, 1131-1137.

[79] Engle, M., Li, Y., Woese, C. R., Wiegel, J., Isolation and characterization of a novel alkalitolerant thermophile, *Anaerobranca horikoshii* gen. nov., sp. nov., *Int. J. Syst. Bacteriol.* **1995**, 45, 454-461.

[80] Rainey, F. A., Donnison, A. M., Janssen, P. H., Saul, D., Rodrigo, A., Bergquist, P. L., Daniel, R. M., Stackebrandt, E., Morgan, H. W., Description of *Caldicellulosiruptor saccharolyticus* gen. nov., sp. nov.: an obligate anaerobic, extremely thermophilic, cellulolytic bacterium, *FEMS Microbiol. Lett.* **1994**, 120, 263-266.

[81] Rainey, F. A., Stackebrandt, E., *Thermoanaerobacter acetoethylicus* comb. nov. (*Thermobacteroides acetoethylicus* Approved List No 9, 1982) and designation of *Coprothermobacter proteolyticus* gen. nov., *Int. J. Syst. Bacteriol.* **1993**, 43, 857-859.

[82] Saiki, T., Kobayashi, Y., Kawagoe, K., Beppu, T., *Dictyoglomus thermophilum* gen. nov., sp. nov., a chemoorganotrophic, anaerobic, thermophilic bacterium, *Int. J. Syst. Bacteriol.* **1985**, 35, 253-259

[83] Magot, M., Ravot, G., Campaignolle, X., Ollivier, B., Patel, B. K. C., Fardeau, M.-L., Thomas, P., Crolet, J.-L., Garcia, J.-L., *Dethiosulfovibrio peptidovorans* gen. nov., sp. nov., a new anaerobic slightly halophilic, thiosulfate-reducing bacterium from corroding offshore oil wells, *Int. J. Syst. Bacteriol.* **1997**, 47, 818-824.

[84] Strömpl, C., Tindall, B. J., Jarvis, G. N., Lünsdorf, H., Moore, E. R. B., Hippe, H., A re-evaluation of the taxonomy of the genus *Anaerovibrio*, with the reclassification of *Anaerovibrio glycerini* as *Anaerosinus glycerini* gen. nov., comb. nov., and *Anaerovibrio burkinabensis* as *Anaeroarcus burkinensis* gen. nov. comb. nov., *Int. J. Syst. Bacteriol.* **1999**, 49, 1861-1872.

[85] Lai, C.-H., Males, B. M., Dougherty, P. A., Berthold, P., Listgarten, M. A., *Centipeda periodontii* gen. nov., sp. nov. from human periodontal lesions, *Int. J. Syst. Bacteriol.* **1983**, 33, 628-635

[86] Baena, S., Fardeau, M.-L., Woo, T. H. S., Olliver, B., Labat, M., Patel, B. K. C., Phylogenetic relationship of three amino-acid-utilizing anaerobes, *Selenomonas acidaminovorans*, *"Selenomonas acidaminophila"* and *Eubacterium acidaminophilum*, as inferred from partial 16S rDNA nucleotide sequences, and proposal of *Thermanaerovibrio*

acidaminovorans gen. nov., comb. nov. and *Anaeromusa acidaminophila* gen. nov., comb. nov., *Int. J. Syst. Bacteriol.* **1999**, 49, 969-974.

[87] Janssen, P. H., O'Farrel, K. A., *Succinispira mobilis* gen. nov., sp. nov., a succinate-decarboxylating anaerobic bacterium, *Int. J. Syst. Bacteriol.* **1999**, 49, 1009-1013.

[88] Krumholz, L. R., Bryant, M. P., Brulla, W.J., Vicini, J. L., Clark, J. H., Stahl, D. A., Proposal of *Quinella ovalis* gen. nov., sp. nov., based on phylogenetic analysis, *Int. J. Syst. Bacteriol.* **1993**, 43, 293-296.

[89] Willems, A., Collins, M. D., Phylogenetic placement of *Dialister pneumosintes* (formerly *Bacteroides pseumosintes*) within the *Sporomusa* subbranch of the *Clostridium* subphylum of the Gram-positive bacteria, *Int. J. Syst. Bacteriol.* **1995**, 45, 403-405.

[90] Kane, M. D., Breznak, J. A., *Acetonema longum* gen. nov., sp. nov., an H_2/CO_2 acetogenic bacterium from the termite, *Pterotermes occidentis, Arch. Microbiol.* **1991**, 156, 91-98.

[91] Van Gylswyk, N. O., *Succiniclasticum ruminis* gen. nov., sp. nov., a ruminal bacterium converting succinate to propionate as the sole energy-yielding mechanism, *Int. J. Syst. Bacteriol.* **1995**, 45, 297-300.

[92] Van Gylswyk, N. O., Hippe, H., Rainey, F. A., *Schwartzia succinivorans* gen. nov., sp. nov., another ruminal bacterium utilizing succinate as the sole energy source, *Int. J. Syst. Bacteriol.* **1997**, 47, 155-159.

[93] Utkin, I., Woese, C. R., Wiegel, J., Isolation and characterization of *Desulfitobacterium dehalogenans* gen. nov., sp. nov., an anaerobic bacterium which reductively dechlorinates chlorophenolic compounds, *Int. J. Syst. Bacteriol.* **1994**, 44, 612-619.

[94] Friedrich, M., Springer, N., Ludwig, W., Schink, B., Phylogenetic positions of *Desulfofustis glycolicus* gen. nov., sp. nov., and *Syntrophobotulus glycolicus* gen. nov., sp. nov., two new strict anaerobes growing with glycolic acid, *Int. J. Syst. Bacteriol.* **1996**, 46, 1065-1069.

[95] Stams, A. J. M., Hansen, T. A., Fermentation of glutamate and other compounds by *Acidaminobacter hydrogenoformans* gen. nov. sp. nov., an obligate anaerobe isolated from black mud. Studies with pure cultures and mixed cultures with sulfate-reducing and methanogenic bacteria, *Arch. Microbiol.* **1984**, 137, 329-337.

[96] Collins, M. D., Facklam, R. R., Rodrigues, U. M., Ruoff, L K., Phylogenetic analysis of some *Aerococcus*-like organisms from clinical sources: description of *Helcococcus kunzii* gen. nov., sp. nov., *Int. J. Syst. Bacteriol.* **1993**, 43, 425-429.

[97] Greening, R. C. J., Leedle, A. Z., Enrichment and isolation of *Acetitomaculum ruminis* gen. nov.: acetogenic bacteria from the bovine rumen, *Arch. Microbiol.* **1989**, 151, 399-406.

[98] Moore, L. V. H., Moore, W. E. C., *Oribaculum catoniae* gen. nov., sp. nov.; *Catonella morbi* gen. nov., sp. nov.; *Hallella seregens* gen. nov., sp. nov.; *Johnsonella ignava* gen. nov., sp. nov.; and *Dialister pneumosintes* gen. nov., sp. nov., nom rev., anaerobic Gram-negative bacilli from the human gingival crevice, *Int. J. Syst. Bacteriol.* **1994**, 44, 187-192.

[99] Krumholz, L. R., Bryant, M. P., *Syntrophococcus sucromutans* sp. nov., gen nov. uses carbohydrates as electron donors and formate, methoxymonobenzoids or *Methanobrevibacter* as electron acceptor systems, *Arch. Microbiol.* **1986**, 143, 313-318.

[100] Van Gylswyk, N. O., Hippe, H., Rainey, F. A., *Pseudobutyrivibrio ruminis* gen. nov., sp. nov, a butyrate-producing bacterium from the rumen that closely resembles *Butyrivibrio fibrisolvens* in phenotype, *Int. J. Syst. Bacteriol.* **1996**, 46, 559-563.

[101] Broda, D. M., Saul, D. J., Bell, R. G., Musgrave, D. R., *Clostridium algidixylanolyticum* sp. nov., a new aerobic thermophilic xylan-degrading bacterium isolated from farm soil, *Int. J. Syst. Evol. Microbiol.* **2000**, 50, 623-631.

[102] Wilkinson, S. R., Young, M., Goodacre, R., Morris, J. G., Farrow, J. A. E., Collins, M. D., Phenotypic and genotypic differences between certain strains of *Clostridium acetobutylicum, FEMS Microbiol. Lett.* **1995**, 125, 199-204.

[103] Tamburini, E., Daly, S., Steiner, U., Vandini, C., Mastromei, G., *Clostridium felsineum* and *Clostridium acetobutylicum* are two distinct species that are phylogenetically closely related *Int. J. Syst. Evol. Microbiol.* **2001**, 51, 963-966.

[104] Hutson, R. A., Thompson, D. E., Collins, M. D., Genetic interrelationships of saccharolytic *Clostridium botulinum* types B, E and F and related clostridia as revealed by small subunit rRNA gene sequences, *FEMS Microbiol. Lett.* **1993**, 108, 103-110.

[105] Ecklund, M. W., Poysky, F. T., Meyers, J. A., Pelroy, G. A., Interspecies conversion of *Clostridium botulinum* type C to *Clostridium novyi* type A by bacteriophage, *Science* **1974**, 186, 456-458.

[106] Hutson, R. A., Thompson, D. E., Lawson, P. A., Schocken-Itturino, R. P., Böttger, E. C., Collins, M. D., Ant. van Leeuwenhoek, Genetic interrelationships of proteolytic *Clostridium botulinum* types A, B, and F and and other members of the *Clostridium botulinum* complex as revealed by small subunit rRNA gene sequences, *Antonie van Leeuwenhoek J. Microbiol.* **1993**, 64, 273-283.

[107] McClane, B. C., ASM First International Conference on the Molecular Genetics and Pathogenesis of the Clostridia, *ASM News* **1995**, 61, 465-468.

[108] Rönner, S. G. E., Stackebrandt, E., Further evidence for the genetic heterogeneity of *Clostridium botulinum* as determined by 23S rDNA oligonucleotide probing, *Syst. Appl. Microbiol.* **1994**, 17, 180-188.

[109] Rönner, S. G. E., Stackebrandt, E., Identification of *Clostridium perfringens* by 16S and 23S rDNA oligonucleotide probes, *Syst. Appl. Microbiol.* **1994**, 17, 425-4323.

[110] Ecklund, M. W., Poysky, F. T., Interconversion of type C and D strains of *Clostridium botulinum* by specific bacteriophages, *Appl. Microbiol.* **1974**, 27, 251-258.

[111] Oguma, K., Iida, H., Shiozaki, M., Inoue, K., Antigenicity of converting phages obtained from *Clostridium botulinum* types C and D strains, *Infect. Immun.* **1976**, 13, 855-860.

[112] Zoetendal, E. G., Akkermans, D. L., de Vos, W. M., Temperature gradient gel electrophoresis analysis of 16S rDNA from human fecal samples reveals stable and host-specific communities of active bacteria, *Appl. Environ. Microbiol.* **1998**, 64, 3854-3859.

[113] Nuesslein, K., Tiedje, J. M., Characterization of the dominant and rare members of a young Hawaiian soil bacterial community with small-subunit ribosomal DNA amplified from DNA fractionated on the basis of its guanine and cytosine composition *Appl. Environ. Microbiol.* **1988**, 64, 1283-1289.

[114] Bowman, J. P., Rea, S. M., McCammon, S. A., McMeekin, T. A., Diversity and community structure within anoxic sediment from marine salinity meromictic lakes and a coastal meromictic marine basin, Vestfold Hills, Eastern Antarctica, *Environ. Microbiol.* **2000**, 2, 227-237.

[115] Tindall, B. J., Brambilla, E., Steffen, M., Neumann, R., Pukall, R., Kroppenstedt, R. M., Stackebrandt, E., Cultiveable microbial diversity: gnawing at the Gordian knot, *Environ. Microbiol.* **2000**, 2, 310-318.

[116] Brambilla, E., Hippe, H., Hagelstein, A., Tindall, B. J., Stackebrandt, E., 16S rDBA diversity of cultured and uncltured prokaryotes of a mat sample from Lake Fryxell, McMurdo Dry Valleys, Antarctica, *Extremophiles,* **2001**, 5, 23-33.

3
Biology and Physiology

Wilfrid J. Mitchell

3.1
Introduction

The clostridia are a heterogeneous group of bacteria which share a small number of common features: they are anaerobic, Gram-positive, endospore-forming rods without the capacity for dissimilatory sulfate reduction. As a result, the genus is one of the largest in the prokaryote kingdom and a wide range of metabolic and physiological diversity is encountered. The species which have attracted most attention are those which are perceived as having biotechnological potential, particularly in the area of conversion of renewable biomass to commodity chemicals, and those which produce potent toxins which are causative agents of disease.

The clostridia are unremarkable bacteria in some respects. Many species grow as typical fermentative anaerobes, deriving energy from the incomplete oxidation of organic molecules; however, the genus also includes species which can grow autotrophically. Some metabolically specialized types exhibit an extremely narrow substrate range, and possess enzymes with unusual characteristics. It has been noted that the study of clostridia has been responsible for the development of understanding of several important aspects of metabolic biochemistry [1]. Genetic organization is typical of bacteria, with genes of related function clustered and expressed in a coordinated fashion. Furthermore, although information in many cases is incomplete, it appears that clostridia employ similar molecular devices to other bacteria (repressors, activators, two-component systems) to control gene expression and to adapt to changing environments.

The purpose of this chapter is to present a summary of the main features of clostridial physiology and metabolism, appreciation of which is essential for exploitation of their biological diversity. The survey includes reference to *Thermoanaerobacter* and *Thermoanaerobacterium* strains which were formerly considered to be clostridia [2].

3.2
Cell structure

Cells of most clostridial strains occur as straight or curved rods with rounded, blunt or tapering ends. Cells typically measure 0.5 to 2 μm in diameter and can be up to 30 μm in length, but cell size may vary with age and the composition of the culture medium [3]. The majority of strains are motile with peritrichous flagella. In conditions which are unfavorable for growth, vegetative cells differentiate to form metabolically dormant endospores, which are resistant to a variety of stresses and harsh conditions (see section 3.6). Endospores may be oval or spherical and have a central, subterminal or terminal location within the mother cell, depending on the species, and they usually distend the cell [3].

The clostridial cell wall contains peptidoglycan; in the majority of cases this is of the directly-linked meso-diaminopimelic acid type, although there are a number of exceptions [4, 5]. Teichoic acids have been found in some strains, as have other phosphate-containing polymers [6]. In many species Gram-positive cells are only seen in young cultures, while some show Gram-negative behavior at all times [3]. Nevertheless, electron micrographs indicate that the cells have a single-layered cell wall as found typically in Gram-positive bacteria. Atypical staining behavior has been suggested to be due to possession of a thin peptidoglycan layer [7-10].

In common with many other bacteria, the outermost layer of the cell envelope of a significant number of clostridia has been shown to be a 2-D crystalline array of regularly ordered protein or glycoprotein subunits [11, 12]. This surface layer, or S-layer, is usually composed of a single, strongly acidic and nonpolar polypeptide species, although in the case of several strains of *C. difficile* two protein subunits have been identified [13, 14]. S-layers show either hexagonal, square or oblique symmetry. Although varying lattice structures have been observed even within a single species, S-layer type was found to represent a characteristic taxonomic feature in the highly thermophilic strains, *C. thermohydrosulfuricum* and *C. thermosaccharolyticum* [15, 16]. The S-layer protomers appear to be anchored non-covalently to the underlying cell wall via regions referred to as "surface-layer homologous" (SLH) domains [17, 18]. The function of the S-layer is not completely understood. Since it covers the entire cell surface, it seems likely that it can act as a molecular sieve, preventing large molecules from entering or leaving the cell. Other roles which have been ascribed to bacterial S-layers include protection of the cell from predation and provision of attachment sites for exoenzymes [19].

Analyses of the lipid content of clostridial membranes, mostly confined to the *C. butyricum* group, have indicated the presence of glycerophospholipids which exist in both diacyl and plasmalogen (1-alk-1′-enyl-2-acyl) forms. An unusual glycerol acetal of the latter ether lipids also is present, often in significant quantities (for a summary see [20]). Membrane lipid composition has been shown to be a useful tool in classification [21-23]. Change in growth temperature is accompanied by changes in both acyl and alkyl chains and in lipid class composition, as the bacteria attempt to ensure that the membrane remains in the liquid crystalline phase [24, 25]. Variations in lipid class composition are also observed during growth in media lacking biotin but containing fatty acid supplements, conditions which allow control of the fatty acyl content of the membrane [26-28]. These changes are presumably designed to maintain the membrane in an optimal state for the functioning of membrane-bound proteins such as solute transporters and the ATPase complex (see sections 3.5.1.2 and 3.5.2). It is also well established that solvents have an effect on membrane structure, and this is likely to be an important determinant of product toxicity in solventogenic strains. Butanol increases the fluidity of the *C. acetobutylicum* membrane, an effect which is countered by an increase in the ratio of saturated to unsaturated acyl chains [29, 30]. A mutant with increased butanol tolerance was shown to have an increased capacity to modify membrane lipid composition, confirming that this ability is an important protective mechanism [31].

3.3 Effects of oxygen

As obligate anaerobes, clostridia are unable to grow in the presence of high concentrations of oxygen but nevertheless do show varying degrees of tolerance to oxygen. A number of different factors may contribute to oxygen toxicity [32]. In a study of the effects of oxygen on growth of *C. beijerinckii* (formerly *C. acetobutylicum*) NCIMB 8052, O'Brien and Morris demonstrated that aeration of a culture was accompanied by an increase in the redox potential (E_h) from − 400 to + 150 mV, and growth was inhibited [33]. However, under anaerobic conditions, using the artificial electron acceptor potassium ferricyanide, the redox potential could be poised at a value as high as + 370 mV without affecting growth and metabolism. Under conditions of low oxygen concentration and E_h of − 50 mV, *C. acetobutylicum* grew as well as in anaerobic conditions, but if E_h of an aerobic culture was lowered to − 50 mV by addition of dithiothreitol, growth was inhibited. It was therefore apparent that oxygen, or one or more products derived from it, rather than increased redox potential was responsible for inhibition of growth.

Oxygen derivatives such as the superoxide anion, hydrogen peroxide, and hydroxyl radicals are well-known to damage biological molecules, and a popular explanation for growth inhibition in the presence of oxygen is the lack of systems to adequately eliminate these species. However, while clostridia in general lack catalase, the enzymes superoxide dismutase (SOD) and NADH/NADPH peroxidase are apparently distributed widely [34, 35] indicating an ability to remove superoxide and peroxides. SOD activity was found to be elevated in *C. perfringens*, which was more tolerant to oxygen than other strains tested [34] but a direct correlation between SOD activity and oxygen tolerance has not been established. It has also been proposed that O_2-dependent drainage of reducing power (NADH) by oxidase activity may be an important factor in growth inhibition as a result of interference with intermediary metabolism. Soluble NADH and NADPH oxidase activities have in fact been detected in a range of clostridial strains [35]. The specific activity of NADH oxidase in *C. acetobutylicum* was found to be increased about 6-fold during growth in the presence of oxygen, and changes in the pattern of fermentation end products formed (increased acetate and reduced butyrate) were consistent with a decreased availability of NADH in the cell [33]. The oxidase activities are likely to make a significant contribution to the ability of *C. acetobutylicum* and *C. butyricum* to consume oxygen without any lasting inhibitory effects [33, 35]. However, whether they are the critical determinant of growth inhibition by oxygen remains a matter of conjecture. Oxygenation of cultures of *C. beijerinckii* also resulted in an immediate fall in intracellular ATP concentration, the basis of which is not understood [33].

3.4 Growth conditions and nutritional requirements

As mentioned above, the clostridia display a considerable diversity, and this is reflected in a wide range of optimal growth conditions [3, 36]. The majority of clostridia are neutrophiles with a pH optimum in the range 6.5 to 7.0. However, some strains are adapted to higher or lower pH in tune with their metabolic capabilities. For example, purinolytic strains liberate considerable amounts of ammonia through degradation of nitrogenous, heterocyclic compounds and have an optimal pH of 8.0 [37, 38]. On the other hand, acid-forming clostridia such as *C. acetobutylicum, C. beijerinckii, C. butyricum,* and *C. thermoaceticum* lower the pH to between 4 and 5 during growth. The pH can affect the ability to utilize sugars, and this can lead to misinterpretation of properties due to the conditions under which tests are performed [39].

The range of temperature optima among clostridia is extremely wide, with the complete spectrum from psychrophilic to thermophilic species represented [38]. The majority are mesophilic, growing optimally between 30 and 40 °C. A number of thermophilic strains, with an optimum growth temperature of 60 to 70 °C, have attracted attention due to interest in thermostable enzymes and the potential for developing fermentation processes which operate at elevated temperature, while *C. thermocellum* is a model organism for the study of cellulose degradation (see section 3.5.1.1). *C. thermohydrosulfuricum* and *C. thermosulfurogenes,* which were the focus of studies on polysaccharide-degrading enzymes [40, 41] have recently been reclassified as strains of *Thermoanaerobacter* and *Thermoanaerobacterium* [2]. Psychrophilic strains are fewer in number; both *C. estertheticum* and *C. putrefaciens* can be regarded as true psychrophiles, with growth optima below 25 °C and no growth at 22 °C and above 30 °C, respectively [42, 43].

The vast majority of clostridia are heterotrophic. They may be saccharolytic or proteolytic, neither or both. Some strains may be regarded as specialized with respect to a restricted or preferred substrate range. Complex media containing yeast extract, meat extract and peptone, with a fermentable sugar, have long been used for routine culture and are available commercially. However, synthetic media have been devised, and in the case of strains with specialized requirements may be the preferred means of cultivation. An optimal medium for a particular strain should be devised empirically, since the optimal composition, both qualitative and quantitative, may reflect the nuances of metabolism of the strain in question [39]. Salt, trace elements, and vitamins such as p-aminobenzoate, biotin, and thiamine are commonly required, but the need for amino acids, purines, and pyrimidines is variable. Some strains, such as *C. bifermentans*, are particularly fastidious with respect to a requirement for amino acids [38]. The nutritional requirements of clostridia have been reviewed in detail by Hippe et al. [36].

3.5
Metabolic properties

Clostridia exhibit the ability to metabolize an extremely wide range of organic molecules, including sugars and other carbohydrates, organic acids, alcohols, aromatic compounds, amino acids, amines, purines, and pyrimidines. This general metabolic diversity has been reviewed in detail [38, 39], while other articles have concentrated on factors relevant to the fermentative production of acids and solvents [44, 45].

3.5.1
Metabolism of carbohydrates

An extensive range of carbohydrates, from monosaccharides to large polymers, can be fermented by clostridia to provide carbon compounds and energy for growth. In common with other fermentations, a considerable amount of the carbon available in the substrate is converted to fermentation end products which are characteristic of the organism. The acetone-butanol (AB) fermentation of *C. acetobutylicum* was an important industrial process during the first half of the twentieth century, although it has subsequently declined as a result of inability to compete with petroleum-based chemical synthesis of these solvents [46]. The proven industrial record, and recognized potential, of clostridial fermentations has underpinned the study of carbon metabolism throughout the last 30 years.

3.5.1.1 Degradation of polysaccharides

Polymers such as cellulose, xylan, starch, and pectin are widely distributed in plant biomass and can be important substrates for clostridia in nature. They are degraded by extracellular enzymes or enzyme complexes, with the resulting low molecular weight compounds being assimilated by the cell. Many clostridia produce depolymerase enzymes, and interest in the polymers as cheap and renewable substrates for fermentation has stimulated research in this area.

Cellulose is found in plant cell walls as sheets of parallel chains of β-1,4-linked glucose residues. Hydrolysis of crystalline cellulose is dependent on the synergistic action of two types of cellulase enzymes: β-1,4-endoglucanases which cleave chains randomly, and β-1,4-exoglucanses (cellobiohydrolases) which remove cellobiose units from the non-reducing end [47]. Strains such as *C. acetobutylicum* produce endoglucanase but cannot degrade crystalline cellulose [48, 49], while those which produce an appropriate spectrum of enzymes do so efficiently. Additional enzymes, cellodextrinases and β-glucosidase, which show optimal activity against short-chain cellodextrins and cellobiose, respectively, may contribute to efficient hydrolysis by removing the initial digestion products. *C. thermocellum* accumulates cellobiose and cellodextrins [50] and metabolizes them intracellularly by phosphorylitic cleavage [51, 52], and the fact that *C. stercorarium* also produces cellobiose and cellodextrin phosphorylases [53] suggests a similar ability to assimilate the products of cellulose breakdown.

In *C. thermocellum*, which has become a model bacterial species for the study of cellulose digestion, cellulase activity is found predominantly in a large, exocellular multienzyme complex called the cellulosome [54-56]. Similar complexes are formed by *C. cellobioparum* [57], C. *cellulovorans* [57, 58], *C. cellulolyticum* [59] and *C. josui* [60], while several different high-molecular-weight multiprotein cellulase complexes have been identified in *C. papyrosolvens* [61, 62]. The major structural component of the cellulosome, variously referred to as CipA, CipC, or CbpA, is a multidomain protein which anchors the complex to the cell surface, binds cellulose, and acts as a scaffolding protein which binds the catalytic components and holds them in an active structural configuration [63]. These functions are the properties of identifiable domains within the protein. *Cohesin* domains, about 150 amino acids in length [64], interact with conserved, duplicated amino acid segments (*dockerins*; 22-24 amino acids) found in cellulosomal enzymes, usually at the C-terminus [65]. The precisely controlled spatial relationship between the catalytic components is believed to be important for effective cellulose breakdown. However, the cellulose binding domain of the cellulosome also is important in enhancing cellulase activity, perhaps by modifying the cellulose surface and thereby increasing accessibility to cellulolytic enzymes [66, 67]. The CipA protein of *C. thermocellum* contains a dockerin (of a different type) which serves to anchor the cellulosome to the cell surface via interaction with the SbdA protein. SbdA contains an N-terminal cohesin, while at the C-terminus is a triplicated segment of "surface layer homology" (SLH) repeats which probably mediate anchoring to the cell surface [68-70]. On the other hand, CipC of *C. cellulolyticum* does not contain a dockerin, implying a different mechanism of attachment to the cell [71].

In contrast to the complex architecture of the cellulosome, the cellulase system of the thermophile *C. stercorarium* has been resolved into just five enzymatic components; two Avicelases, two β-cellobiosidases, and a β-glucosidase. Avicelase I exhibited both endo- and exo-glucanase activities, while Avicelase II is an exoglucanase with activity also against cellodextrins [72, 73]. This cellulase system clearly lacks the structural complexity found in others. Nevertheless, the Avicelases do act synergistically in degradation of crystalline cellulose, although no synergism was apparent with amorphous substrates [74, 75].

Xylan is a branched polymer, principally of β-1,4-linked xylose residues but also containing other constituents such as arabinose, glucose, galactose, and glucuronate. It is a major component of hemicellulose which constitutes up to around

30% of the mass of plant cell walls. Xylan is degraded by the synergistic action of endoxylanases and β-xylosidases, along with enzymes which remove side groups, with the degree of substrate degradation determined by the activities of the enzymes produced [76-79]. Several clostridia synthesize xylanases [65, 80]. The cellulosome of *C. thermocellum* contains several xylanases, but no β-xylosidase activity [81]. As a result, the bacterium does not grow on xylan, apparently due to an inability to accumulate the digestion products efficiently [81-83]. The physiological role of the xylanases may be to increase accessibility of cellulose to the degradative enzymes in the cellulosome.

The cellulosome of *C. thermocellum* contains at least 16 enzymic components (12 cellulases, 3 xylanases, and a lichenase). These components are not easily dissociated, but the study of individual enzymes has been advanced by molecular biological approaches involving gene cloning and sequencing, and expression in *Escherichia coli*. Over 20 genes from *C. thermocellum* encoding glucanases, xylanases, and related enzymes have been sequenced, and the enzymes have been found to belong to several families of glycosyl hydrolases [47, 65]. The genes are widely distributed around the chromosome [84], which is reflected in the predominantly monocistronic mode of expression [85]. Major cellulase genes are, however, clustered in *C. cellulovorans, C. cellulolyticum,* and *C. josui* [60, 86, 87]. Despite the availability of a large number of cellulase and xylanase genes, little is known regarding regulation of their expression. Studies of the effects of soluble sugars on cellulase synthesis and activity in *C. thermocellum* have yielded variable results, most likely attributable to strain differences and different assay conditions, meaning that different enzymes were being monitored. Cellobiose has been shown to inhibit degradation of crystalline cellulose by crude cellulase, the purified cellulosome and the isolated CelS exoglucanase [88-91] but is not an effective inhibitor of endoglucanase activity [92-95]. It has generally been reported that synthesis of "true" cellulase is influenced by growth conditions, in particular the energy status of the cells and the presence of soluble sugars [96-99]. However, no evidence has been found for a specific inducer or repressor of gene expression. Individual endoglucanases have been shown to be differentially expressed in *C. thermocellum* [85]. None of the genes studied, *celA, celD, celC,* and *celF* were expressed in early exponential phase, perhaps indicating that they are catabolite repressed. mRNAs encoding *celA, celD,* and *celF* were detected in late exponential and early stationary phase cells, while *celC,* which encodes a non-cellulosomal endoglucanase, was expressed almost exclusively in stationary phase. Both the *celA* and *celD* genes could apparently be transcribed from two start sites with promoters resembling the consensus of *Bacillus subtilis* σ^{28} and *B. subtilis* σ^{54}/E. coli σ^{70}, respectively [85, 100]. This implies that control of gene expression may be achieved through changes in the σ subunit of RNA polymerase, although there is currently no direct evidence for this. In accord with the changing levels of synthesis of individual enzymes, the composition of the cellulosome has been shown to vary with growth conditions [71, 101, 102]. The fact that the cohesins of CipA do not discriminate between different enzymes [103, 104] is compatible with a variable cellulosome composition determined by the complement of enzymes available at different times.

Starch is a cheap and abundant polymer with two components: amylose, which is a linear polymer of α-1,4-linked glucose residues, and amylopectin, a branched polymer, in which the individual chains are connected via α-1,6-linkages. Degradation of the chains is achieved by endo-acting (α-amylase) and exo-acting (β-amylase and glucoamylase) enzymes, while α-1,6-bonds in amylopectin are hydrolyzed by the "debranching" enzyme pullulanase. α-Glucosidase hydrolyzes maltose and short oligosaccharides which are produced during primary digestion. All these enzymes have been found among the saccharolytic clostridia, and because of their biotechnological potential those in thermophilic strains have attracted particular attention [40, 41, 45]. Synergistic action between pullulanase and amylase (cyclodextrinase) enzymes has been observed [105, 106]. In addition, novel amylopullulanases, which exhibit activity against both α-1,4 and α-1,6-glucosyl linkages, have been isolated from strains of *C. thermohydrosulfuricum* (now reclassified as *Thermoanaerobacter ethanolicus* and *T. thermohydrosulfuricus*) and *C. thermosulfurogenes* (now *Thermoanaerobacterium thermosulfurigenes*) [2, 107-109]. Other enzymes which hydrolyze both types of glucosyl linkages are the glucoamylase of *C. thermosaccharolyticum* [110], and an α-glucosidase of *C. beijerinckii* which showed optimum activity against isomaltose [111].

The growth conditions resulting in optimal synthesis of different starch-degrading enzymes in clostridia and related bacteria are variable. In general, enzyme synthesis is induced by starch, dextrin, and maltose, and in the majority of cases is subject to catabolite repression by glucose or other readily metabolized sugars (reviewed in detail in [45]). The mechanisms of induction and repression are not known. Potential regulatory sequences were identified in the promoter regions of the genes encoding amylopullulanase in *T. thermohydrosulfuricus, T. ethanolicus,* and *T. thermosulfurigenes* [112-114] and β-amylase in *T. thermosulfurigenes* [115], but their physiological relevance has not been demonstrated. A regulatory gene termed *regA*, which encoded a protein showing 40% identity to the catabolite control protein (CcpA) of *B. subtilis,* was identified in *Clostridium* sp. P262 (now named *C. saccharobutylicum*; [116]). This gene complemented a *ccpA* mutant of *B. subtilis* with respect to repression of amylase synthesis, although repression was not relieved in the absence of glucose. In addition, when *regA* was expressed in *E. coli* together with a gene encoding a starch degrading enzyme (*staA*) from *C. beijerinckii,* starch degradation was inhibited [117]. These findings implicate the RegA protein, which contains a helix-turn-helix DNA binding motif, as a transcriptional regulator and suggest that catabolite repression in clostridia may conform to the general model described for low G-C Gram-positive bacteria [118]. Catabolite-derepressed mutants have been isolated following NTG mutagenesis and selection for resistance to the glucose analogue 2-deoxyglucose [119-121]. The mutants displayed various phenotypes, but have not been characterized in detail with respect to loss of repression of amylase synthesis.

In common with enzymes secreted by other prokaryotes, amylases and pullulanases produced by clostridia and related bacteria are synthesized in a precursor form with an N-terminal signal sequence which is removed during the secretion process [112, 115, 122-124]. While these enzymes are predominantly extracellular, the amylase (cyclodextrinase) and amylopullulanase of *T. thermosulfurigenes* EM1 are cell-bound. The amylopullulanase is believed to be anchored to the cell surface via three "surface-layer homologous" (SLH) domains at the C-terminus [124]. Under certain growth conditions, loss of enzyme into the extracellular medium is accompanied by the disintegration of the outer layers of the cell [125, 126]. Two further enzymes of *T. thermosulfurigenes* EM1, a xylanase and a polygalacturonate hydrolase, contain similar SLH domains [127]. Although no such domains are found in the amylase (cyclodextrinase) of this strain, the extracellular form of the enzyme is smaller than predicted from the encoding *amyA* gene sequence, perhaps due to removal of an anchoring segment during release of the enzyme from the cell [122, 128].

Pectins are composed of a backbone of galacturonic acid residues carrying small substituents such as methanol, acetate, and monosaccharides attached through ester linkages; up to 90% of the galacturonic acid residues may be esterified. Pectin fermentation, resulting in dissolution of the middle lamella of the cell wall and disintegration of plant tissues, is the basis of the retting of flax and hemp by bacteria such as *C. felsinium*, *C. laniganii*, and *C. flavum* [129]. Pectin degrading clostridia are readily isolated from soil, sediments, or decaying vegetable matter [130-132]. Degradation of pectin and polygalacturonic acid is generally achieved via endo- or exo-acting lyase and hydrolase enzymes, together with esterases which remove substituent groups [133-135]. *C. multifermentans* produces an unusual exopectate lyase complex which exhibits coordinated pectin esterase and polygalacturonate lyase activities, and releases unsaturated digalacturonic acid residues from the reducing end of the substrate molecule [136-138]. *C. thermosaccharolyticum* similarly produces an extracellular complex incorporating two coordinated activities, but in this case digalacturonic acid is released from the non-reducing end of the substrate by pectin esterase and polygalacturonate hydrolase [139]. In this strain, hydrolysis of the degradation products, digalacturonate and trigalacturonate, appears to occur intracellularly [140]. In general, pectinolytic enzymes are inducible and repressed by glucose [129, 130, 135, 140], but the underlying mechanisms have not been described.

3.5.1.2 Uptake and metabolism of carbohydrates

Bacteria employ a number of mechanisms to accumulate low molecular weight solutes from their surroundings. The predominant mechanism of carbohydrate uptake in clostridia is the phosphoenolpyruvate (PEP)-dependent phosphotransferase system (PTS) which is characterized by the comcomitant uptake and phosphorylation of its substrates (Figure 1; [45]). The simultaneous accumulation and derivatization of the substrate represents an energetically economical mechanism of uptake which is wide-spread in obligate and facultative anaerobes, and the clostridia are typical anaerobes in this respect. The PTS comprises a multiprotein phosphoryl transfer chain which transfers phosphate from the donor, PEP, to the sugar which is phosphorylated as it enters the cell. In most cases, the chain has been shown to comprise two general proteins, enzyme I and HPr, which participate in all the phosphotransferases in the cell, together with two hydrophilic domains (IIA and

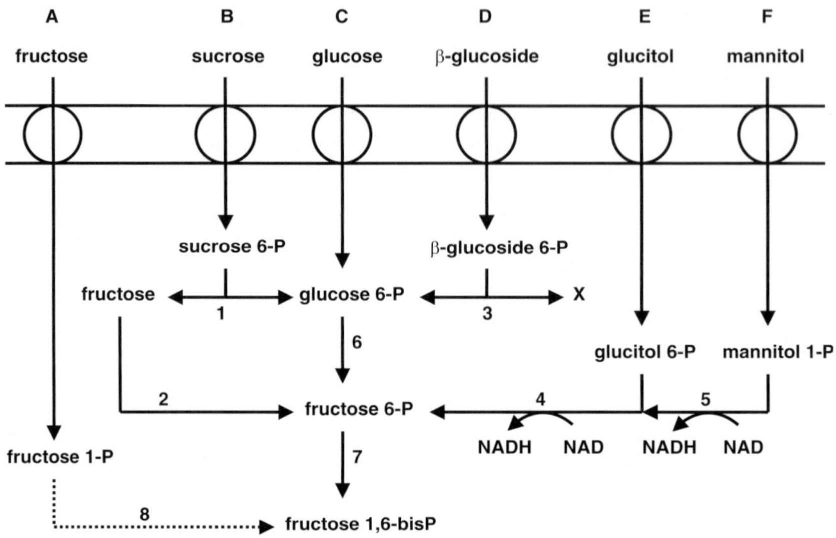

Figure 1. Uptake and metabolism of carbohydrates in clostridia. All uptake systems shown are PEP-dependent phosphotransferase systems which both transport and phosphorylate the substrate. (A) fructose in *C. beijerinckii* [158]; (B) sucrose in *C. beijerinckii* and *C. acetobutylicum* [146, 147, 157]; (C) glucose in *C. beijerinckii* [143, 158]; (D) β-glucoside in *C. longisporum* [145]; (E) glucitol in *C. beijerinckii* [144]; (F) mannitol in *C. acetobutylicum* [156]. 1, sucrose 6-phosphate hydrolase; 2, fructokinase; 3, phospho-β-glucosidase; 4, glucitol 6-phosphate dehydrogenase; 5, mannitol 1-phosphate dehydrogenase; 6, hexokinase; 7, phosphofructokinase; 8, putative fructose 1-phosphate kinase. X, aglycon group of β-glucosides arbutin or salicin.

IIB) of a substrate-specific, membrane associated, “enzyme II” complex. The third enzyme II domain, IIC, is hydrophobic and integral to the membrane, and serves to mediate entry of the substrate [45, 141, 142]. The components of the clostridial PTS have been shown to interact functionally with components from other bacteria to phosphorylate glucose in a cell-free assay system [143], and the close relationship between the PTS in clostridia and in other bacteria also is apparent from sequences of protein components. Thus, the glucitol PTS of *C. beijerinckii* [144], a β-glucoside PTS of *C. longisporum* [145] and the sucrose PTS of both *C. beijerinckii* [146] and *C. acetobutylicum* [147] are each homologous to their counterparts in other bacteria. The two sucrose systems serve to illustrate that the clostridial PTS exhibits the variable domain architecture identified among other bacterial phosphotransferases [141], since the respective *scrA* genes encode $IICB^{scr}$ and $IICBA^{scr}$ proteins. The *scr* operon of *C. beijerinckii* does not in fact encode a IIA domain. Since the IIA domain of the *C. acetobutylicum* sucrose PTS is homologous to IIA^{glc} (glucose) domains in other bacteria, it has been suggested that the sucrose PTS of *C. beijerinckii* may “hi-jack” the IIA domain of the glucose PTS, as has been shown to occur in *B. subtilis* [147, 148].

Not all sugars are accumulated by clostridia via a PTS. In general, inability to detect PEP-dependent phosphorylation of a substrate by cell extracts is taken to indicate that its uptake is by an alternative mechanism. In most such cases, the dependence of accumulation on either ATP or a transmembrane proton gradient has been inferred from the effects of metabolic inhibitors [45]. However, in *C. pasteurianum* direct evidence has been presented for uptake of both galactose and gluconate via a H^+-symport [149], utilizing the proton gradient generated by the membrane-bound, H^+-translocating ATPase [150, 151]. Also, the presence of sugar-binding proteins in *Thermoanaerobacter* and *Thermoanaerobacterium* strains [114, 152, 153] is indicative of the operation of high affinity, ATP-dependent transport by systems belonging to the ATP-binding cassette (ABC) transporter family [154].

Following transport into the cytoplasm, carbohydrate substrates are metabolized to pyruvate via the Embden-Meyerhof-Parnas (EMP) pathway (Figure 2A; [44, 155]). Glucose, which is converted to glucose 6-phosphate during PTS-mediated uptake, enters the pathway directly and its metabolism yields 2 mol of ATP and 2 mol of NADH from each mol hexose. Other substrates taken up by the PTS must be further modified to form a pathway intermediate(s). For example glucitol 6-phosphate and mannitol 1-phosphate, the products of the glucitol PTS and mannitol PTS, respectively, are oxidized to fructose 6-phosphate [144, 156], while sucrose 6-phosphate is hydrolyzed to glucose 6-phosphate and fructose, which is subsequently phosphorylated by fructokinase [157]. Fructokinase is required specifically for sucrose metabolism, since fructose itself is taken up by a fructose 1-phosphate-forming PTS (Figure 1; [158]). As in other bacteria, genes encoding the converting enzymes are found in association with genes encoding the respective uptake systems, allowing for coordinate control of their expression, and it is clear that the general principles of operon structure, induction of gene expression by substrate and repression by a more readily metabolized source of carbon apply to the clostridia [159]. Details of the underlying mechanisms are just beginning to emerge. However, it is already apparent that a variety of regulatory strategies are employed based on transcriptional repressors and activators and antiterminator proteins, while the mechanism of catabolite repression seems likely to show similarities to that in other low G-C Gram-positive bacteria [45, 144-147, 156]. Gene clusters encoding enzymes of the EMP pathway have also been identified. The key regulatory enzymes phosphofructokinase and pyruvate kinase are encoded by adjacent genes in *C. acetobutylicum* ATCC 824 and appear to be cotranscribed [160], while a cluster encoding glyceraldehyde 3-phosphate dehydrogenase, phosphoglycerate kinase, triosephosphate isomerase and phosphoglycerate mutase has been isolated from *C. acetobutylicum* DSM 792. The lack of transcriptional signals between these genes suggests that they also are coordinately expressed [161].

Pentoses are metabolized intracellularly by a combination of phosphorylation, isomerization, and epimerization, and the resulting phosphorylated intermediates are converted to fructose 6-phosphate and glyceraldehyde 3-phosphate by transaldolase and transketolase (Figure 2B; [44, 162]. Corresponding enzymes and genes have been found in clostridia and related strains [163-168, 455]. The conversion of 3 mol of pentose to pyruvate yields 5 mol ATP and 5 mol NADH. The phosphoketolase pathway incorporating phosphorolytic cleavage of xylulose 5-phosphate is

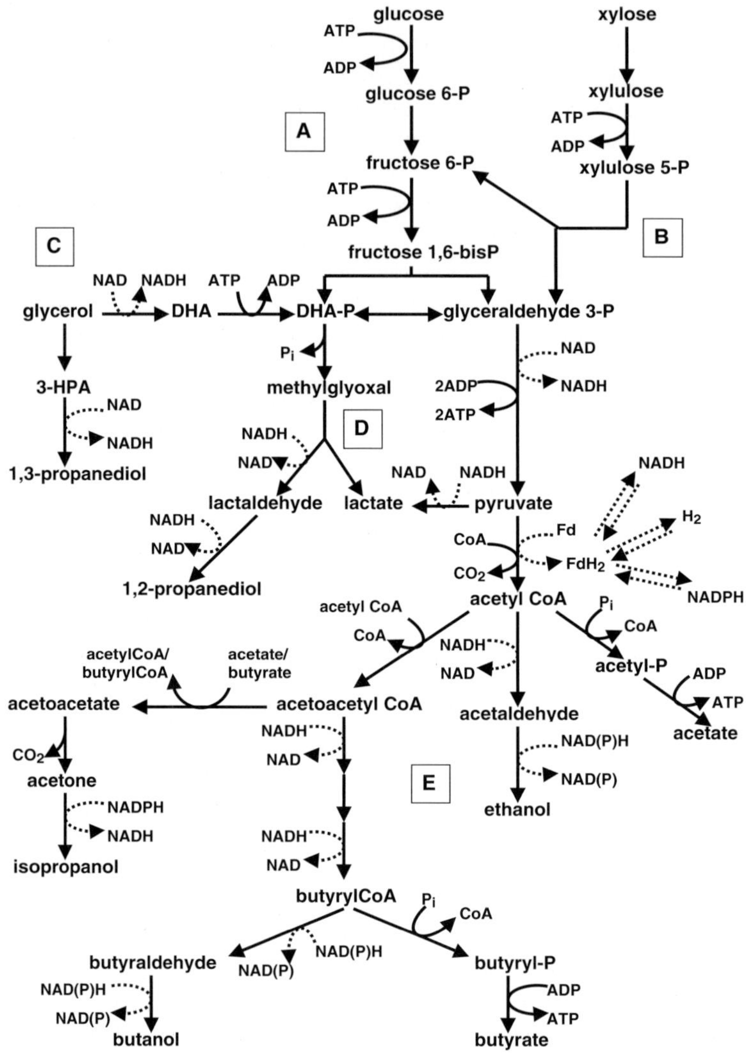

Figure 2. Sugar fermentation pathways in heterotrophic clostridia. In all fermentations, carbon metabolism is coordinated with electron transfer pathways to ensure an oxidation/reduction balance. Electron transferring reactions are shown by broken lines. Transfer of electrons between FdH_2, NADH, and NADPH is a key determinant of the fermentation profile. (A) glucose breakdown by the Embden-Meyerhof-Parnas pathway [44]; (B) xylose breakdown by the pentose phosphate pathway [44]; (C) glycerol dismutation by *C. acetobutylcium*, *C. pasteurianum* and *C. butyricum* [170, 172, 173]; (D) methylglyoxal by-pass in *C. sphenoides* [201]; (E) production of acids and alcohols by acidogenic and solventogenic strains [44, 45, 451].

not found in clostridia. Gluconate is degraded by several clostridia via a modified Entner-Doudoroff pathway, in which the first enzymatic step is catalyzed by gluconate dehydratase [169], to yield 2 mol pyruvate, 1 mol ATP, and 1 mol NADH. Glycerol fermentation is accomplished by oxidation to dihydroxyacetone, which is then phosphorylated and metabolized via the latter steps of the EMP pathway yielding 2 mol NADH and 1 mol ATP per mol glycerol. However, part of the glycerol is converted to 3-hydroxypropionaldehyde by a dehydratase, and then reduced to 1,3-propanediol (Figure 2C; [170-173]). The genes encoding enzymes of this reductive pathway, glycerol dehydratase and 1,3-propanediol dehydrogenase, in *C. pasteurianum* are clustered, but separated by four ORFs (open reading frames) of unknown function. The enzymes themselves exhibit high similarity to corresponding enzymes of enteric bacteria [174, 175].

It is evident that the common end product in degradation of the above substrates, pyruvate, is a key intermediate in clostridial metabolism. Different clostridial species are, however, characterized by different routes of pyruvate metabolism (Figure 2E). Some lactate may be formed by reduction of pyruvate, but under most conditions pyruvate is predominantly cleaved by pyruvate-ferredoxin (Fd) oxidoreductase to form acetyl CoA, CO_2, and reduced ferredoxin (FdH_2). Both the oxidoreductase enzyme and the electron acceptor Fd are iron-sulfur proteins which contain 4Fe-4S clusters [176]. Fd, which appears to be present in all clostridia, occupies a crucial position in metabolism. Electrons from reduced Fd can be used in the reduction of protons to form H_2, or can be transferred to NAD(P) and thereby used in the formation of alternative reduced fermentation products. The fate of acetyl CoA is dependent on both the species and growth conditions. A variety of different products can result from clostridial fermentations (Figure 2E; [38, 177]) and many of the enzymes involved have been purified and characterized [44, 45, 178, 179]. Formation of either acetate or butyrate allows for the production of one further mol ATP for each mol product (equivalent to 2 ATP per mol glucose metabolized in the case of acetate).

Fermentations which result in a spectrum of products are poised to ensure a balance between oxidative reactions during glycolysis and reductive reactions in the latter fermentation steps. Three oxidoreductases can transfer electrons between Fd and a donor or acceptor, and the nature of the fermentation depends to a large extent on their relative activities. Hydrogenase (Fd-H_2 oxidoreductase) forms hydrogen, while the other two enzymes transfer electrons between Fd and pyridine nucleotides. NADPH-Fd oxidoreductase may play a significant role in the generation of NADPH for biosynthesis, particularly since many clostridia appear to lack glucose 6-phosphate dehydrogenase [44]. Nevertheless, the existence of some NADPH-specific dehydrogenases indicates a role also in product formation [179]. NADH-Fd oxidoreductase may transfer electrons from NADH to Fd, but this reaction involves an unfavorable redox change and requires acetyl CoA as an activator. The reverse reaction is inhibited by NADH [180-182]. The relative activities of this enzyme, in response to the concentrations of intracellular metabolites, may therefore be an important determinant of the fermentation. Indeed, when *C. butyricum* was grown on glycerol, hydrogenase activity and the intracellular concentrations of NADH and acetyl CoA were lower than in cells grown on glucose. The predicted effect on Fd-NADH oxidoreductase, resulting in substantial electron transfer from Fd to NADH, is consistent with the observation that the yield of 1,3-propanediol was higher (and the amount of H_2 evolved was lower) than expected from the amount of NADH generated during glycerol metabolism [173, 183].

It is the loss of electrons in H_2 formation which allows the cell to form acetate by a non-reductive pathway. However, despite the energetic advantage, acetate cannot be formed exclusively unless glycolytic NADH is re-oxidized via NADH-Fd oxidoreductase and hydrogenase. Under conditions in which H_2 evolution is prevented or limited, additional electrons must be directed towards the formation of reduced end products. Therefore, since H_2 accumulation itself results in inhibition of hydrogenase, most fermentations must produce some reduced compounds. Butyrate fermentations typically result in the formation of both acetate and butyrate, which represents a balance between energetic efficiency and the need to reoxidize NADH during metabolism of pyruvate [155].

It is clear that growth conditions affect the fermentation pattern. Fermentation of reduced substrates results in a greater production of reduced end products, as would be expected if an oxidation-reduction balance is maintained. *C. acetobutylicum* grown in a chemostat on a mixture of glucose and glycerol produced alcohols and considerably less H_2 than during growth on glucose alone, associated with changes in activity of Fd-NAD oxidoreductase and alcohol-forming enzymes [184]. Even a species such as *C. pasteurianum,* which is regarded as an acid producer, can form butanol or other alcohols when either mannitol or glycerol is the substrate [172, 183, 185]. Conversely, neither *C. rubrum, C. butyricum, C. thermosulfurogenes,* nor *C. puniceum* formed significant quantities of alcohols during pectin fermentation (except for methanol released by the action of pectin esterase), consistent with the substrate being in a more oxidized state than hexose sugar [130, 132, 186, 187].

In batch culture, the acetone-butanol fermentation of *C. acetobutylicum* occurs in two phases [188, 189]. In the first, acidogenic phase corresponding to exponential growth, the principal products are acetate, butyrate, H_2, and CO_2. As growth slows at the onset of the stationary phase, there is a switch to solventogenesis; butanol, acetone, and ethanol become the main products (in a ratio of approximately 6:3:1), H_2 production decreases, and some of the acids are reassimilated. Growth under phosphate or sulfate limitation was found to be effective for production of acetone and butanol from glucose in a chemostat at a pH between 4 and 5 [190, 191]. Interestingly, under sulfate-limited conditions, lactate was formed above pH 5 and became the major product as the pH was further raised [191]. Similarly, lactate was the main product in an iron-limited chemostat above pH 5.1 [192]. These effects are likely due to alterations in the activity of enzymes and the availability of intermediates of the fermentation pathways. Activity of pyruvate-Fd oxidoreductase and hydrogenase, both of which are iron-sulfur proteins, is likely to be reduced in response to iron limitation [193, 194], while lactate dehydrogenase is virtually inactive at pH 5 and below [195]. Thus, NADH is reoxidized via the most appropriate route, which is determined by the growth environment.

An appreciation of the interplay between carbon and electron flow has underpinned numerous further attempts to stimulate butanol production by *C. acetobutylicum* by altering culture conditions. For example, when hydrogenase activity was inhibited by increasing the amount of dissolved H_2 in the culture medium or by addition of CO, formation of butanol by *C. acetobutylicum* was stimulated [196-198]. The pattern of electron flow can also be altered by addition of artificial electron carriers such as methyl viologen, apparently due to diversion of electrons from reduced Fd to form NADH rather than H_2 [194, 199, 200]. In both cases, additional NADH is reoxidized in butanol formation. Varying culture conditions has a similar effect on product formation by *C. pasteurianum*. Under phosphate limitation in a chemostat, glucose was fermented almost exclusively to acetate and butyrate, whereas lactate was produced under iron limitation. CO gassing resulted in formation of significant amounts of lactate, ethanol, and butanol, while production of acetate, butyrate, and CO_2 was markedly decreased, again indicating that formation of reduced end products is favored when hydrogenase is inhibited [172].

When grown under phosphate limitation in a chemostat, *C. sphenoides* was found to produce two new products, 1,2-propanediol and lactate, in addition to ethanol, acetate, H_2, and CO_2 . These were apparently synthesized via the methylglyoxal by-pass, which provides an alternative route for the metabolism of dihydroxyacetone phosphate (Figure 2D; [201]). Inhibition of methylglyoxal synthase by phosphate allows for control of the pathway in response to intracellular phosphate concentration. Other species tested did not produce propanediol under phosphate limitation, although *C. oroticum* and *C. indolis* did form lactate. A variant of this pathway was found in *C. thermosaccharolyticum* growing on some sugars [202]. Thus, in addition to ethanol, lactate, and acetate, cultures grown on glucose, xylose, mannose, and cellobiose formed 1,2-propanediol. This was proposed to be the result of reduction of methylglyoxal via acetol. In this case, phosphate did not affect propanediol production; however, the strains used had been subjected to extensive mutation and selection, which could have affected regulation of the pathway.

3.5.2
Homoacetogens and autotrophic growth

The homoacetogens are an exception to the general considerations described above, in that they are capable of fermenting sugars to form acetate as the only metabolic product. However, this acetate is produced by both a fermentative (oxidative) and a synthetic (reductive) route (Figure 3). The yield of acetate is almost 3 mol per mol glucose or fructose metabolized; 2 mol are generated through fermentative reactions, which also yield 2 mol CO_2, 4 mol ATP and reducing equivalents which are used to reduce the CO_2 to a third mol of acetate. Saccharolytic acetogens include both mesophiles (*C. aceticum, C. formicoaceticum, C. magnum,* and *C. mayombei*) and thermophiles (*C. thermoaceticum* and *C. thermoautotrophicum*). The substrate for homoacetogenic growth is not restricted to hexoses, with some species able to utilize pentoses, organic acids and alcohols [177, 203, 204].

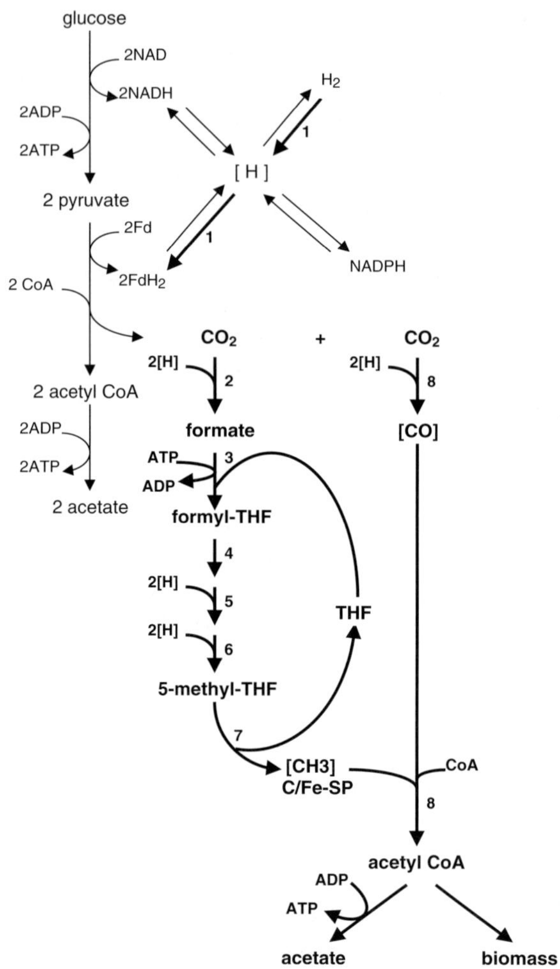

Figure 3. Acetate formation by homoacetogenic clostridia. Acetate is formed either from hexoses such as glucose or autotrophically from CO_2 [177, 205]. The autotrophic pathway is shown by bold lines. Under autotrophic conditions, electrons for CO_2 reduction are derived from H_2. Square brackets indicate an enzyme-bound intermediate. 1, hydrogenase; 2, formate dehydrogenase; 3, formyl-tetrahydrofolate (THF) synthetase; 4, methenyl-THF cyclohydrolase; 5, methylene-THF dehydrogenase; 6, methylene-THF reductase; 7, methyltransferase; 8, CO dehydrogenase/acetyl CoA synthase.

The homoacetogens *C. thermoaceticum, C. thermoautotrophicum,* and *C. aceticum* are capable of autotrophic growth on a mixture of H_2 plus CO_2, but this ability has not been demonstrated for *C. formicoaceticum* [205]. *C. magnum* can grow with H_2 plus CO_2 provided the medium contains a low concentration (0.025 %) of yeast extract [206]. One-carbon (C-1) compounds (CO, formate, and methanol) can also be used, as can the O-methyl groups of molecules such as vanillate and syringate [207-209]. Following a period of adaptation, growth can occur in media essentially free of organic nutrients other than small amounts of vitamins [210, 211]; a number of trace metals are essential, being constituents of enzymes which constitute the pathway of acetate formation from CO_2. This pathway, now referred to as the Wood or Wood-Ljungdahl pathway, was first characterized in *C. thermoaceticum,* but it is now established that it, or variations of it, operate in many anaerobic bacteria including acetogens, methanogens, and sulfate reducers. A number of detailed reviews of the pathway, have been published [177, 203, 205, 212, 213].

Carbon dioxide is metabolized by two routes (Figure 3). One mol CO_2 is reduced to formate and subsequently converted to 5-methyltetrahydrofolate (5-methyl-THF) by a series of tetrahydrofolate enzymes, while the other is reduced to carbon monoxide by CO dehydrogenase. Under autotrophic conditions, the reducing equivalents required are provided by the action of hydrogenase which reduces Fd and hence NAD(P) at the expense of molecular H_2. Following transfer of the methyl group of 5-methyl-THF to a corrinoid/iron sulfur protein (C/Fe-SP) mediated by a methyltransferase, condensation with CoA and enzyme-bound CO is catalyzed by the CO dehydrogenase [214]. The latter enzyme, a nickel/Fe-S/zinc protein, therefore is a multifunctional CO dehydogenase/acetyl CoA synthase. Production of acetyl CoA from CO, 5-methyl-THF and CoA has been achieved using a combination of purified CO dehydrogenase, C/Fe-SP, methyltransferase and Fd [215]. *In vivo,* the resulting acetyl CoA may be used in synthesis of cellular biomass [205] or converted to acetate with the formation of ATP.

Inspection of the autotrophic pathway reveals that there is no net synthesis of ATP, and accordingly there must be another mechanism of energy conservation. This is also indicated by unusually high growth yields under heterotrophic conditions. Evidence has been presented for a chemiosmotic mechanism of ATP synthesis in *C. thermoaceticum* and *C. thermoautotrophicum.* These strains contain an anaerobic, membrane-bound electron transport chain comprising b-type cytochromes, menaquinone, and other redox carriers including ferredoxin, rubredoxin, and a flavoprotein, together with dehydrogenase and reductase enzymes. Menaquinone appears to be involved in electron transfer between low potential cyt b_{559} and high potential cyt b_{554} [216, 217]. Oxidation of CO and other electron donors (formaldehyde, methanol, H_2, and NADH) has been shown to result in reduction of the membrane-bound electron carriers [217, 218], and results in the formation of a transmembrane proton gradient (protonmotive force) which is capable of driving the accumulation of amino acids by membrane vesicles [219, 220]. The chemiosmotic circuit of ATP synthesis is completed by a membrane-bound F_1F_0 ATP synthase. This complex has been purified from both *C. thermoautotrophicum* and *C. thermoaceticum* [221, 222]. The number of subunits, four (α,β,γ, and ε) in the

F_1 part and two (δ and c) in the F_0 part, is fewer than in *E. coli*, but the purified ATPase from *C. thermoautotrophicum* has been shown to catalyze ATP-dependent proton translocation across proteoliposome membranes [222]. The *atp* operon of *C. thermoaceticum* in fact comprises nine genes encoding proteins equivalent to all components of the *E. coli* ATP synthase. Since all the genes are expressed, a post-transcriptional regulation of assembly of the complex is implied [221].

Under heterotrophic conditions, in which the fixation of CO_2 is not obligatory, the acetogenic clostridia can use alternative electron acceptors including fumarate and nitrate [223, 224]. Fumarate dissimilation was also found to support growth of *C. formicoaceticum* and *C. aceticum* on methanol and H_2, respectively, under CO_2-limited conditions, although CO_2 was the preferred electron sink when both CO_2 and fumarate were present [223]. Nitrate, on the other hand, is a preferred electron acceptor, in the presence of which the yield of acetate is reduced by one third, corresponding to loss of the acetate produced by CO_2 reduction. Nitrate also inhibits autotrophic growth and the use of C-1 compounds as sole carbon and energy source [225]. The basis of preference for nitrate in *C. thermoaceticum* appears to lie in coordinate regulation of synthesis of the enzymes of the Wood-Ljungdahl pathway. The genes encoding the enzymes of the acetyl CoA synthesis pathway (CO dehydrogenase, C/Fe-SP, and methyltransferase) have been shown to be clustered on the *C. thermoaceticum* chromosome [213, 226, 227]. CO_2 appears to induce expression of the Wood-Ljungdahl pathway, and it has been suggested that nitrate may inhibit CO_2-mediated induction. The mechanism(s) involved are not understood at present [228].

3.5.3 Metabolism of organic acids, alcohols, and aromatic compounds

It is well documented that the clostridia exhibit a wide variety of metabolic processes, and an extensive range of compounds can be metabolized. Various strains have been shown to metabolize mono-, di-, or tri-carboxylic acids [38, 39] which can serve as sole carbon and energy source for growth. Citrate can support the growth of a number of strains with citrate lyase apparently the key degradative enzyme [229], while tartrate is a selective substrate for "*C. tartarivorum*", now classified as *C. thermosaccharolyticum* [230]. Several acids can be metabolized by different routes; for example, metabolism of glycerate, lactate, and pyruvate is a property found among acetate-, propionate- and butyrate-forming strains (see [38] and references therein).

The alcohols methanol and ethanol can be metabolized by the homoacetogenic clostridia, although differences in the ability to use these compounds have been observed between species [38]; bicarbonate (CO_2) is used as electron acceptor, and the only product is acetate (see section 3.5.2). Ethanol, propanol, and butanol can also be metabolized by *C. thermoaceticum* in the presence of dimethylsulfoxide or thiosulfate as electron acceptors, the alcohols being oxidized to the corresponding acids, and these electron acceptors also stimulate growth on methanol [231]. Fermentation of alcohols may be dependent on the presence of a co-substrate.

C. neopropionicum can ferment ethanol to acetate, propionate, and some propanol and can also metabolize propanol in combination with acetate [232]. *C. kluyveri* metabolizes ethanol plus acetate to form butyrate and caproate [233]; ethanol may be replaced by propanol, and acetate by one of several alternative acids, with consequent formation of additional products [234, 235]. The characteristics of succinate metabolism were consistent with splitting to two molecules of acetate, with reducing equivalents provided by ethanol [235]. *C. mayombei*, an acetogenic strain isolated from a termite gut, can grow on succinate as a sole carbon source, generating propionate and CO_2 [236]. Consideration of the associated free energy change has led to the suggestion that this fermentation may involve a membrane-bound, Na^+-translocating decarboxylase as found in *Propionigenium modestum* [38, 237], but this has not been confirmed.

Diols can also be used as carbon and energy sources. Ethylene glycol (1,2-ethanediol) is fermented by *C. aceticum* and *C. glycolicum*, the latter employing a radical mechanism with the production of an aldehyde [238]. *C. glycolicum* also grows on 1,2-propanediol. Longer chain alcohols, such as 2,3-butanediol and acetoin are utilized by the homoacetogenic species *C. aceticum* and *C. magnum* [239]. *C. thermoaceticum* and *C. thermoautotrophicum* are capable of growing at the expense of aromatic compounds such as syringate and vanillate, as a result of utilization of methoxy groups released by an inducible O-demethylating enzyme which is repressed by glucose and methanol [207-209]. *C. thermoaceticum* also is able to decarboxylate vanillate, but not syringate, to generate CO_2-equivalents for growth [240]. Attack of aromatic rings has seldom been reported. However, both *C. scindens* and *C. orbiscindens* have been found to cleave flavonoid rings [241, 242], while purine, pyrimidine, and pyranine rings can be degraded by some species (see section 3.5.5.3).

3.5.4 Homeostasis

With the exception of the autotrophic homoacetogens, clostridia are reliant on substrate-level phosphorylation to generate ATP. However, it is clear that, in common with other bacteria, the heterotrophic species and acetogenic species growing heterotrophically do maintain transmembrane ion gradients [150, 243, 244]. Clostridia exhibit a limited capacity to regulate cytoplasmic pH, which decreases during growth due to the production of acidic fermentation products. These acids, in their undissociated form, act as uncouplers which effectively render the membrane permeable to protons and thus lower the cytoplasmic pH [245]. Growth of *C. thermoaceticum* on glucose was found to be inhibited when the internal pH became lower than 5.5 [243]. The production of acetone and butanol by *C. acetobutylicum*, which occurs at low pH and is accompanied by the reassimilation of acetate and butyrate formed in the early stages of fermentation, can be viewed as a detoxification mechanism ([244], reviewed in [45]).

Establishment of the transmembrane proton gradient is dependent on the membrane-bound ATPase [150, 151, 243, 246]. *C. fervidus*, in which energy transduction is exclusively dependent on Na^+ rather than H^+ ions (see section 3.5.5.2), cannot maintain a pH gradient and consequently growth is limited to within the pH range

6.3 to 7.7 [247]. The ATPase of *C. pasteurianum* was reported to consist of only four different subunits; therefore, although the complex exhibits similar general properties to the ATPase complexes of other bacteria, it has a simpler molecular structure [151]. The purified *C. pasteurianum* ATPase was shown to be capable of ATP synthesis although the kinetic characteristics appeared to favor ATP hydrolysis, consistent with the function of the complex in the growing cell [248]. Nevertheless, the *atp* operon has been found to encode nine subunits and has a composition and organization identical to that found in many bacteria [249]. Therefore, as for the homoacetogens *C. thermoautotrophicum* and *C. thermoaceticum*, some as yet undefined discrimination between subunits must be exerted at the level of assembly of the complex.

Ion fluxes, in particular of K^+ and Na^+, are known to play a vital role in bacterial pH regulation [250], while modulation of the K^+ content of the cytoplasm also is a primary response of bacterial cells to osmotic stress [251]. Secondary ion translocation systems are wide-spread, facilitating the flux of ions and the maintenance of transmembrane ion gradients which are ultimately dependent on either respiration or ATP hydrolysis. Na^+-H^+ antiport activity has been demonstrated in *C. thermoaceticum*, although no such activity was found in *C. acetobutylicum* [252]. The latter species does contain a *kdp* system, inducible by low concentrations of potassium, which encodes a complex analogous to the high affinity K^+ transport ATPase of *E. coli*. Synthesis of the complex is inducible by low concentrations of potassium, and is controlled by a two-component, sensor kinase/response regulator system [253-255]. Elements of potentially important physiological regulatory systems have therefore been identified in clostridial species, but the extent of their distribution and their role in pH and osmotic homeostasis remain to be established.

3.5.5 Nitrogen metabolism

3.5.5.1 Assimilation of ammonium ions

In common with a large number of heterotrophic bacteria, clostridia are able to grow using ammonium as the sole source of nitrogen. Assimilation of NH_4^+ is generally dependent on three enzymes, glutamine synthetase (GS), glutamate synthetase (glutamine-2-oxoglutarate aminotransferase, GOGAT), and glutamate dehydrogenase (GDH); these enzymes are present in clostridia, although the full complement has not been found in all cases [256-259]. Enzyme activities are controlled in response to the nitrogen status of the medium. The key enzyme in assimilation of NH_4^+ under nitrogen-limited conditions is GS which catalyzes the reaction:

$$\text{L-glutamate} + NH_4^+ + \text{ATP} \rightarrow \text{L-glutamine} + \text{ADP} + \text{Pi}$$

The GS of *C. acetobutylicum* P262 (now classified as *C. saccharobutylicum*; [116]) is a member of the class I group found in the majority of bacteria [260]. A detailed study of the *glnA* gene encoding the enzyme has revealed a novel mechanism of control of gene expression. The gene is expressed from two upstream promoter sequences. Downstream is a third promoter (P3) oriented towards *glnA*, which directs the transcription of an antisense RNA complementary to 43 bases of the 5′ end of *glnA* mRNA including the ribosome binding site and the start codon. A role for this antisense RNA in regulation of GS synthesis is indicated by the variation in cellular RNA and GS activity in response to the nitrogen content of the medium. In nitrogen-rich medium, there was a 1.6-fold excess of antisense transcripts over *glnA* mRNA, and GS activity was low. Conversely, under conditions of nitrogen limitation, the number of antisense RNA transcripts was approximately sixfold lower, the cellular content of *glnA* mRNA exceeded antisense mRNA by fivefold, and GS activity increased [261]. It has also been suggested that the antisense RNA may be involved in regulation of expression of a putative regulatory gene which overlaps the antisense coding region and encodes a product which exhibits homology with both response regulators of two-component systems and transcriptional antiterminators. Expression of the antisense molecule, mediated by activation of P3 in response to a cellular signal reflecting a nitrogen-rich environment, would therefore exert a dual control over *glnA* expression [262].

3.5.5.2 **Metabolism of amino acids**

There have been few studies of amino acid uptake by clostridia. Membrane vesicles prepared from *C. acetobutylicum* were found to accumulate a number of amino acids by a H^+ symport mechanism [263]. Similarly, uptake of glycine, alanine, and serine by vesicles of *C. thermoautotrophicum* and *C. thermoaceticum* driven by the oxidation of CO showed the characteristics expected of H^+ symport [219, 220]. On the other hand, amino acid uptake in the peptidolytic thermophile *C. fervidus* showed no dependence on H^+ but rather occurs by Na^+ symport [247, 264]. Uptake of 12 amino acids, including neutral, acidic, basic, and aromatic types, was similarly dependent on Na^+ ions. The transmembrane Na^+ gradient required to support accumulation of the amino acids is generated by a Na^+-translocating V-type ATPase in the cell membrane [265, 266]. This bacterium therefore appears to rely on sodium ion circulation for membrane-associated, energy-dependent processes.

All the essential amino acids can be metabolized by clostridia [267, 268]; in fact, almost all known organisms capable of fermenting amino acids are clostridia or related bacteria. The range of specific pathways known to be involved in amino acid metabolism by clostridia has been reviewed in detail by Barker [269], Andreesen et al. [39] and Bahl and Dürre [38]. Metabolism involves both oxidative and reductive reactions, and in general the products are ammonia, CO_2, hydrogen, and short-chain fatty acids. Oxidative metabolism is coupled to ATP-generating reactions which are essential for growth. Single amino acids are fermented but the well-known Stickland reaction, in which oxidation of one amino acid is coupled to reduction of another, is typical for a number of species. Oxidative reactions typically involve deaminations, transaminations or eliminations, which are similar to reactions found in other bacteria with the production of a 2-oxoacid. An acyl CoA may be formed by oxidative decarboxylation of the 2-oxoacid, with electrons transferred to Fd; acetyl CoA is then formed by a CoA transferase and this is converted to acetate with consequent production of ATP (see Figure 2E). 2-Oxoacids may also be metabolized by reduction to 2-hydroxyacid, and subsequent activation to a CoA-derivative, dehydration to enoyl CoA and reduction to a saturated fatty acid. Arginine is metabolized by the deiminase pathway which produces ornithine, ammonia and CO_2 with the formation of ATP in the terminal reaction catalyzed by carbamate kinase; ornithine can be further oxidized via the Stickland reaction. In other specialized pathways, glutamate is metabolized via coenzyme B_{12}-dependent carbon-carbon rearrangement to form β-methylaspartate, while lysine metabolism involves shifts in the positions of the amino groups. In keeping with the general inability of clostridia to cleave aromatic rings, aromatic amino acids are only partially degraded. Branched-chain amino acids also appear to be incompletely metabolized. A study of proteolytic clostridia demonstrated a correlation between the membrane complement of fatty acids and ability to oxidize valine, leucine, and isoleucine to the corresponding branched-chain, volatile fatty acids from which membrane lipids could be derived [270].

Proline and glycine can be metabolized by reductive cleavage of the C_α-N bond to form acetate or 5-aminovalerate, respectively, the latter being further metabolized by a Stickland-type reaction. *C. purinolyticum* is capable of growth on glycine as a sole substrate. The key enzymes are glycine decarboxylase and glycine reductase; 1 mol glycine is oxidized to CO_2, while 3 mol are reduced to acetate using the electrons released. Both oxidative and reductive pathways are associated with the formation of ATP, through conversion of formyl-tetrahydrofolate (FTHF) to formate and tetrahydrofolate (catalyzed by FTHF synthetase) and glycine reductase, respectively [271]. Glycine reduction has attracted considerable interest, since it was recognized that the process is linked to ATP synthesis [272], which occurs via the formation of acetyl phosphate [273]. Protein components of glycine reductase have been purified from *C. sticklandii*, *C. purinolyticum*, and *C. litorale* (reviewed in [274]) and genes encoding the enzyme have been cloned and sequenced. Both P_A and P_B proteins, which comprise glycine reductase, contain selenocysteine specified by the codon TGA [275-277], which explains the absolute requirement of glycine reduction for selenium. The *grdA* and *grdB* genes in *C. litorale*, encoding P_A and P_B, respectively, form an operon and are adjacent to the genes *trxB* and *trxA* encoding the thioredoxin system which is the natural electron donor to glycine reductase [277, 278].

While glycine reduction to acetate is linked to ATP formation, proline reduction in *C. sporogenes* has been shown to be associated with vectorial H^+ translocation across the cell membrane. This suggests a chemiosmotic mechanism of free energy conservation, and is consistent with increased growth yields when proline was added to cells growing in a glucose-limited chemostat [279]. Although D-proline is the substrate of the reductase, similar activity was observed with L-proline due to the presence of an active proline racemase. Proline reductase in *C. sticklandii* has recently been isolated as a soluble selenoenzyme, but association of the enzyme with the cell membrane *in vivo* has not been excluded [280].

Many clostridial strains can ferment other amine compounds such as betaine, choline, creatine, creatinine and ethanolamine, either as single or co-substrates

[281]. Betaine can act as an electron acceptor in the Stickland reaction, being cleaved to triethanolamine and acetate. While sarcosine metabolism is not a common property, it can be fermented together with alanine by *C. litorale* [283, 284].

3.5.5.3 **Metabolism of heterocyclic compounds**

A small number of clostridia are able to degrade heterocyclic nitrogenous compounds. Three species, *C. acidiurici, C. cylindrosporum,* and *C. purinolyticum* are obligately purinolytic, fermenting only purines and some of the degradation products. The latter organism has the wider substrate range, and is the only one capable of growth on adenine [285]. Purines are converted to xanthine by deaminases and xanthine dehydrogenase, a seleno-flavoprotein containing both an Fe-S centre and a molybdenum cofactor. This key enzyme catalyzes a variety of oxidation/reduction reactions in purine degradation pathways [286]. Xanthine is the start point for ring cleavage by a series of hydrolytic, deamination, and decarboxylation reactions with the formation of acetate, formate, ammonia, and CO_2, with a stoichiometry which reflects the oxidation/reduction state of the substrate [285]. ATP is generated via formyl-THF synthetase, and by reduction of glycine which is formed in the degradative pathway [287]. Under selenium starvation conditions, *C. purinolyticum* degrades uric acid by an alternative pathway involving pyrimidine derivatives, forming glycine as an additional product [288].

The pyrimidines uracil and orotic acid are completely degraded to β-alanine or aspartate, ammonia and CO_2 by *C. glycolicum* and *C. oroticum* (formerly known as *Zymobacterium oroticum*), respectively, via reductive, hydrolytic pathways. They are, however, poor growth substrates [289, 290]. Among the proteolytic species, *C. sporogenes* and some types of *C. botulinum* were found to deaminate cytosine and to reduce uracil and thymine to the dihydro forms, but further metabolism was not observed [291]. *C. barkeri* ferments nicotinic acid to acetate, propionate, ammonia, and CO_2 in a pathway which involves hydroxylation, reduction, and hydrolytic cleavage of the pyridine ring [292]. The first step is catalyzed by an NADP-coupled nicotinic acid hydroxylase, another enzyme which contains selenium, molybdenum, flavin and Fe-S clusters [293].

3.5.5.4 **Dinitrogen fixation**

Dinitrogen fixation is a common property of both mesophilic and thermophilic clostridial strains [256-258]. Nitrogenase is a metalloenzyme that catalyses ATP-dependent reduction of N_2 to yield two molecules of NH_3, using electrons donated by reduced Fd (or in some cases flavodoxin). The enzyme has two components which are referred to as the Fe protein and the MoFe protein reflecting the metal composition of the prosthetic groups. Electron transfer (one electron at a time) from the Fe to the MoFe protein, which is coupled to ATP hydrolysis and is rate-limiting for the reaction, forms a super-reduced molybdoferredoxin which binds N_2 and reduces it stepwise to ammonia [294]. An extremely large amount of ATP, estimated at 2 ATP per electron or 16 mol per mol N_2 reduced, is required. In general, nitrogenase activity is strongly repressed in the presence of NH_4^+, as might be expected for a reaction which consumes such a considerable amount of energy [256-258].

C. pasteurianum was the first free-living nitrogen-fixing organism isolated in pure culture, and among the clostridia it has received most attention. The properties and primary structures of the Fe and MoFe proteins are highly conserved among nitrogen-fixing bacteria, but the *C. pasteurianum* enzyme does show significant differences from the others [295-298] and hybrid nitrogenase enzymes with components from *C. pasteurianum* are not highly active [299]. The crystal structures of both components of the *C. pasteurianum* nitrogenase have recently been determined [298, 300]. A [2Fe-2S] ferredoxin, which has been shown to interact with the nitrogenase MoFe protein, has been implicated in nitrogen fixation but whether it is involved in electron donation (normally associated with 2[4Fe-4S] ferredoxin) or has some other function has not been established [301].

The nitrogen-fixation (*nif*) genes of *C. pasteurianum* are clustered on the chromosome, and are apparently arranged into transcriptional units according to their function; one operon (*nifH1nifDnifK*) comprising structural nitrogenase genes, and two involved in molybdenum transport and synthesis of the molybdenum-iron cofactor [297, 302]. However, the factors involved in regulation of *nif* gene expression are not well defined. *C. pasteurianum* in fact contains six *nifH*-like genes [295, 303]. One of these, *nifH2,* is located upstream of *nifH1,* but is not part of the nitrogenase operon. Another, *nifH3,* is found some distance upstream of three genes (*anfDGK*) which appear to encode components of an alternative Fe-only nitrogenase [304, 305]. Biochemical evidence has in fact been presented for the existence of a Mo-independent nitrogenase in *C. pasteurianum* [306], and this may be encoded by *nifH3* together with these genes. This possibility is supported by the homology observed between the protein encoded by *nifH3* and AnfH of the *A. vinelandii* alternative nitrogenase [307].

3.6
Spores and sporulation

3.6.1
Conditions for sporulation

One of the characteristic determinants of the genus *Clostridium* is the ability to form endospores, and this represents the most dramatic response of the bacteria to their environment. In conditions which are unfavorable for growth, vegetative cells differentiate to form metabolically dormant spores which are resistant to harsh treatments such as heat, chemicals and radiation (Figure 4). The precise conditions which are conducive to sporulation of different clostridia are highly variable. Most strains will sporulate in commonly used anaerobic media containing protein hydrolysates and yeast extract, although there is a wide variation in sporulation frequency, while in some cases defined media have been developed which are compatible with sporulation (see [308, 309] and references therein). The factors which are important for initiation of sporulation have been reviewed in detail by Woods and Jones [308]. Although bacterial sporulation is generally believed to be

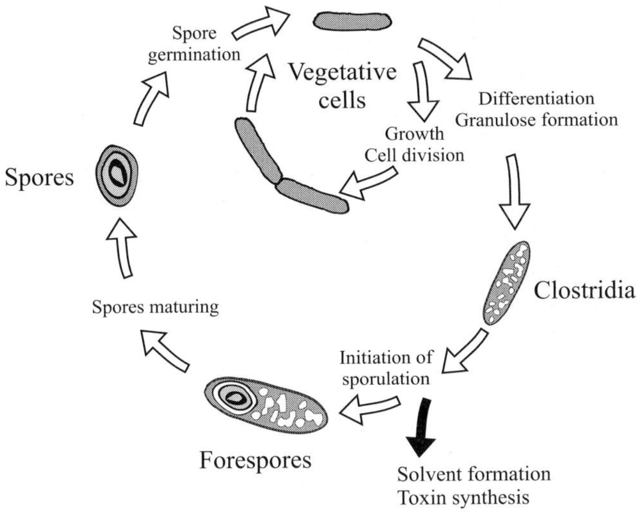

Figure 4. Cell cycle of clostridia. Morphological changes associated with sporulation are illustrated schematically, and the association of solvent formation (in *C. acetobutylicum* and related strains) and toxin formation (in *C. perfringens*) with sporulation is indicated. Adapted from Schuster et al. [452].

initiated in response to a lack of nutrients, clostridia appear to require an exogenous source of carbon and energy throughout the process. Carbon catabolite repression of sporulation in clostridia is well documented. However, in strains which exhibit catabolite repression, sporulation occurs provided the repressing carbon source is fed continuously in limiting amounts. In a study of the effects of different carbon sources in *C. thermosaccharolyticum*, a correlation was observed between lowered growth rate and stimulation of sporulation [310]. Similarly, a poorly used carbon source appeared to be important for good sporulation in *C. perfringens* [311, 312]. The mechanisms by which carbohydrates repress sporulation are not understood. However, repression of sporulation and of enzyme synthesis was found to be separable by mutation in *C. thermohydrosulfuricum* and *C. thermosulfurogenes*, indicative of different signal transduction pathways [119, 120]. A nitrogen source may also be essential for spore formation, and the nature of the nitrogen source can affect the frequency of sporulation. Other factors which may affect the ability to sporulate are pH, oxygen, and temperature. In general, sporulation appears to be favored by conditions which result in a decreased growth rate in the presence of adequate energy and carbon source reserves. An increase in substrate consumption is often seen during the onset of sporulation, while in some strains of *C. botulinum*, endogenous carbohydrate reserves appear to be utilized during spore formation [313, 314].

In *B. subtilis*, sporulation is always associated with depletion of the cellular pool of the guanine nucleotides GTP and GDP [315]. Sporulation of *C. perfringens* was found to be stimulated by methylxanthines such as caffeine and theophylline which lowered the growth rate, their effects being dependent on the nature of the carbohydrate energy source in some strains [316-318]. Since methylxanthines are purine analogues, it was considered that they may act through interference with purine metabolism, a conclusion consistent with stimulation of growth rate and inhibition of sporulation by guanosine [312]. While it was noted that methylxanthines also inhibit phosphodiesterases, the apparent absence of cAMP from *C. perfringens* appears to exclude this nucleotide as an effector of sporulation [319]. Therefore, although methylxanthines were not effective in all *C. perfringens* strains tested, the findings do suggest a comparable relationship between purines and sporulation in clostridia and bacilli.

3.6.2
Spore properties

Spore development involves an unequal cell division, the smaller cell (forespore) being engulfed by the larger one so that the *endo*spore develops *inside* the mother cell. In many species, the cell is distended by the spore [3]. Spore formation, which takes several hours, is accompanied by morphological, physiological, and biochemical changes, and the resulting refractile spore is structurally very different from a vegetative cell [320-322]. Within the outer proteinaceous, sac-like exosporium, are layers of spore coat proteins, separated by a membrane from the thick underlying cortex composed of peptidoglycan. The innermost spore protoplast comprises a germ cell wall, a membrane and the spore core which contains DNA, ribosomes, and enzymes as well as significant quantities of dipicolinic acid and Ca^{2+} ions, but has a low water content. Small acid soluble proteins (SASP) are present which bind to DNA and play a major role in spore UV resistance. Clostridial spores have been shown to contain the α/β-type of SASP [323, 324]. SASP alter the UV photochemistry of DNA, with the result that formation of pyrimidine dimers is greatly diminished while a different spore photoproduct (5-thyminyl-5,6-dihydrothymine) is formed with greater frequency [325]. These photoproducts are repaired efficiently during the early minutes of spore germination. Some clostridial spores have appendages of unknown function, the morphology of which is variable [326-329].

Most studies of the properties of clostridial spores have concentrated on strains which are pathogenic or of potential economic importance. In heat-treatment of low-acid canned foods, the primary goal is to inactivate spores of *C. botulinum*, which have a D-value (time to achieve a ten-fold reduction of viable organisms) of 7–30 min at 100 °C and 0.1–0.2 min at 120 °C. Spores of some other strains, including *C. sporogenes* and *C. thermosaccharolyticum*, are even more resistant [330]. Spores formed by thermophilic strains are generally more heat-resistant than those from mesophiles or psychrophiles. In a study of three thermophilic clostridia, *C. thermocellum*, *C. thermosulfurogenes*, and *C. thermohydrosulfuricum*, Hyun et al. observed a correlation between a higher degree of spore heat resistance and thickness of the spore cortex layer, implying that the cortex is a major determining factor in heat stability [331]. The most resistant spores of *C. thermohydrosulfuricum* exhibited a D value at 121 °C of 11 min, indicating that a proportion can survive a standard autoclave treatment for 15-20 min. While the compositional features which contribute to the heat resistance of spores are not well understood, the potential role of the cortex in maintaining the central protoplast in a dehydrated state has been recognized [332]. Mineralization also is an important factor, although the relationship with heat resistance is complex. Resistance of spores of *C. botulinum* and *C. perfringens* is lower following removal of mineral ions by acid treatment [333, 334]. Since recovery of survivors was improved by treatment with lysozyme, and spore lytic enzyme can be stabilized under some ionic conditions, it was suggested that sensitization of enzymes required for germination may be the cause of thermal inactivation [334]. By contrast, lysozyme had no effect on the recovery of *Bacillus* spores [335]. *C. botulinum* has been shown to accumulate high concentrations of transition metals, particularly zinc, during sporulation, and heat resistance was found to vary in spores produced in media supplemented with different metal ions [337]. The underlying reasons for the effect of metal ions have not been determined.

3.6.3 Mechanism of spore formation

Spore formation has been divided into stages (0 to VII) based on electron microscope observations, and physical and biochemical analyses [320, 330, 338, 339]. However, the order of events is not the same in all cases; several clostridia have been observed to lay down coat protein layers prior to the cortex, rather than the generally observed pattern of cortex first [321, 322, 329, 340-342]. The molecular events associated with sporulation have been extensively characterized in the aerobic bacterium *B. subtilis*, and this organism serves as the model system of differentiation in prokaryotes [338, 339]. Initiation of spore formation is controlled by Spo0A, a transcription factor which modulates gene expression during the transition from the exponential to the stationary phase. Spo0A is the response regulator of a two-component, signal transduction regulatory system, and in growing cells exists predominantly in the dephosphorylated state. Under conditions in which sporulation is initiated, it is phosphorylated by a phosphorelay involving a number of kinases and is thereby able to activate or repress gene expression by binding to specific DNA targets (referred to as "0A boxes") found upstream of regulated genes [343]. Subsequently, changes in gene expression are controlled by the synthesis and activation of alternative σ factors which associate with RNA polymerase and alter the promoter specificity of the enzyme. Five σ factors are known to be produced at various stages. Early in sporulation, the amount of σ^H, a minor σ factor in exponential cells, increases. At later stages, two σ factors (σ^E and σ^K) direct gene expression in the mother cell, while σ^F and σ^G are active in the forespore. The differentiation process is therefore a tightly controlled sequence of events culminating in production of the endospore and its release from the cell.

By comparison with *B. subtilis*, sporulation in clostridia is not well understood, although some details of the process are now beginning to emerge. Homologues of Spo0A were found in six strains of *Clostridium* representing phylogenetically diverse groups, and it was also observed that a homologue of the *B. subtilis spoIVB* gene was present upstream of *spo0A* in each case, demonstrating a similar gene organization [344]. Analysis of the Spo0A homologues identified three highly conserved regions in the effector domain, one of which was proposed to be the DNA-binding surface of the protein. In the same study, a mutant of *C. beijerinckii* NCIMB 8052, in which the *spo0A* gene was inactivated by site-specific integration, was shown to be completely asporogenous, confirming the involvement of the gene in control of spore formation.

The isolation of drug-resistant RNA polymerase mutants of *C. acetobutylicum* which also exhibited sporulation defects [345, 346] argued that altered RNA polymerase specificity may be involved in regulation of sporulation. Purified RNA polymerases from *C. perfringens* and *C. acetobutylicum* consist of a complex of α, β, β' and σ subunits as in other bacteria [347, 348], and genes encoding homologues of vegetative and sporulation-specific σ factors have been cloned from *C. acetobutylicum* or identified by hybridization. The *sigA* (vegetative σ^A) gene is expressed as an operon with *dnaE*, the product of which shows homology to bacterial primases, while the *sigE* (σ^E) and *sigG* (σ^G) genes form a cluster with *spoIIGA* which encodes a sporulation specific protease [349, 350]. While *sigA* was transcribed constitutively, expression of the sporulation-specific sigma factors was shown to occur sequentially in the same sequence (E-G-K) as demonstrated for *B. subtilis*. Furthermore, the promoter sequences identified were consistent with the role of the σ factors in directing gene expression [351]. The presence of a σ^F consensus promoter sequence upstream of *sigG* provided indirect evidence for the presence of this σ factor.

Recently the *spoIIG* (*sigE*) gene of *C. acetobutylicum*, encoding σ^E, was shown to restore sporulation in a *spoIIG* mutant of *B. subtilis*, although temporal expression of the *spoIIG* gene itself appeared to differ in the two bacteria [352]. Therefore, the available evidence indicates that, despite the taxonomic distance between *Bacillus* and *Clostridium* and the fact that nutritional conditions leading to sporulation are markedly different in the two genera, control of the sporulation process is fundamentally similar at the molecular level.

3.6.4 Spore germination

Under appropriate conditions, a spore can revert back to a vegetative cell relatively quickly. This conversion is referred to as germination, and can be defined as a series of inter-related degradation events, triggered by specific germinants, which leads to irreversible loss of typical dormant spore properties [353]. The germination process involves three steps: activation, germination and outgrowth. Activation, which increases the rate and extent of germination, can be achieved by brief heating to elevated but sub-lethal temperatures. Clostridial spores also respond to both pH and alcohols [354, 355]. Germination of clostridial spores can be initiated by various nutrients including amino acids (in particular L-alanine), sugars, lactate, and nicotinamide, and by non-nutrient chemicals such as bicarbonate (summarized in [356]), but is inhibited by oxygen [357]. The mechanisms responsible are, however, ill-defined. Morphological changes associated with germination have been monitored in a number of studies [321, 358-361]. During germination, spores lose their refractility, and the core is transformed into a stainable cytoplasm with ribosomes and a nuclear area. The outgrowth period which follows is characterized by ordered synthesis of RNA, protein, membrane, cell wall, and DNA. Clostridial spore germination and outgrowth occur within the sporangium, the resulting vegetative cell being released through the exosporium.

Cortex hydrolysis, eliminating the osmoregulatory constraint exerted by this layer on the core, would appear to be the key event in spore germination. It is accompanied by imbibition of water, swelling, and release of dipicolinic acid, Ca^{2+} and mucopeptide. Two enzymes involved in cortex degradation have been identified in spores of *C. perfringens* S40. One, named SCLE (spore cortex-lytic enzyme), is an N-acetylmuramyl-L-alanine amidase which acts on intact spore cortex [362], while the other, CFLE (cortical fragment-lytic enzyme), is a muramidase which acts on spore cortex fragments, but has minimal activity on decoated spores [363]. Both enzymes are synthesized at the time of forespore formation, and have been shown to be located at the outside of the cortex of dormant spores implying that degradation proceeds inwards from the periphery [363-365]. SCLE is synthesized as an unusual precursor comprising an N-terminal pre-sequence, N-terminal pro-sequence and C-terminal pro-sequence in addition to the mature enzyme. The N-terminal pre- and C-terminal pro-sequences likely play a role in targeting the enzyme to its correct location, and are removed during spore development. The resulting pro-enzyme is then activated by proteolytic removal of the N-terminal pro-sequence during germination [365, 366]. On the other hand, CFLE is synthesized as the mature enzyme [363]. Activation of SCLE is believed to be the critical step in the commitment to the germination process [353]; this enzyme will produce fragments which can be further acted on by CFLE, a substrate for which is not available until germination is under way.

3.6.5 Events associated with sporulation and stress

It is well established that the transition from exponential to stationary phase is associated with physiological changes other than those directly involved in endospore formation. There are reports of increased production of enzymes such as amylase in *C. perfringens* and protease in *C. sporogenes* [367, 368], and in *C. cellulolyticum* protease activity was correlated with the level of sporulation observed [369]. However, considerably greater emphasis has been placed on understanding the factors controlling the onset of solvent formation in *C. acetobutylicum* and related strains, and toxin production by pathogenic clostridia.

3.6.5.1 Solvent formation

The acetone-butanol fermentation of *C. acetobutylicum* was operated on an industrial scale in the past and, while the technology has been proven, the fermentation is currently unable to compete with chemical synthesis from petroleum products.

Nevertheless interest in the fermentation as an alternative technology, based on renewable fermentation substrates, continues. An understanding of the mechanisms which control the metabolic switch from acid to solvent formation (see section 3.5.1.2), which occurs at the onset of stationary phase, is essential for development of a more effective process.

A number of factors including pH, butyric acid and other metabolic intermediates, NAD(P)H, ATP/ADP ratio, and limitation of certain nutrients have been implicated in the change of metabolism to solvent formation (reviewed recently in [45]). It is clear that changes in gene expression are important, with synthesis of enzymes involved in the pathways leading to acetone and butanol formation being induced [370-373], but little is known of the mechanisms by which this is achieved. The promoters directing solvent gene expression resemble the σ^A promoter consensus, consistent with the fact that the RNA polymerase holoenzyme purified from exponential phase (acidogenic) and solventogenic cells is identical [374], and arguing against a change in σ factor being responsible for induction of solvent synthesis. In addition, the putative *adc* (acetoacetate decarboxylase) promoter of *C. acetobutylicum* has been shown to be functional in both acidogenic and solventogenic phases of growth [375], implying the existence of specific induction or repression mechanisms in regulation of solvent formation. There are several reports of isolation of mutants which show a general loss of solvent formation. Some mutants generated by transposon mutagenesis were shown to carry a single copy of the transposon integrated into the chromosome, indicative of pleiotropically acting elements concerned with regulation of solvent gene expression [376-378]. However, with the exception of a mutant which appeared to have a defect in translation due to poor expression of a tRNA for threonine [379], these mutants have not been characterized. A putative regulatory gene (*solR*) has been located upstream of the *sol* operon encoding aldehyde/alcohol dehydrogenase and CoA transferase in *C. acetobutylicum* ATCC 824. Overexpression of *solR* resulted in repression of the solvent genes *aad* (aldehyde/alcohol dehydrogenase) and *adc* (acetoacetate decarboxylase), while inactivation of *solR* stimulated gene expression and solvent production [380]. Transcription of *solR* itself was strong in exponential phase and declined in stationary phase cells. Thus SolR, which exhibits features of a DNA-binding protein, was proposed to act as a transcriptional repressor of the onset of solvent formation. The mechanism of action of this protein as a transcriptional regulator has, however, recently been questioned [381].

It has long been recognized that sporulation is required for maintenance of good solvent production, with high solvent-producing strains selected and maintained through cycles of sporulation, heat exposure, germination, and outgrowth [189]. Furthermore, repeated batch subculture or growth in continuous culture often results in the loss of the ability to form both solvents and spores, a phenomenon referred to as strain degeneration [382]. Nevertheless, some asporogenous mutants retain the capacity for solventogenesis [383-385]. The industrial strain *C. acetobutylicum* P262 (now named *C. saccharobutylicum*; [116]) shows clearly defined morphological changes associated with the transition to stationary phase (Figure 4). At the end of exponential growth, cells accumulate granulose and vegetative cells convert to swollen, phase-bright clostridial forms and synthesize a capsule. Mutants were isolated which had a pleiotropic defect in these properties together with solvent formation and sporulation, suggestive of common regulatory mechanisms, but some mutants were also obtained which were defective in individual pathways [345, 346]. A correlation was observed between clostridial form and solvent production; sporulation mutants which were unable to form the clostridial stage (*cls*, blocked before stage II) were unable to form solvents and granulose, while those blocked after this (*spo*) entered the clostridial stage and produced granulose, a capsule and normal levels of solvents [345]. Butanol production has in fact been ascribed to stage III of the sporulation process [386]. Inhibitors of chromosome replication, which is necessary for initiation of sporulation-specific events, did not prevent induction of clostridial stage, granulose, capsule and solvent formation [346]. On the other hand, the shift to solventogenesis appeared to be essential for spore production, since cultures which had not undergone transition to the clostridial stage did not form spores [387]. This is consistent with a number of studies investigating the relationship between solventogenesis and sporulation in which solvent negative mutants which could still sporulate were never isolated [377, 378, 385, 388]. Mutants with these properties are invariably defective for enzymes of the solvent-forming pathways [389, 390]. The solventogenic stage also appears to be an essential step in sporulation of *C. thermosaccharolyticum* [391].

Evidence for a direct regulatory link between sporulation and solventogenesis was provided by the observation that, at a frequency indicative of a single mutational event, a pleiotropic (sporulation, granulose, and solvent negative) mutant of *C. beijerinckii* reverted to regain all three traits [385]. *C. acetobutylicum* mutants carrying a single chromosomal transposon insertion were also found to be defective in both sporulation and solvent formation, and hybridization analysis suggested that multiple regulatory elements may contribute to control of these processes [377].

Recently, investigations have centered on the role of the transition state regulator, Spo0A, in control of solvent formation. An integrative *spo0A* mutant of *C. beijerinckii* was shown to be asporogenous, and in addition was unable to form solvents or to synthesize granulose [344, 378, 381]. Sequences closely matching the consensus “0A box” have been found upstream of the solvent genes *adc*, *adhE* (*aad*) and *bdh* (butanol dehydrogenase) in *C. acetobutylicum*, as well as *ptb* (phosphotransbutyrylase) in *C. beijerinckii* [378]. The latter gene is repressed at the onset of solventogenic metabolism [375]. A combination of *in vitro* Spo0A-binding and *in vivo* expression assays, employing wild-type and modified promoters, implicated Spo0A-0A box interaction in regulation of expression of these genes [381]. Thus, solvent formation and sporulation appear to share a common Spo0A-dependent mechanism of initiation.

3.6.5.2 Toxin formation

Several clostridia produce potent toxins, and an understanding of regulation of their synthesis will be important in prevention of serious infections. Some toxins appear to be produced in association with sporulation, as indicated by the isolation of mutants which are negative for both sporulation and toxin production and the recovery of both properties in revertants [392]. The production of *C. perfringens* enterotoxin (CPE) is clearly linked to sporulation, a fact consistent with spore formation being essential for food poisoning which is believed to be caused by the toxin [393-395]. Maximal formation of the toxin was observed to occur after the appearance of heat-resistant spores, and it was released together with mature spores when the sporangium lysed [396]. The toxin is, in fact, synthesized without an N-terminal signal peptide, indicating that it is not actively secreted [397]. Prior to mother cell lysis, CPE can accumulate to 15% of total cell protein in food poisoning isolates, and may be sequestered in inclusion bodies [397a].

Analysis of sporulation mutants indicated that those blocked at stage 0 could not express CPE, while those blocked at stages III, IV, and V did so, but at a reduced level. It was also found that revertants of a stage 0 mutant also regained the ability to synthesize CPE, implying that a single locus affected both spore and toxin formation [398]. Control of CPE synthesis is at the level of transcription, as shown by mRNA analysis and functional studies of the putative *cpe* promoter which became active about 1 hour after entry into stationary phase [399, 400]. Remarkably, similar regulation of *cpe* expression was observed when the gene was transformed into naturally *cpe*-negative hosts. This suggests that, although the vast majority of *C. perfringens* strains do not carry the *cpe* gene, associated transcriptional factors are present which likely have other regulatory functions in the cell [400]. Although details of transcriptional regulation are not available, perfect consensus binding sites for the transition state regulator Hpr have been identified in regions flanking the *cpe* gene [401]. In *B. subtilis,* Hpr is a repressor of gene expression, the concentration of which is controlled by the level of Spo0A-P in the cell [402]. Therefore, it is reasonable to suggest that the expression of *cpe* may be under indirect control by Spo0A. In addition, promoter sequences resembling the consensus of both SigK and SigE have been identified upstream of the *cpe* gene [403]. These are the mother cell-specific σ factors in *B. subtilis,* consistent with the location of CPE synthesis in *C. perfringens.* It seems likely that *cpe* expression is subject to various controls in the sporulating cell.

C. perfringens strains also produce a range of extracellular toxins associated with gas gangrene or clostridial myonecrosis, but these are expressed in exponential cells and have no relationship to sporulation. Toxin production is regulated at the level of transcription by two proteins, VirR and VirS, which constitute a two-component system of the type known to regulate a variety of environmental responses in bacteria [404, 405]. The response regulator of the system, VirR, is believed to stimulate expression of toxin genes either directly or via stimulation of expression of intermediary regulators. Although a low molecular weight compound, "substance A", has been implicated in the activation of the Vir proteins, the nature of the environmental stimulus is unknown (reviewed in [405, 406]).

C. difficile is the major causative agent of pseudomembranous colitis and is also responsible for antibiotic-associated diarrhea. Two related toxins (Toxin A and Toxin B) are produced, the expression of which is repressed during exponential growth and increased substantially as the cells enter stationary phase [407, 408].

However, despite the growth phase-dependent induction, studies investigating the relationship between toxin formation and sporulation in *C. difficile* have yielded conflicting results. Ketley et al. reported that toxin A synthesis was not associated with production of spores or their constituents [409], while a later study found a correlation between the two events [410]. Toxin expression has been shown to be influenced by the nutritional status of the growth medium; in particular, certain amino acids appear to be important for toxin production while synthesis is repressed by glucose and other rapidly metabolized sugars [408, 411-413].

The *toxA* and *toxB* genes, together with three small ORFs, form a small pathogenicity island with gene order *txe1*(or *txeR*), *toxB*, *txe2*, *toxA*, *txe3*, also referred to as *tcdDBEAC* [414]. Control of *tox* gene expression appears complex, with both polycistronic and gene-specific transcripts observed [408, 415]. The *txeR* gene encodes a *trans*-acting protein which positively regulates transcription of the *tox* genes while *txe3*, which is transcribed in the opposite orientation to the other genes, is highly expressed in exponentially growing cells and has been implicated as a repressor of *tox* gene expression [407, 416]. It has been suggested that glucose repression may be targeted at expression of *txeR*. The TxeR protein shows similarity to σ factors of the extracellular function (ECF) family, which control a range of functions including expression of heat shock genes [417]. It is therefore possible that TxeR is a factor which activates toxin synthesis in response to stress. Indeed, stress-associated toxin production has been observed [418, 420], but nothing is known of the signal transduction pathway which might activate TxeR.

C. botulinum and *C. tetani* produce potent neurotoxins which are related at the level of amino acid sequence. Seven (A-G) serologically distinct botulinum toxins (BoNTs) have been recognized, while a single form of tetanus toxin (TeNT) is known. *C. botulinum* strains are divided into four distinct groups, all of which contain both toxigenic and non-toxigenic types (group IV *C. botulinum* strains are sometimes referred to a *C. argentinense*). BoNT is also produced by strains of *C. butyricum* and *C. barati* [421-423]. BoNTs are produced in association with nontoxic proteins to form complexes known as progenitor toxins, of which three forms are recognized. These progenitor toxins exhibit greater stability to high temperature and extremes of pH compared to purified BoNT, which may be rationalized with the food-borne route of toxin transmission. Genes representing all forms of the toxin have been isolated. In each case the gene encoding BoNT is part of a transcriptional unit which also includes genes encoding any associated nontoxic, non-hemagglutinating protein (NTNH) and hemagglutinins (HA). In general, genes encoding NTNH and BoNT are contiguous and arranged divergently with an operon encoding HA components (for review see [422]), although different gene arrangements have been identified in some strains [424].

BoNT expression is activated by a transcriptional factor, BotR. BotR/A (found in *C. botulinum* type A strains) has been shown to bind to the promoter regions upstream of the NTNH/BoNT and HA operons [425], and homologous proteins in other types presumably act in a similar fashion. Toxin formation has been shown to be affected by the nitrogen status of the medium, being repressed by arginine and tryptophan and stimulated by casein and casein hydrolysate [426-428], but the associated environmental and cellular signals have not been identified. Toxin expression appears to be maximal in stationary phase, and the highest levels of toxin production are generally found in cell populations which undergo rapid autolysis and do not sporulate [428, 429]. Only in the case of type C_2 toxin has an association between toxin production and sporulation been reported [430].

The tetanus toxin is not associated with nontoxic proteins, and the gene encoding TeNT in *C. tetani* is expressed monocistronically. Again, toxin formation is affected by nitrogenous compounds [431-433] and, like botulinum toxin, TeNT is produced in highest quanitities by non-sporulating cultures [429]. Expression of TeNT is under the control of the transcriptional activator TetR, which is related to BotR, and both BotR/A and BotR/C have been shown to be capable of stimulating TeNT synthesis in *C. tetani*. TetR and BotR are members of a family of regulatory genes found in *Clostridium* species which includes TxeR of *C. difficile* [434].

A strain of *C. bifermentans* has been found to produce larvicidal toxins which are active against mosquitoes and blackflies. Production of these toxins is associated with sporulation, as is the case in the larvicidal bacilli, *B. thuringiensis israelensis* and *B. sphaericus*. Larvicidal activity reaches a maximum between 5 and 8 hours after initiation of sporulation, but decreases considerably upon lysis of the mother cell sporangium, apparently due to inactivation by proteases [435-437]. The mechanism of control of toxin synthesis has not been described.

3.6.5.3 The heat shock response

When subjected to several environmental stresses, all organisms respond by inducing the synthesis of a set of protective proteins; these are known as heat shock proteins (hsp), since they are characteristically produced following a temperature upshift. Following a sudden temperature shift from 28-30 °C to above 40 °C, up to 15 proteins were shown to be transiently synthesized by two strains of *C. acetobutylicum*. In batch culture, some hsp were also induced by exposure to solvents or air, or at the end of the exponential growth phase [438], while the same was observed in continuous culture by lowering the pH or the growth rate [439]. The parallels between the response to heat shock and conditions favoring solvent formation suggested that hsp might be involved in the metabolic shift.

The two most prominent hsp, which were also present in non-stressed cells, were identified as the highly conserved molecular chaperones GroEL and DnaK by immuno-reaction to antibodies raised against the corresponding proteins from *E. coli* [438, 439]. Subsequently, genes encoding further hsp have been isolated and characterized and shown to be organized within the *dnaK* operon comprising seven genes, *orfA, grpE, dnaK, dnaJ, orfB, orfC,* and *orfD* [440, 441] and the *groE* operon comprising *groES* and *groEL* [442]. In addition, a gene termed *hsp18* was found to constitute a third, monocistronic operon [443]. The *groE* operon of *C. thermocellum* has recently been sequenced and shown to have a similar organization to that in *C. acetobutylicum* [444].

The proteins DnaK, DnaJ, and GrpE from *C. acetobutylicum* DSM 1731 were recently shown to have the capacity to refold guanidinium-denatured firefly luciferase, and to prevent aggregation of OrfA. Furthermore, ATPase activity of *E. coli*

DnaK was stimulated by DnaJ from both *C. acetobutylicum* and *E. coli* [441]. Therefore, the DnaK system appears to be functionally equivalent as a chaperone system to its counterpart in *E. coli*. The GroEL protein of *C. thermocellum* has also been shown to be an ATPase and, like GroEL of *E. coli*, is inhibited by *E. coli* GroES. The release of GroEL and GroES from cell pellets by a buffer wash similar to that which has been shown to release the cellulosome has led to speculation that these chaperones may play a role in regulating translocation and assembly of cellulosomal components on the cell surface [445].

Hsp gene expression in *C. acetobutylicum* and *C. thermocellum* is controlled at the level of transcription, with maximum levels of mRNA apparent 15 minutes after temperature upshift from 30 to 42 °C or from 60 to 70 °C, respectively, but there is no evidence for the involvement of an alternative σ factor as is the case in *E. coli* [444, 446]. On the contrary, regulation has been proposed to be mediated by binding of a repressor (putatively OrfA in *C. acetobutylicum*) to a conserved palindromic sequence located between the transcription and translation start sites [441, 444, 446]. Further work will be required to elucidate the details of the regulatory mechanism, the nature of the signal transduction system responsible for activation, and the interaction between the heat shock response and solvent formation in *C. acetobutylicum*.

3.7 Conclusion

The genus *Clostridium* comprises a large number of species which, although sharing a number of general properties, exhibit a diverse range of biological and physiological characteristics. Study of the metabolic principles which underpin this diversity has proved a fertile area for investigation, leading to significant advances in understanding of the biochemistry, energetics and growth of the clostridia and other anaerobes. This has been accompanied by an appreciation of the potential for exploiting the metabolic capabilities of the bacteria. In areas such as cellulose degradation, fermentation for production of fuels and chemicals, and stereospecific biotransformations the clostridia and their enzymes are being viewed with interest by biotechnologists, encouraged by the successful operation of the *C. acetobutylicum* acetone-butanol fermentation in the past [447]. Even the lethal *C. botulinum* neurotoxin is being used therapeutically in the treatment of human dystonias [429].

As described in this chapter the principal pathways of metabolism of a wide range of compounds by clostridia have been established, and the process of differentiation to form endospores is well studied from a morphological perspective. Although understanding of molecular control mechanisms is in most cases more rudimentary, the advent of procedures for gene cloning, manipulation and analysis has contributed significantly to advances in this area over the last few years [448]. The recent development of reporter gene systems [375, 408, 428, 449, 450] and the availability of complete clostridial genome sequences, will now act as a stimulus for further investigations of the control of expression of clostridial genes. The study of the biology and physiology of the clostridia promises to be a productive pursuit for some time to come.

Acknowledgement
I am grateful to the Biotechnology and Biological Sciences Research Council for financial support.

References

[1] Morris, J. G. (1993), History and future potential of the clostridia in biotechnology, in: *The Clostridia and Biotechnology* (Woods, D. R., Ed.), pp. 1-23, Stoneham: Butterworth-Heinemann.

[2] Lee, Y-E., Jain, M. K., Lee, C., Lowe, S. E., Zeikus, J. G. (1993), Taxonomic distinction of saccharolytic thermophilic anaerobes: description of *Thermoanaerobacterium xylanolyticum* gen. nov., sp. nov., and *Thermoanaerobacterium saccharolyticum* gen. nov., sp. nov.; reclassification of *Thermoanaerobacterium brockii, Clostridium thermosulfurogenes*, and *Clostridium thermohydrosulfuricum* E100-69 as *Thermoanaerobacter brockii* comb. nov., *Thermoanaerobacterium thermosulfurigenes* comb. nov., and *Thermoanaerobacter thermohydrosulfuricus* respectively; and transfer of *Clostridium thermohydrosulfuricum* 39E to *Thermoanaerobacter ethanolicus, Int. J. Syst. Bacteriol.* **43,** 41-51.

[3] Cato, E. P., George, W. L., Finegold, S. M. (1986), Genus *Clostridium* Prazmowski 1880, in: *Bergey"s Manual of Systematic Bacteriology Vol.2* (Sneath, P. H. A., Mair, N. S., Sharpe, E. S., Holt, J. G., Eds.), pp. 1141-1200, Baltimore: Williams and Williams.

[4] Cummins, C. S., Johnson, J. L. (1971), Taxonomy of the clostridia: wall composition and DNA homologies in *Clostridium butyricum* and other butyric acid producing clostridia, *J. Gen. Microbiol.* **67,** 33-46.

[5] Schleifer, K. H., Kandler, O. (1972), Peptidoglycan types of bacterial cell walls and their taxonomic implications, *Bacteriol. Rev.* **36,** 407-477.

[6] Poxton, I. R., Cartmill, T. D. I. (1982), Immunochemistry of the cell-surface carbohydrate antigens of *Clostridium difficile, J. Gen. Microbiol.* **128,** 1365-1370.

[7] Hollaus, F., Sleytr, U. B. (1972), On the taxonomy and fine structure of some hyperthermophilic saccharolytic clostridia, *Arch. Microbiol.* **86,** 129-146.

[8] Van Gylswyck, N. O., Morris, E. J., Els, H. J. (1980), Sporulation and cell wall structure of *Clostridium polysaccharolyticum* comb. nov. (formerly *Fusobacterium polysaccharolyticum*), *J. Gen. Microbiol.* **121,** 491-493.

[9] König, H., Buckel, W., Langworthy, T. A. (1985), Ultrastructure of the cell envelope and amino acid composition of the murein of *Clostridium symbiosum, FEMS Microbiol. Lett.* **30,** 283-288.

[10] Rogers, G. M., Messner, P. (1992), Improved description of the cell wall architecture of the xylanolytic eubacterium *Clostridium xylanolyticum, Int. J. Syst. Bacteriol.* **42,** 492-493.

[11] Sleytr, U. B., Messner, P. (1983), Crystalline surface layers on bacteria, *Ann. Rev. Microbiol.* **37,** 311-339.

[12] Messner, P., Sleytr, U. B. (1992), Crystalline bacterial cell-surface layers, *Adv. Microb. Physiol.* **33,** 213-275.

[13] Kawata, T., Takeoka, A., Takumi, K., Masuda, K. (1984), Demonstration and preliminary characterization of a regular array in the cell wall of *Clostridium difficile, FEMS Microbiol. Lett.* **24,** 323-328.

[14] Takeoka, A., Takumi, K., Koga, T., Kawata, T. (1991), Purification and characterization of S layer proteins from *Clostridium difficile* GAI 0714, *J. Gen. Microbiol.* **137,** 261-267.

[15] Sleytr, U. B., Glauert, A. M. (1976), Ultrastructure of the cell walls of two closely related clostridia that possess different regular

arrays of surface subunits, *J. Bacteriol.* **126,** 869-882.

[16] Sára, M., Kalsner, I., Sleytr, U.B. (1988), Surface properties from the S-layer of *Clostridium thermosaccharolyticum* D120-70 and *Clostridium thermohydrosulfuricum* L111-69, *Arch. Microbiol.* **149,** 527-533.

[17] Lemaire, M., Miras, I., Gounon, P., Béguin, P. (1998), Identification of a region responsible for binding to the cell wall within the S-layer protein of *Clostridium thermocellum, Microbiology* **144,** 211-217.

[18] Brechtel, E., Matuschek, M., Hellberg, A., Egelseer, E.M., Schmid, R., Bahl, H. (1999), Cell wall of *Thermoanaerobacterium themosulfurigenes* EM1: isolation of the components and attachment of the xylanase XynA, *Arch. Microbiol.* **171,** 159-165.

[19] Sára, M., Sleytr, U.B. (2000), S-layer proteins, *J. Bacteriol.* **182,** 859-868.

[20] Goldfine, H. (1993), Phospholipid biosynthetic enzymes of butyric acid – producing clostridia, in: *Genetics and Molecular Biology of Anaerobic Bacteria* (Sebald, M., Ed.), pp. 354-362, New York: Springer-Verlag.

[21] Johnston, N.C., Goldfine, H. (1983), Lipid composition in the classification of the butyric acid-producing clostridia, *J. Gen. Microbiol.* **129,** 1075-1081.

[22] Ghanem, F.M., Ridpath, A.C., Moore, W.E.C., Moore, L.V.H. (1991), Identification of *Clostridium botulinum, Clostridium argentinense,* and related organisms by cellular fatty acid synthesis, *J. Clin. Microbiol.* **29,** 1114-1124.

[23] Johnston, N.C., Goldfine, H., Fischer, W. (1994), Novel polar lipid composition of *Clostridium innocuum* as the basis for an assessment of its taxonomic status, *Microbiology* **140,** 105-111.

[24] Khuller, G.K., Goldfine, H. (1974), Phospholipids of *Clostridium butyricum.* V. Effects of growth temperature on fatty acid, alk-1-enyl ether group, and phospholipid composition, *J. Lipid Res.* **15,** 500-507.

[25] Goldfine, H., Khuller, G.K., Bone, R.P., Silverman, B., Selick, H., Johnston, N.C., Vanderkooi, J.M., Horwitz, A.F. (1977), Effects of growth temperature and supplementation with exogenous fatty acids on some physical properties of *Clostridium butyricum* phospholipids, *Biochim. Biophys. Acta* **488,** 341-352.

[26] Johnston, N.C., Goldfine, H. (1985), Phospholipid aliphatic chain composition modulates lipid class composition, but not lipid asymmetry in *Clostridium butyricum, Biochim. Biophys. Acta* **813,** 10-18.

[27] Goldfine, H., Johnston, N.C., Mattai, J., Shipley, G.G. (1987), Regulation of bilayer stability in *Clostridium butyricum*: studies on the polymorphic phase behaviour of the ether lipids, *Biochemistry* **26,** 814-822.

[28] Johnston, N.C., Goldfine, H. (1992), Replacement of the aliphatic chains of *Clostridium acetobutylicum* by exogenous fatty acids: regulation of phospholipid and glycolipid composition, *J. Bacteriol.* **174,** 1843-1853.

[29] Vollherbst-Schneck, K., Sands, J.A., Montenecourt, B.S. (1984), Effect of butanol on lipid composition and fluidity of *Clostridium acetobutylicum* ATCC 824, *Appl. Environ. Microbiol.* **47,** 193-194.

[30] Lepage, C., Fayolle, F., Hermann, M., Vandecasteele, J.-P. (1987), Changes in membrane lipid composition of *Clostridium acetobutylicum* during acetone-butanol fermentation: effects of solvents, growth temperature and pH, *J. Gen. Microbiol.* **133,** 103-110.

[31] Baer, S.H., Blaschek, H.P., Smith, T.L. (1987), Effect of butanol challenge and temperature on lipid composition and membrane fluidity of butanol-tolerant *Clostridium acetobutylicum, Appl. Environ. Microbiol.* **53,** 2854-2861.

[32] Morris, J.G. (1975), The physiology of obligate anaerobiosis, *Adv. Microb. Physiol.* **12,** 169-246.

[33] O'Brien, R.W., Morris, J.G. (1971), Oxygen and the growth and metabolism of *Clostridium acetobutylicum, J. Gen. Microbiol,* **68,** 307-318.

[34] Hewitt, J., Morris, J.G. (1975), Superoxide dismutase in some obligately anaerobic bacteria. *FEBS Lett.* **50,** 315-318.

[35] Kawasaki, S., Nakagawa, T., Nishiyama, Y., Benno, Y., Uchimura, T., Komagata, K., Kozaki, M, Niimura, Y. (1998), Effect of oxygen on the growth of *Clostridium butyricum* (type species of the genus *Clostridium*), and the distribution of enzymes for oxygen and for active oxygen species in clostridia, *J. Ferment. Bioeng.* **86,** 368-372.

[36] Hippe, H., Andreesen, J.R., Gottschalk, G. (1992). The genus *Clostridium*-nonmedical, in: *The Prokaryotes* (Balows, A., Trüper, H.G.,

Dworkin, M., Harder, W., Schleifer, K-H., Eds.), 2nd Ed., pp. 1800-1866, New York: Springer-Verlag.

[37] Schiefer-Ullrich, H., Wagner, R., Dürre, P., Andreesen, J. R. (1984), Comparative studies on physiology and taxonomy of obligately purinolytic clostridia, *Arch. Microbiol.* **138,** 345-353.

[38] Bahl, H., Dürre, P. (1993), Clostridia, in: *Biotechnology* (Rehm, H.-J., Reed, G., Pühler, A., Stadler, P., Eds.), pp. 285-323, Weinheim: VCH.

[39] Andreesen, J. R., Bahl, H., Gottschalk, G. R. (1989), Introduction to the physiology and biochemistry of the genus *Clostridium,* in: *Biotechnology Handbooks Vol.3: Clostridia* (Minton, N. P., Clarke, D. J., Eds.), pp. 27-62, New York: Plenum Press.

[40] Saha, B. C., Lamed, R., Zeikus, J. G. (1989), Clostridial enzymes, in: *Biotechnology Handbooks Vol.3: Clostridia* (Minton, N. P., Clarke, D. J., Eds.), pp. 227-263, New York: Plenum Press.

[41] Antranikian, G. (1990), Physiology and enzymology of thermophilic anaerobic bacteria degrading starch, *FEMS Microbiol. Rev.* **75,** 201-218.

[42] Roberts, T. A., Hobbs, G. (1968), Low temperature growth characteristics of clostridia, *J. Appl. Bacteriol.* **31,** 75-78.

[43] Collins, M. D., Rodrigues, U. M., Dainty, R. H., Edwards, R. A., Roberts, T. A. (1992), Taxonomic studies on a psychrophilic *Clostridium* from vacuum-packed beef: Description of *Clostridium estertheticum* sp. nov., *FEMS Microbiol. Lett.* **96,** 235-240.

[44] Rogers, P. (1986), Genetics and biochemistry of *Clostridium* relevant to development of fermentation processes, *Adv. Appl. Microbiol.* **31,** 1-60.

[45] Mitchell, W. J. (1998), Physiology of carbohydrate to solvent conversion by clostridia, *Adv. Microb. Physiol.* **39,** 31-130.

[46] Jones, D. T., Woods, D. R. (1986), Acetone-butanol fermentation revisited, *Microbiol. Rev.* **50,** 484-524.

[47] Tomme, P., Warren, R. A. J., Gilkes, N. R. (1995), Cellulose hydrolysis by bacteria and fungi, *Adv. Microb. Physiol.* **37,** 1-81.

[48] Allcock, E. R., Woods, D. R. (1981), Carboxymethyl cellulase and cellobiase production by *Clostridium acetobutylicum* in an industrial fermentation medium, *Appl. Environ. Microbiol.* **41,** 539-541.

[49] Lee, S. F., Forsberg, C. W., Gibbins, L. N. (1985), Cellulolytic activity of *Clostridium acetobutylicum, Appl. Environ. Microbiol.* **50,** 220-228.

[49] Sauer, U., Treuner, A., Buccholz, M., Santangelo, J. D., Dürre, P. (1994), Sporulation and primary sigma factor homologous genes in *Clostridium acetobutylicum, J. Bacteriol.* **176,** 6572-6582.

[50] Strobel, H. J., Caldwell, F. C., Dawson, K. A. (1995), Carbohydrate transport by the anaerobic thermophile *Clostridium thermocellum* LQR1, *Appl. Environ. Microbiol.* **61,** 4012-4015.

[51] Alexander, J. K. (1968), Purification and specificity of cellobiose phosphorylase from *Clostridium thermocellum, J. Biol. Chem.* **243,** 2899-2904.

[52] Sheth, K., Alexander, J. K. (1969), Purification and properties of β-1,4-oligoglucan: orthophosphate glucosyltransferase from *Clostridium thermocellum, J. Biol. Chem.* **244,** 457-464.

[53] Reichenbecher, M., Lottspeich, F., Bronnenmeier, K. (1997), Purification and properties of a cellobiose phosphorylase (CepA) and a cellodextrin phosphorylase (CepB) from the cellulolytic thermophile *Clostridium stercorarium, Eur. J. Biochem.* **247,** 262-267.

[54] Lamed, R., Bayer, E. A. (1988), The cellulosome of *Clostridium thermocellum, Adv. Appl. Microbiol.* **33,** 1-46.

[55] Felix, C. R., Ljungdahl, L. G. (1993), The cellulosome: the exocellular organelle of *Clostridium, Ann. Rev. Microbiol.* **47,** 791-819.

[56] Béguin, P., Alzari, P. M. (1998), The cellulosome of *Clostridium thermocellum, Biochem. Soc. Trans.* **26,** 178-185.

[57] Lamed, R., Naimark, J., Morgenstern, E., Bayer, E. A. (1987), Specialised cell surface structures in cellulolytic bacteria, *J. Bacteriol.* **169,** 3792-3800.

[58] Doi, R. H., Goldstein, M., Hashida, S., Park, J. S., Tagaki, M. (1994), The *Clostridium cellulovorans* cellulosome, *CRC Crit. Rev. Microbiol.* **20,** 87-93.

[59] Gal, L., Pages, S., Gaudin, C., Belaich, A., Reverbel-Leroy, C., Tardif, C., Belaich, J-P. (1997), Characterization of the cellulolytic complex (cellulosome) produced by *Clostridium cellulolyticum, Appl. Environ. Microbiol.* **63,** 903-909.

[60] Kakiuchi, M., Isui, A., Suzuki, K., Fujino, T., Fujino, E., Kimura, T., Karita, S., Saki, K.,

Ohmiya, K. (1998), Cloning and DNA sequencing of the genes encoding *Clostridium josui* scaffolding protein CipA and cellulase CelD and identification of their gene products as major components of the cellulosome, *J. Bacteriol.* **180**, 4303-4308.

[61] Pohlschröder, M., Leschine, S. B., Canale-Parola, E. (1994), Multicomplex cellulase-xylanase system of *Clostridium papyrosolvens*, *J. Bacteriol.* **176**, 70-76.

[62] Pohlschröder, M., Canale-Parola, E., Leschine, S. B. (1995), Ultrastructural diversity of the cellulase complexes of *Clostridium papyrosolvens* C7, *J. Bacteriol.* **177**, 6625-6629.

[63] Bayer, E. A., Morag, E., Lamed, R. (1994), The cellulosome – a treasure trove for biotechnology, *Trends Biotechnol.* **12**, 379-386.

[64] Gerngross, U. T., Romaniec, M. P. M., Kobayashi, T., Huskisson, N. S., Demain, A. L. (1993), Sequencing of a *Clostridium thermocellum* gene (*cipA*) encoding the cellulosomal S(L) protein reveals an unusual degree of internal homology, *Mol. Microbiol.* **8**, 325-334.

[65] Hazlewood, G. P., Gilbert, H. J. (1993), Xylan and cellulose utilization by the clostridia, in: *The Clostridia and Biotechnology* (Woods, D. R., Ed.), pp. 311-341, Stoneham: Butterworth-Heinemann.

[66] Kataeva, I., Guglielmi, G., Béguin, P. (1997), Interaction between *Clostridium thermocellum* endoglucanase CelD and polypeptides derived from the cellulosome-integrating protein CipA: stoichiometry and cellulolytic activity of the complexes, *Biochem. J.* **326**, 617-624.

[67] Pagès, S., Gal, L., Bélaïch, A., Gaudin, C., Tardif, C., Bélaïch, J-P. (1997), Role of scaffolding protein CipC of *Clostridium cellulolyticum* in cellulose degradation, *J. Bacteriol.* **179**, 2810-2816.

[68] Leibovitz, E., Ohayon, H., Gounon, P., Béguin, P. (1997), Characterization and subcellular localization of the *Clostridium thermocellum* scaffoldin dockerin binding protein SdbA, *J. Bacteriol.* **179**, 2519-2523.

[69] Chavaux, S., Matuschek, M., Béguin, P. (1999), Distinct affinity of binding sites for S-layer homologous domains in *Clostridium thermocellum* and *Bacillus anthracis* cell envelopes, *J. Bacteriol.* **181**, 2455-2458.

[70] Tamaru, Y., Doi, R. H. (1999), Three surface layer homology domains at the N terminus of the *Clostridium cellulovorans* major cellulosomal subunit EngE, *J. Bacteriol.* **181**, 3270-3276.

[71] Mohand-Oussaid, O., Payot, S., Guedon, E., Gelhaye, E., Youyou, A., Petitdemange, H. (1999), The extracellular xylan degradative system in *Clostridium cellulolyticum* cultivated on xylan: evidence for cell-free cellulosome production, *J. Bacteriol.* **181**, 4035-4040.

[72] Bronnenmeier, K., Staudenbauer, W. L. (1988), Resolution of *Clostridium stercorarium* cellulase by fast protein liquid chromatography (FPLC), *Appl. Microbiol. Biotechnol.* **27**, 432-436.

[73] Bronnenmeier, K., Staudenbauer, W. L. (1990), Cellulose hydrolysis by a highly thermostable endo-1,4-β-glucanase (Avicelase I) from *Clostridium stercorarium*, *Enzyme Microb. Technol.* **12**, 431-436.

[74] Riedel, K., Ritter, J., Bronnenmeier, K. (1997), Synergistic interaction of the *Clostridium stercorarium* cellulases Avicelase I (CelZ) and Avicelase II (CelY) in the degradation of microcrystalline cellulose, *FEMS Microbiol. Lett.* **147**, 239-243.

[75] Riedel, K., Bronnenmeier, K. (1999), Active-site mutations which change the substrate specificity of the *Clostridium stercorarium* cellulase CelZ. Implications for synergism, *Eur. J. Biochem.* **262**, 218-223.

[76] Schwarz, W. H., Adelsberger, H., Jauris, S. (1990), Xylan degradation by the thermophile *Clostridium stercorarium*: cloning and expression of xylanase, β-D-xylosidase, and α-L-arabinofuranosidase genes in *Escherichia coli*, *Biochem. Biophys. Res. Commun.* **170**, 368-374.

[77] Schwarz, W. H., Bronnenmeier, K., Krause, B., Lottspeich, F., Staudenbauer, W. L. (1995), Debranching of arabinoxylan: properties of the thermoactive recombinant α-L-arabinofuranosidase from *Clostridium stercorarium* (ArfB), *Appl. Microbiol. Biotechnol.* **43**, 856-860.

[78] Zverlov, V. V., Liebl, W., Bachleitner, M., Schwarz, W. (1998), Nucleotide sequence of *arfB* of *Clostridium stercorarium*, and prediction of catalytic residues of α-L-arabinofuranosidases based on local similarity with several families of glycosyl hydrolases, *FEMS Microbiol. Lett.* **164**, 337-343.

[79] Fernandes, A. C., Fontes, C. M. G. A., Gilbert, H. J., Hazlewood, G. P., Fernandes, T. H., Ferreira, L. M. A. (1999), Homologous xylanases from *Clostridium thermocellum*: evidence for bi-functional activity, synergism

between xylanase catalytic modules and the presence of xylan-binding domains in enzyme complexes, *Biochem. J.* **342**, 105-110.

[80] Bronnenmeier, K., Staudenbauer, W. L. (1993), Molecular biology and genetics of substrate utilization in clostridia, in: *The Clostridia and Biotechnology* (Woods, D. R., Ed.), pp. 261-309, Stoneham: Butterworth-Heinemann.

[81] Morag, E., Bayer, E. A., Lamed, R. (1990), Relationship of cellulosomal and noncellulosomal xylanases of *Clostridium thermocellum* to cellulose-degrading enzymes, *J. Bacteriol.* **172**, 6098-6105.

[82] Garcia-Martinez, D. V., Shinmyo, A., Madia, A., Demain, A. L. (1980), Studies on cellulase production by *Clostridium thermocellum*, *Eur. J. Appl. Microbiol. Biotechnol.* **9**, 189-197.

[83] Wiegel, J., Mothershed, C. P., Puls, J. (1985), Differences in xylan degradation by various non-cellulolytic thermophilic anaerobes and *Clostridium thermocellum*, *Appl. Environ. Microbiol.* **49**, 656-659.

[84] Guglielmi, G., Béguin, P. (1998), Cellulase and hemicellulase genes of *Clostridium thermocellum* from five independent collections contain few overlaps and are widely scattered across the chromosome, *FEMS Microbiol. Lett.* **161**, 209-215.

[85] Mishra, S., Béguin, P., Aubert, J-P. (1991), Transcription of *Clostridium thermocellum* endoglucanase genes *celF* and *celD*, *J. Bacteriol.* **173**, 80-85.

[86] Tamaru, Y., Doi, R. H. (2000), The *engL* gene cluster of *Clostridium cellulovorans* contains a gene for cellulosomal *manA*, *J. Bacteriol.* **182**, 244-247.

[87] Bagnara-Tardif, C., Gaudin, C., Belaich, A., Hoest, P., Citard, T., Belaich, J-P. (1992), Sequence analysis of a gene cluster encoding cellulases from *Clostridium cellulolyticum*, *Gene* **119**, 17-28.

[88] Johnson, E. A., Reese, E. T., Demain, A. L. (1982), Inhibition of *Clostridium thermocellum* cellulase by end products of cellulolysis, *J. Appl. Bacteriol.* **4**, 64-71.

[89] Lamed, R., Kenig, R., Setter, E., Bayer, E. A. (1985), Major characteristics of the cellulolytic system of *Clostridium thermocellum* coincide with those of the purified cellulosome, *Enzyme Microb. Technol.* **7**, 37-41.

[90] Morag, E., Halevy, I., Bayer, E. A., Lamed, R. (1991), Isolation and properties of a major cellobiohydrolase from the cellulosome of *Clostridium thermocellum*, *J. Bacteriol.* **173**, 4155-4162.

[91] Kruus, K., Andreacchi, A., Wang, W. K., Wu, J. H. D. (1995), Product inhibition of the recombinant CelS, an exoglucanase component of the *Clostridium thermocellum* cellulosome, *Appl. Microbiol. Biotechnol.* **44**, 399-404.

[92] Shinmyo, A., Garcia-Martinez, D. V., Demain, A. L. (1979), Studies on the extracellular cellulolytic enzyme complex produced by *Clostridium thermocellum*, *J. Appl. Biochem.* **1**, 202-209.

[93] Pètre, J., Longin, R., Millet, J. (1981), Purification and some properties of an endo-β-1,4-glucanase from *Clostridium thermocellum*, *Biochimie* **63**, 629-639.

[94] Schwarz, W. H., Gräbnitz, F., Staudenbauer, W. L. (1986), Properties of a *Clostridium thermocellum* endoglucanase produced in *Escherichia coli*, *Appl. Environ. Microbiol.* **51**, 1293-1299.

[95] Golovchenko, N. P., Singh, R. N., Velikodvorskaya, G. A., Akimenko, V. K. (1993), Isolation and characterization of a lichenan-degrading hydrophobic endoglucanase of *Clostridium thermocellum*, *Appl. Microbiol. Biotechnol.* **39**, 74-79.

[96] Johnson, E. A., Bouchot, F., Demain, A. L. (1985), Regulation of cellulase formation in *Clostridium thermocellum*, *J. Gen. Microbiol.* **131**, 2303-2308.

[97] Nochur, S. V., Roberts, M. F., Demain, A. L. (1993), True cellulase production by *Clostridium thermocellum* grown on different carbon sources, *Biotechnol. Lett.* **15**, 641-646.

[98] Halliwell, G., Phillips, T. M., Halliwell, N. (1995), Microcrystalline forms of cellulose as substrates for strains of *Clostridium thermocellum* and cellulase formation, *Proc. Biochem.* **30**, 243-250.

[99] Blair, B. G., Anderson, K. L. (1999), Regulation of cellulose-inducible structures of *Clostridium cellulovorans*, *Can. J. Microbiol.* **45**, 242-249.

[100] Béguin, P., Rocancourt, M., Chebrou, M.-C., Aubert, J-P. (1986), Mapping of mRNA encoding endoglucanase A from *Clostridium thermocellum*, *Mol. Gen. Genet.* **202**, 251-254.

[101] Bayer, E. A., Setter, E., Lamed, R. (1985), Organization and distribution of the cellulosome in *Clostridium thermocellum*, *J. Bacteriol.* **163**, 552-559.

[102] Bhat, S., Goodenough, P. W., Owen, E., Bhat, M. K. (1993), Cellobiose – a true inducer of cellulosome in different strains of *Clostridium thermocellum*, *FEMS Microbiol. Lett.* **111**, 73-78.
[103] Yaron, S., Morag, E., Bayer, E. A., Lamed, R., Shoham, Y. (1995), Expression, purification and subunit-binding properties of cohesins 2 and 3 of the *Clostridium thermocellum* cellulosome, *FEBS Lett.* **360**, 121-124.
[104] Lytle, S., Myers, C., Kruus, K., Wu, J. H. D. (1996), Interactions of the CelS binding ligand with various receptor domains of the *Clostridium thermocellum* cellulosomal scaffolding protein, CipA, *J. Bacteriol.* **178**, 1200-1203.
[105] Spreinat, A., Antranikian, G. (1992), Analysis of the amylolytic enzyme system of *Clostridium thermosulfurogenes* EM1: purification and synergistic action of pullulanases and maltohexaose forming α-amylase, *Starch* **44**, 305-312.
[106] Wind, R. D., Liebl, W., Buitelaar, R. M., Penninga, D., Spreinat, A., Dijkhuizen, L., Bahl, H. (1995), Cyclodextrin formation by the thermostable α-amylase of *Thermoanaerobacterium thermosulfurigenes* EM1 and reclassification of the enzyme as a cyclodextrin glycosyltransferase, *Appl. Environ. Microbiol.* **61**, 1257-1265.
[107] Melasniemi, H. (1988), Purification and some properties of the extracellular α-amylase-pullulanase produced by *Clostridium thermohydrosulfuricum*, *Biochem. J.* **250**, 813-818.
[108] Saha, B. C., Mathupala, S. P., Zeikus, J. G. (1988), Purification and characterization of a highly thermostable novel pullulanase from *Clostridium thermohydrosulfuricum*, *Biochem. J.* **250**, 343-348.
[109] Spreinat, A., Antranikian, G. (1990), Purification and properties of a thermostable pullulanase from *Clostridium thermosulfurogenes* EM1 which hydrolyzes both α-1,6 and α-1,4-glycosidic linkages, *Appl. Microbiol. Biotechnol.* **33**, 511-518.
[110] Specka, U., Mayer, F., Antranikian, G. (1991), Purification and properties of a thermoactive glucoamylase from *Clostridium thermosaccharolyticum*, *Appl. Environ. Microbiol.* **57**, 2317-2323.
[111] Albasheri, K. A., Mitchell, W. J. (1995), Identification of two α-glucosidase activities in *Clostridium acetobutylicum* NCIB 8052, *J. Appl. Bacteriol.* **78**, 149-156.
[112] Melasniemi, H., Paloheimo, M., Hemiö, L. (1990), Nucleotide sequence of the α-amylase-pullulanase gene from *Clostridium thermohydrosulfuricum*, *J. Gen. Microbiol.* **136**, 447-454.
[113] Mathupala, S. P., Lowe, S. E., Podkovyrov, S. M., Zeikus, J. G. (1993), Sequencing of the amylopullulanase (*apu*) gene of *Clostridium thermohydrosulfuricum* 39E and identification of the active site by site-directed mutagenesis, *J. Biol. Chem.* **268**, 16332-16344.
[114] Sahm, K., Matuschek, M., Müller, H., Mitchell, W. J., Bahl, H. (1996), Molecular analysis of the *amy* gene locus of *Thermoanaerobacterium thermosulfurigenes* EM1 encoding starch-degrading enzymes and a binding protein-dependent maltose transport system, *J. Bacteriol.* **178**, 1039-1046.
[115] Kitamoto, N., Yamagata, H., Kato, T., Tsukagoshi, N., Ukada, S. (1988), Cloning and sequencing of the gene encoding thermophilic β-amylase of *Clostridium thermosulfurogenes*, *J. Bacteriol.* **170**, 5848-5854.
[116] Shaheen, R., Shirley, M., Jones, D. T. (2000), Comparative fermentation studies of industrial strains belonging to four species of solvent-producing clostridia, *J. Mol. Microbiol. Biotech.* **2**, 115-124.
[117] Davison, S. P., Santangelo, J. D., Reid, S. J., Woods, D. R. (1995), A *Clostridium acetobutylicum* regulator gene (*regA*) affecting amylase production in *Bacillus subtilis*, *Microbiology* **141**, 989-996.
[118] Saier, M. H. Jr., Chavaux, S., Cook, G. M., Deutscher, J., Paulsen, I. T., Reizer, J., Ye, J.-J. (1996), Catabolite repression and inducer control in Gram-positive bacteria, *Microbiology* **142**, 217-230.
[119] Hyun, H. H., Zeikus, J. G. (1985a), Regulation and genetic enhancement of glucoamylase and pullulanase production in *Clostridium thermohydrosulfuricum*, *J. Bacteriol.* **164**, 1146-1152.
[120] Hyun, H. H., Zeikus, J. G. (1985b), Regulation and genetic enhancement of β-amylase production in *Clostridium thermosulfurogenes*, *J. Bacteriol.* **164**, 1162-1170.
[121] Annous, B. A., Blaschek, H. P. (1991), Isolation and characterization of *Clostridium acetobutylicum* mutants with enhanced amylolytic activity, *Appl. Environ. Microbiol.* **57**, 2544-2548.

[122] Bahl, H., Burchhardt, G., Spreinat, A., Haeckel, K., Wienecke, A., Schmidt, B., Antranikian, G. (1991), α-Amylase of *Clostridium thermosulfurogenes* EM1: nucleotide sequence of the gene, processing of the enzyme, and comparison to other α-amylases, *Appl. Environ. Microbiol.* **57**, 1554-1559.

[123] Verhasselt, P., Vanderleyden, J. (1992), Cloning and expression of *Clostridium acetobutylicum* genes involved in carbohydrate utilization, in: *Genetics and Molecular Biology of Anaerobic Bacteria* (Sebald, M., Ed.), pp. 301-316, New York: Springer-Verlag.

[124] Matuschek, M., Burchhardt, G., Sahm, K., Bahl, H. (1994), Pullulanase of *Thermoanaerobacterium thermosulfurigenes* EM1 (*Clostridium thermosulfurogenes*): molecular analysis of the gene, composite structure of the enzyme and a common model for its attachment to the cell surface, *J. Bacteriol.* **176**, 3295-3302.

[125] Antranikian, G., Herzberg, C., Gottschalk, G. (1987a), Production of thermostable α-amylase, pullulanase and α-glucosidase in continuous culture by a new *Clostridium* isolate, *Appl. Environ. Microbiol.* **53**, 1668-1673.

[126] Antranikian, G., Herzberg, C., Mayer, F., Gottschalk, G. (1987b), Changes in cell-envelope structure during massive production of amylase and pullulanase, *FEMS Microbiol. Lett.* **41**, 193-197.

[127] Matuschek, M., Sahm, K., Zibat, A., Bahl, H. (1996), Characterization of genes from *Thermoanaerobacterium thermosulfurigenes* EM1 that encode two glycosyl hydrolases with conserved S-layer-like domains, *Mol. Gen. Genet.* **252**, 493-496.

[128] Haeckel, K., Bahl, H. (1989), Cloning and expression of the thermostable α-amylase from *Clostridium thermosulfurogenes* (DSM 3896) in *Escherichia coli*, *FEMS Microbiol. Lett.* **60**, 333-338.

[129] Lanigan, G.W. (1959), Studies on the pectinolytic anaerobes *Clostridium flavum* and *Clostridium laniganii*, *J. Bacteriol.* **77**, 1-9.

[130] Ng, H., Vaughn, R.H. (1963), *Clostridium rubrum* sp. n. and other pectinolytic clostridia from soil, *J. Bacteriol.* **85**, 1104-1113.

[131] Lund, B.M. (1972), Isolation of pectic clostridia from potatoes, *J. Appl. Bacteriol.* **35**, 609-614.

[132] Schink, B., Zeikus, J.G. (1982), Microbial ecology of pectin decomposition in anoxic lake sediments, *J. Gen. Microbiol.* **128**, 393-404.

[133] Lund, B.M., Brocklehurst, T.F. (1978), Pectic enzymes of pigmented strains of *Clostridium*, *J. Gen. Microbiol.* **104**, 59-66.

[134] Schink, B., Zeikus, J.G. (1983a), Characterization of pectinolytic enzymes of *Clostridium thermosulfurogenes*, *FEMS Microbiol. Lett.* **17**, 295-298.

[135] Nakajima, N., Ishihara, K., Tanabe, M., Matsubara, K., Matsuura, Y. (1999), Degradation of pectic substances by two pectate lyases from a human intestinal bacterium, *Clostridium butyricum-beijerinckii* group, *J. Biosci. Bioeng.* **88**, 331-333.

[136] Macmillan, J.D., Vaughn, R.H. (1964), Purification and properties of a polygalacturonic acid-*trans*-eliminase produced by *Clostridium multifermentans*, *Biochemistry* **3**, 564-572.

[137] Macmillan, J.D., Phaff, H.J., Vaughn, R.H. (1964), The pattern of action of an exopolygalacturonic acid-*trans*-eliminase from *Clostridium multifermentans*, *Biochemistry* **3**, 572-578.

[138] Sheiman, M.I., Macmillan, J.D., Miller, L., Chase, T. Jr. (1976), Coordinated action of pectinesterase and polygalacturonate lyase complex of *Clostridium multifermentans*, *Eur. J. Biochem.* **64**, 565-572.

[139] Vanrijssel, M., Gerwig, G.J., Hansen, T.A. (1993a), Isolation and characterization of an extracellular glycosylated protein complex from *Clostridium thermosaccharolyticum* with pectin methylesterase and polygalacturonate hydrolase activity, *Appl. Environ. Microbiol.* **59**, 828-836.

[140] Vanrijssel, M., Smidt, M.P., Van Kouwen, G., Hansen, T.A. (1993b), Involvement of an intracellular oligogalacturonate hydrolase in metabolism of pectin by *Clostridium thermosaccharolyticum*, *Appl. Environ. Microbiol.* **59**, 837-842.

[141] Saier, M.H. Jr., Reizer, J. (1992), Proposed uniform nomenclature for the proteins and protein domains of the bacterial phosphoenolpyruvate: sugar phosphotransferase system, *J. Bacteriol.* **174**, 1433-1438.

[142] Postma, P.W., Lengeler, J.W., Jacobson, G.R. (1993), Phosphoenolpyruvate: carbohydrate phosphotransferase systems of bacteria, *Microbiol. Rev.* **57**, 543-594.

[143] Mitchell, W.J., Shaw, J.E., Andrews, L. (1991), Properties of the glucose phospho-

transferase system of *Clostridium acetobutylicum* NCIB 8052, *Appl. Environ. Microbiol.* **57**, 2534-2539.

[144] Tangney, M., Brehm, J.K., Minton, N.P., Mitchell, W.J. (1998a), A gene system for glucitol transport and metabolism in *Clostridium beijerinckii* NCIMB 8052, *Appl. Environ. Microbiol.* **64**, 1612-1619.

[145] Brown, G.D., Thomson, J.A. (1998), Isolation and characterisation of an aryl-beta-D-glucoside uptake and utilisation system (*abg*) from the gram-positive ruminal *Clostridium* species *C. longisporum*, *Mol. Gen. Genet.* **257**, 213-218.

[146] Reid, S.J., Rafudeen, M.S., Leat, N.G. (1999), The genes controlling sucrose utilization in *Clostridium beijerinckii* NCIMB 8052 constitute an operon, *Microbiology* **145**, 1461-1472.

[147] Tangney, M., Mitchell, W.J. (2000), Analysis of a catabolic operon for sucrose transport and metabolism in *Clostridium acetobutylicum* ATCC 824, *J. Mol. Microbiol. Biotechnol.* **2**, 71-80.

[148] Sutrina, S., Reddy, P., Saier, M.H. Jr., Reizer, J. (1990), The glucose permease of *Bacillus subtilis* is a single polypeptide chain that functions to energise the sucrose permease, *J. Biol. Chem.* **265**, 18581-18589.

[149] Booth, I.R., Morris, J.G. (1975), Protonmotive force in the obligately anaerobic bacterium *Clostridium pasteurianum*: a role in galactose and gluconate uptake, *FEBS Lett.* **59**, 153-157.

[150] Riebeling, V., Thauer, R.K., Jungermann, K. (1975), The internal alkaline pH gradient, sensitive to uncoupler and ATPase inhibitor in growing *Clostridium pasteurianum*, *Eur. J. Biochem.* **55**, 445-453.

[151] Clarke, D.J., Fuller, F.M., Morris, J.G. (1979), The proton-translocating adenosine triphosphatase of the obligately anaerobic bacterium *Clostridium pasteurianum*. 1. ATP phosphohydrolase activity, *Eur. J. Biochem.* **98**, 597-612.

[152] Erbeznik, M., Strobel, H.J., Dawson, K.A., Jones, C.R. (1998b), The D-xylose-binding protein, XylF, from *Thermoanaerobacter ethanolicus* 39E: cloning, molecular analysis, and expression of the structural gene, *J. Bacteriol.* **180**, 3570-3577.

[153] Jones, C.R., Ray, M., Dawson, K.A., Strobel, H.J. (2000), High-affinity maltose binding and transport by the thermophilic anaerobe *Thermoanaerobacter ethanolicus* 39E, *Appl. Environ. Microbiol.* **66**, 995-1000.

[154] Higgins, C.F. (1992), ABC transporters: from microorganisms to man, *Ann. Rev. Cell Biol.* **8**, 67-113.

[155] Thauer, R.K., Jungermann, K., Decker, K. (1977), Energy conservation in chemotrophic anaerobic bacteria, *Bacteriol. Rev.* **41**, 100-180.

[156] Behrens, S., Mitchell, W.J., Bahl, H. (2001), Molecular analysis of the mannitol operon of *Clostridium acetobutylicum* encoding a phosphotransferase system and a putative PTS-modulated regulator, *Microbiology* **147**, 75-86.

[157] Tangney, M., Rousse, C., Yazdanian, M., Mitchell, W.J. (1998b), Sucrose transport and metabolism in *Clostridium beijerinckii* NCIMB 8052, *J. Appl. Microbiol.* **84**, 914-919.

[158] Mitchell, W.J. (1996), Carbohydrate uptake and utilization by *Clostridium beijerinckii* NCIMB 8052, *Anaerobe* **2**, 379-384.

[159] Mitchell, W.J., Albasheri, K.A., Yazdanian, M. (1995), Factors affecting utilization of carbohydrates by clostridia, *FEMS Microbiol. Rev.* **17**, 317-329.

[160] Belouski, E., Watson, D.E., Bennett, G.N. (1998), Cloning, sequence and expression of the phosphofructokinase gene of *Clostridium acetobutylicum* ATCC 824 in *Escherichia coli*, *Curr. Microbiol.* **37**, 17-22.

[161] Schreiber, W., Dürre, P. (1999), The glyceraldehyde-3-phosphate dehydrogenase of *Clostridium acetobutylicum*: isolation and purification of the enzyme, and sequencing and localization of the *gap* gene within a cluster of other glycolytic genes, *Microbiology* **145**, 1839-1847.

[162] Rosenberg, S.L. (1980), Fermentation of pentose sugars to ethanol and other neutral products by microorganisms, *Enzyme Microb. Technol.* **2**, 185-193.

[163] Cynkin, M.A., Delwiche, E.A. (1958), Metabolism of pentoses by clostridia. I. Enzymes of ribose dissimilation in extracts of *Clostridium perfringens*, *J. Bacteriol.* **75**, 331-334.

[164] Aduse-Opoku, J., Mitchell, W.J. (1988), Diauxic growth of *Clostridium thermosaccharolyticum* on glucose and xylose, *FEMS Microbiol. Lett.* **50**, 45-49.

[165] Lee, C., Bagdasarian, M., Meng, M., Zeikus, J.G. (1990), Catalytic mechanism of xylose (glucose) isomerase from *Clostridium thermosulfurogenes*. Characterization of the

structural gene and function of active site histidine, *J. Biol. Chem.* **265**, 19082-19090.

[166] Dekker, K., Yamagata, H., Sakaguchi, K., Udaka, S. (1991), Xylose (glucose) isomerase gene from the thermophile *Clostridium thermohydrosulfuricum*: cloning, sequencing and expression in *Escherichia coli*, *Agric. Biol. Chem.* **55**, 221-227.

[167] Meaden, P. G., Aduse-Opoku, J., Reizer, J., Reizer, A., Lanceman, Y. A., Martin, M. F., Mitchell, W. J. (1994), The xylose isomerase-encoding gene (*xylA*) of *Clostridium thermosaccharolyticum*: cloning, sequencing and phylogeny of XylA enzymes, *Gene* **141**, 97-101.

[168] Erbeznik, M., Dawson, K. A., Strobel, H. J. (1998a), Cloning and characterization of transcription of the *xylAB* operon in *Thermoanaerobacter ethanolicus*, *J. Bacteriol.* **180**, 1103-1109.

[168a] Cynkin, M. A., Gibbs, M. (1958), Metabolism of pentoses by clostridia. II. The fermentation of C^{14}-labeled pentoses by *Clostridium perfringens, C. beijerinckii*, and *C. butylicum*, *J. Bacteriol.* **75**, 335-338.

[169] Bender, R., Andreesen, J. R., Gottschalk, G. (1971), 2-Keto-3-deoxygluconate, an intermediate in the fermentation of gluconate by clostridia, *J. Bacteriol.* **107**, 570-573.

[170] Forsberg, C. W. (1987), Production of 1,3-propanediol from glycerol by *Clostridium acetobutylicum* and other *Clostridium* species, *Appl. Environ. Microbiol.* **53**, 639-643.

[171] Biebl, H., Marten, S., Hippe, H., Deckwer, W. D. (1992), Glycerol conversion to 1,3-propanediol by newly isolated clostridia, *Appl. Microbiol. Biotechnol.* **36**, 592-597.

[172] Dabrock, B., Bahl, H., Gottschalk, G. (1992), Parameters affecting solvent production by *Clostridium pasteurianum*, *Appl. Microbiol. Biotechnol.* **58**, 1233-1239.

[173] Abbad-Andaloussi, S., Dürr, C., Raval, G., Petitdemange, H. (1996), Carbon and electron flow in *Clostridium butyricum* grown in chemostat culture on glycerol and on glucose, *Microbiology* **142**, 1149-1158.

[174] Luers, F., Seyfried, M., Daniel, R., Gottschalk, G. (1997), Glycerol conversion to 1,3-propanediol by *Clostridium pasteurianum*: cloning and expression of the gene encoding 1,3-propanediol dehydrogenase, *FEMS Microbiol. Lett.* **154**,

[175] Macis, L., Daniel, R., Gottschalk, G. (1998), Properties and sequence of the coenzyme B_{12}-dependent glycerol dehydratase of *Clostridium pasteurianum*, *FEMS Microbiol. Lett.* **164**, 21-28.

[176] Meyer, J. (2000), Clostridial iron-sulphur proteins, *J. Mol. Microbiol. Biotechnol.* **2**, 9-14.

[177] Ljungdahl, L. G., Hugenholtz, J., Wiegel, J. (1989), Acetogenic and acid-producing clostridia, in: *Biotechnology Handbooks Vol.3: Clostridia* (Minton, N. P., Clarke, D. J., Eds.), pp. 227-263, New York: Plenum Press.

[178] Jones, D. T., Woods, D. R. (1989), Solvent production, in: *Biotechnology Handbooks Vol.3: Clostridia* (Minton, N. P., Clarke, D. J., Eds.), pp. 105-144, New York: Plenum Press.

[179] Chen, J.-S. (1993), Properties of acid- and solvent-forming enzymes of clostridia, in: *The Clostridia and Biotechnology* (Woods, D. R., Ed.), pp. 51-76, Stoneham: Butterworth-Heinemann.

[180] Jungermann, K., Rupprecht, C., Ohrloff, R., Thauer, R. K., Dekker, K. (1971), Regulation of the reduced nicotinamide adenine dinucleotide-ferredoxin reductase system in *Clostridium kluyveri*, *J. Biol. Chem.* **246**, 960-962.

[181] Jungermann, K., Thauer, R. K., Liemenstoll, G., Dekker, K. (1973), Function of reduced pyridine nucleotide-ferredoxin oxidoreductases in saccharolytic clostridia, *Biochim. Biophys. Acta* **305**, 268-280.

[182] Petitdemange, H., Cherrier, C., Raval, G., Gay, R. (1976), Regulation of the NADH and NADPH-ferredoxin oxidoreductases in clostridia of the butyric group, *Biochim. Biophys. Acta* **421**, 334-337.

[183] Heyndrickz, M., De Vos, P., Vancanneyt, M., De Ley, J. (1991b), The fermentation of glycerol by *Clostridium butyricum* LMG $1212t_2$ and $1213t_1$ and *Clostridium pasteurianum* LMG 3285, *Appl. Microbiol. Biotechnol.* **34**, 637-642.

[184] Vasconcelos, I., Girbal, L., Soucaille, P. (1994), Regulation of carbon and electron flow in *Clostridium acetobutylicum* grown in chemostat culture at neutral pH on mixtures of glucose and glycerol, *J. Bacteriol.* **176**, 1443-1450.

[185] Heyndrickz, M., De Vos, P., De Ley, J. (1991a), Fermentation characteristics of *Clostridium pasteurianum* LMG 3285 grown on glucose and mannitol, *J. Appl. Bacteriol.* **70**, 52-58.

[186] Schink, B., Zeikus, J. G. (1983b), *Clostridium thermosulfurogenes* sp. nov., a thermo-

phile that produces elemental sulphur from thiosulphate, *J. Gen. Microbiol.* **129**, 1149-1158.

[187] Holt, R. A., Cairns, A. J., Morris, J. G. (1988), Production of butanol by *Clostridium puniceum* in batch and continuous culture, *Appl. Microbiol. Biotechnol.* **27**, 319-324.

[188] Davies, R., Stephenson, M. (1941), Studies on the acetone-butyl alcohol fermentation. I. Nutritional and other factors involved in the preparation of active suspensions of *Cl. acetobutylicum* (Weizmann), *Biochem. J.* **35**, 1320-1331.

[189] Spivey, M. J. (1978), The acetone/butanol/ethanol fermentation, *Proc. Biochem.* **13**, 2-5.

[190] Bahl, H., Andersch, W., Gottschalk, G. (1982), Continuous production of acetone and butanol by *Clostridium acetobutylicum* in a two-stage phosphate limited chemostat, *Eur. J. Appl. Miocrobiol. Biotechnol.* **15**, 201-205.

[191] Bahl, H., Gottschalk, G. (1984), Parameters affecting solvent production by *Clostridium acetobutylicum* in continuous culture, *Biotech. Bioeng. Symp.* **14**, 215-223.

[192] Bahl, H., Gottwald, M., Kuhn, A., Rale, V., Andersch, W., Gottschalk, G. (1986), Nutritional factors affecting the ratio of solvents produced by *Clostridium acetobutylicum*, *Appl. Environ. Microbiol.* **52**, 169-172.

[193] Junelles, A. M., Janati-Idrissi, R., Petitdemange, H., Gay, R. (1988), Iron effect on acetone-butanol fermentation, *Curr. Microbiol.* **17**, 299-303.

[194] Peguin, S., Soucaille, P. (1995), Modulation of carbon and electron flow in *Clostridium acetobutylicum* by iron limitation and methyl viologen addition, *Appl. Environ. Microbiol.* **61**, 403-405.

[195] Freier, D., Gottschalk, G. (1987), L(+) lactate dehydrogenase activity of *Clostridium acetobutylicum* is activated by fructose-1,6-bisphosphate, *FEMS Microbiol. Lett.* **43**, 229-233.

[196] Kim, B. H., Bellows, P., Datta, R., Zeikus, J. G. (1984), Control of carbon and electron flow in *Clostridium acetobutylicum* fermentations: utililzation of carbon monoxide to inhibit hydrogen production and to enhance butanol yields, *Appl. Environ. Microbiol.* **48**, 764-770.

[197] Doremus, M. G., Linden, J. C., Moreira, A. R. (1985), Agitation and pressure effects on acetone-butanol fermentation, *Biotechnol. Bioeng.* **27**, 852-860.

[198] Meyer, C. L., Roos, J. W., Papoutsakis, E. T. (1986), Carbon monoxide gasing leads to alcohol production and butyrate uptake without acetone formation in continuous cultures of *Clostridium acetobutylicum*, *Appl. Microbiol. Biotechnol.* **24**, 159-167.

[199] Rao, G., Mutharasan, R. (1986), Alcohol production by *Clostridium acetobutylicum* induced by methyl viologen, *Biotechnol. Lett.* **8**, 893-896.

[200] Peguin, S., Goma, G., Delorme, P., Soucaille, P. (1994), Metabolic flexibility of *Clostridium acetobutylicum* in response to methyl viologen addition, *Appl. Microbiol. Biotechnol.* **42**, 611-616.

[201] Tran-Dinh, K., Gottschalk, G. (1985), Formation of D(-)-1,2-propanediol and D(-)-lactate from glucose by *Clostridium sphenoides* under phosphate limitation, *Arch. Microbiol.* **142**, 87-92.

[202] Cameron, D. C., Cooney, C. L. (1986), A novel fermentation: the production of R(-)-1,2-propanediol and acetol by *Clostridium thermosaccharolyticum*, *Biotechnology* **4**, 651-654.

[203] Fuchs, G. (1986), CO_2 fixation in acetogenic bacteria: variations on a theme, *FEMS Microbiol. Rev.* **39**, 181-213.

[204] Daniel, S. L., Drake, H. L. (1993), Oxalate and glyoxylate-dependent growth and acetogenesis by *Clostridium thermoaceticum*, *Appl. Environ. Microbiol.* **59**, 3062-3069.

[205] Ljungdahl, L. G. (1986), The autotrophic pathway of acetate synthesis in acetogenic bacteria, *Ann. Rev. Microbiol.* **40**, 415-450.

[206] Bomar, M., Hippe, H., Schink, B. (1991), Lithotrophic growth and hydrogen metabolism by *Clostridium magnum*, *FEMS Microbiol. Lett.* **83**, 347-350.

[207] Daniel, S. L., Wu, Z., Drake, H. L. (1988), Growth of thermophilic acetogenic bacteria on methoxylated aromatic acids, *FEMS Microbiol. Lett.* **52**, 25-28.

[208] Wu, Z., Daniel, S. L., Drake, H. L. (1988), Characterization of a CO-dependent O-demethylating enzyme system from the acetogen *Clostridium thermoaceticum*, *J. Bacteriol.* **170**, 5747-5750.

[209] El Kasmi, A., Rajasekharan, S., Ragsdale, S. W. (1994), Anaerobic pathway for conversion of the methyl group of aromatic methyl ethers to acetic acid by *Clostridium thermoaceticum*, *Biochemistry* **33**, 11217-11224.

[210] Lundie, L. L., Drake, H. L. (1984), Development of a minimally defined medium for the acetogen *Clostridium thermoaceticum*, *J. Bacteriol.* **159**, 700-703.

[211] Savage, M. D., Drake, H. L. (1986), Adaptation of the acetogen *Clostridium thermoautotrophicum* to minimal medium, *J. Bacteriol.* **165**, 315-318.

[212] Wood, H. G. (1991), Life with CO or CO_2 and H_2 as a source of carbon and energy, *FASEB J.* **5**, 156-163.

[213] Morton, T. A., Chou, C-F., Ljungdahl, L. G. (1993), Cloning, sequencing, and expressions of genes encoding enzymes of the autotrophic acetyl-CoA pathway in the acetogen *Clostridium thermoaceticum*, in: *Genetics and Molecular Biology of Anaerobic Bacteria* (Sebald, M., Ed.), pp. 389-406, New York: Springer-Verlag.

[214] Ragsdale, S. W., Wood, H. G. (1985), Acetate biosynthesis by acetogenic bacteria. Evidence that carbon monoxide dehydrogenase is the condensing enzyme that catalyzes the final steps of the synthesis, *J. Biol. Chem.* **260**, 3970-3977.

[215] Roberts, J. R., Lu, W-P., Ragsdale, S. W. (1992), Acetyl-coenzyme A synthesis from methyltetrahydrofolate, CO, and coenzyme A by enzymes purified from *Clostridium thermoaceticum*: attainment of *in vivo* rates and identification of rate-limiting steps, *J. Bacteriol.* **174**, 4667-4676.

[216] Das, A., Hugenholtz, J., Van Halbeek, H., Ljungdahl, L. G. (1989), Structure and function of a menaquinone involved in electron transport in membranes of *Clostridium thermoautotrophicum* and *Clostridium thermoaceticum*, *J. Bacteriol.* **171**, 5823-5829.

[217] Hugenholtz, J., Ljungdahl, L. G. (1990b), Metabolism and energy generation in homoacetogenic clostridia, *FEMS Microbiol. Rev.* **87**, 383-390.

[218] Hugenholtz, J., Ivey, D. M., Ljungdahl, L. G. (1987), Carbon monoxide-driven electron transport in *Clostridium thermoautotrophicum* membranes, *J. Bacteriol.* **169**, 5845-5847.

[219] Hugenholtz, J., Ljungdahl, L. G. (1989), Electron transport and electrochemical proton gradient in membrane vesicles of *Clostridium thermoautotrophicum*, *J. Bacteriol.* **171**, 2873-2875.

[220] Hugenholtz, J., Ljungdahl, L. G. (1990a), Amino acid transport in membrane vesicles of *Clostridium thermoautotrophicum*, *FEMS Microbiol. Lett.* **69**, 117-122.

[221] Das, A., Ljungdahl, L. G. (1997), Composition and primary structure of the F_1F_0 ATP synthase from the obligately anaerobic bacterium *Clostridium thermoaceticum*, *J. Bacteriol.* **179**, 3746-3755.

[222] Das, A., Ivey, D. M., Ljungdahl, L. G. (1997), Purification and reconstitution into proteoliposomes of the F_1F_0 ATP synthase from the obligately anaerobic Gram-positive bacterium *Clostridium thermoautotrophicum*, *J. Bacteriol.* **179**, 1714-1720.

[223] Matthies, C., Freiberger, A., Drake, H. L. (1993), Fumarate dissimilation and differential reductant flow by *Clostridium formicoaceticum* and *Clostridium aceticum*, *Arch. Microbiol.* **160**, 273-278.

[224] Seifritz, C., Daniel, S. L., Gößner, A., Drake, H. L. (1993), Nitrate as a preferred electron sink for the acetogen *Clostridium thermoaceticum*, *J. Bacteriol.* **175**, 8008-8013.

[225] Fröstl, J. M., Seifritz, C., Drake, H. L. (1996), Effect of nitrate on the autotrophic metabolism of the acetogens *Clostridium thermoautotrophicum* and *Clostridium thermoaceticum*, *J. Bacteriol.* **178**, 4597-4603.

[226] Roberts, D. L., James-Hagstrom, J. E., Garvin, D. K., Gorst, C. M., Runquist, J. A., Baur, J. R., Haase, F. C., Ragsdale, S. W. (1989), Cloning and expression of the gene cluster encoding key proteins involved in acetyl CoA synthesis in *Clostridium thermoaceticum*: CO dehydrogenase, the corrinoid/Fe-S protein, and methyltransferase, *Proc. Natl. Acad. Sci. USA* **86**, 32-36.

[227] Ragsdale, S. W., Baur, J. R., Gorst, C. M., Harder, S. R., Lu, W-P., Roberts, D. L., Runquist, J. A., Schiau, I. (1990), The acetyl-CoA synthase from *Clostridium thermoaceticum*: from gene cluster to active-site metal clusters, *FEMS Microbiol. Rev.* **87**, 397-402.

[228] Arendsen, A. F., Soliman, M. Q., Ragsdale, S. W. (1999), Nitrate-dependent regulation of acetate biosynthesis and nitrate respiration by *Clostridium thermoaceticum*, *J. Bacteriol.* **181**, 1489-1495.

[229] Antranikian, G., Friese, C., Quentmeier, A., Hippe, H., Gottschalk, G. (1984), Distribution of the ability for citrate utilization amongst clostridia, *Arch. Microbiol.* **138**, 179-182.

[230] Matteuzzi, D., Hollaus, F., Biavati, B. (1978), Proposal of neotype for *Clostridium thermohydrosulfuricum* and the merging of *Clostridium tartarivorum* with *Clostridium thermosaccharolyticum*, *Int. J. Syst. Bacteriol.* **28**, 528-531.

[231] Beaty, P. S., Ljungdahl, L. G. (1991), Growth of *Clostridium thermoaceticum* on methanol, ethanol, propanol and butanol in medium containing either thiosulfate or dimethylsulfoxide, *Abstr. Ann. Meet. Am. Soc. Microbiol.* **1991**, p. 236, K-131.

[232] Tholozan, J. L., Touzel, J. P., Samain, E., Grivet, J. P., Prensier, G., Albagnac, G. (1992), *Clostridium neopropionicum* sp. nov., a strict anaerobe fermenting ethanol to propionate through acrylate pathway, *Arch. Microbiol.* **157**, 249-257.

[233] Bornstein, B. T., Barker, H. A. (1948a), The nutrition of *Clostridium kluyveri*, *J. Bacteriol.* **55**, 223-230.

[234] Bornstein, B. T., Barker, H. A. (1948b), The energy metabolism of *Clostridium kluyveri* and the synthesis of fatty acids, *J. Biol. Chem.* **172**, 659-669.

[235] Kenealy, W. R., Waselefsky, D. M. (1985), Studies on the substrate range of *Clostridium kluyveri*: the use of propanol and succinate, *Arch. Microbiol.* **141**, 187-194.

[236] Kane, M. D., Brauman, A., Breznak, J. A. (1991), *Clostridium mayombei* sp. nov., an H_2/CO_2 acetogenic bacterium from the gut of the African soil-feeding termite, *Cubitermes speciosus*, *Arch. Microbiol.* **156**, 99-104.

[237] Dimroth, P. (1994), Bacterial sodium ion-coupled energetics, *Antonie van Leeuwenhoek* **65**, 381-395.

[238] Hartmanis, M. G. N., Stadtman, T. C. (1986), Diol metabolism and diol dehydratase in *Clostridium glycolicum*, *Arch. Biochem. Biophys.* **245**, 144-152.

[239] Schink, B. (1984), *Clostridium magnum* sp. nov., a non-autotrophic homoacetogenic bacterium, *Arch. Microbiol.* **137**, 250-255.

[240] Hsu, T., Daniel, S. L., Lux, M. F., Drake, H. L. (1990), Biotransformations of carboxylated aromatic compounds by the acetogen *Clostridium thermoaceticum*: generation of growth-supportive CO_2 equivalents under CO_2-limited conditions, *J. Bacteriol.* **172**, 212-217.

[241] Winter, J., Moore, L. H., Dowell, V. R. Jr., Bokkenheuser, V. D. (1989), C-ring cleavage of flavonoids by human intestinal bacteria, *Appl. Environ. Microbiol.* **55**, 1203-1208.

[242] Winter, J., Popoff, M. R., Grimont, P., Bokkenheuser, V. D. (1991), *Clostridium orbiscindens* sp. nov., a human intestinal bacterium capable of cleaving the flavonoid C-ring, *Int. J. Syst. Bacteriol.* **41**, 355-357.

[243] Baronofsky, J. J., Schreurs, W. J. A., Kashket, E. R. (1984), Uncoupling by acetic acid limits growth of and acetogenesis by *Clostridium thermoaceticum*, *Appl. Environ. Microbiol.* **48**, 1134-1139.

[244] Huang, L., Gibbins, L. N., Forsberg, C. W. (1985), Transmembrane pH gradient and membrane potential in *Clostridium acetobutylicum* during growth under acetogenic and solventogenic conditions, *Appl. Environ. Microbiol.* **50**, 1043-1047.

[245] Kell, D. B., Peck, M. W., Rodger, G., Morris, J. G. (1981), On the permeability to weak acids and bases of the cytoplasmic membrane of *Clostridium pasteurianum*, *Biochem. Biophys. Res. Commun.* **99**, 81-88.

[246] Clarke, D. J., Morley, C. D., Kell, D. B., Morris, J. G. (1982), On the mode of action of the bacteriocin butyricin 7423. Effects on membrane potential and potassium-ion accumulation in *Clostridium pasteurianum*, *Eur. J. Biochem.* **127**, 105-116.

[247] Speelmans, G., Poolman, B., Abee, T., Konings, W. N. (1993), Energy transduction in the thermophilic anaerobic bacterium *Clostridium fervidus* is exclusively coupled to sodium ions, *Proc. Natl. Acad. Sci. USA* **90**, 7975-7979.

[248] Clarke, D. J., Morris, J. G. (1979), The proton-translocating adenosine triphosphatase of the obligately anaerobic bacterium *Clostridium pasteurianum*. 2. ATP synthetase activity, *Eur. J. Biochem.* **98**, 613-620.

[249] Das, A. X., Ljungdahl, L. G. (2000), The primary structure of the *atp* operon encoding the F_1F_0 ATP synthase from *Clostridium pasteurianum*, *Abstr. Ann. Meet. Am. Soc. Microbiol.* **2000**, p. 374, H-116.

[250] Booth, I. R. (1999), The regulation of intracellular pH in bacteria, *Novartis Foundation Symp.* **221**, 19-28.

[251] Booth, I. R., Higgins, C. F. (1990), Enteric bacteria and osmotic stress: intracellular potassium glutamate as a secondary signal of osmotic stress?, *FEMS Microbiol. Rev.* **75**, 239-246.

[252] Terracciano, J. S., Schreurs, W. J. A., Kashket, E. R. (1987), Membrane conductance of *Clostridium thermoaceticum* and *Clostridium acetobutylicum*: evidence for electrogenic Na^+/H^+ antiport in *Clostridium thermoaceticum*, *Appl. Environ. Microbiol.* **53**, 782-786.

[253] Treuner-Lange, A., Dürre, P. (1996), Molecular biological analysis of kdpD/E, a sensor histidine kinase/response regulator system in *Clostridium acetobutylicum*, *Anaerobe* **2**, 351-363.

[254] Treuner-Lange, A., Kuhn, A., Dürre, P. (1997), The *kdp* system of *Clostridium acetobutylicum*: cloning, sequencing, and transcriptional regulation in response to potassium concentration, *J. Bacteriol.* **179**, 4501-4512.

[255] Behrens, M., Dürre, P. (2000), KdpE of *Clostridium acetobutylicum* is a highly specific response regulator controlling only the expression of the *kdp* operon, *J. Mol. Microbiol. Biotechnol.* **2**, 45-52.

[256] Bogdahn, M., Andreesen, J. R., Kleiner, D. (1983), Pathways and regulation of N_2, ammonium and glutamate assimilation by *Clostridium formicoaceticum*, *Arch. Microbiol.* **134**, 167-169.

[257] Bogdahn, M., Kleiner, D. (1986a), N_2 fixation and NH_4^+ assimilation in the thermophilic anaerobes *Clostridium thermosaccharolyticum* and *Clostridium thermoautotrophicum*, *Arch. Microbiol.* **144**, 102-104.

[258] Bogdahn, M., Kleiner, D. (1986b), Inorganic nitrogen metabolism in two cellulose-degrading clostridia, *Arch. Microbiol.* **145**, 159-161.

[259] Kanamori, K., Weiss, R. L., Roberts, J. D. (1989), Ammonia assimilation pathways in nitrogen-fixing *Clostridium kluyveri* and *Clostridium butyricum*, *J. Bacteriol.* **171**, 2148-2154.

[260] Janssen, P., Jones, W. A., Jones, D. T., Woods, D. R. (1988), Molecular analysis and regulation of the *glnA* gene of the Gram-positive anaerobe *Clostridium acetobutylicum*, *J. Bacteriol.* **170**, 400-408.

[261] Fierro-Monti, I. P., Reid, S. J., Jones, D. T., Woods, D. R. (1992), Differential expression of a *Clostridium acetobutylicum* antisense RNA: implications for the regulation of glutamine synthetase, *J. Bacteriol.* **174**, 7642-7647.

[262] Woods, D. R., Reid, S. J. (1995), Regulation of nitrogen metabolism, starch utilisation and the *β-hbd-adh1* gene cluster in *Clostridium acetobutylicum*, *FEMS Microbiol. Rev.* **17**, 299-306.

[263] Driessen, A. J. M., Ubbink-Kok, T., Konings, W. N. (1988), Amino acid transport by membrane vesicles of an obligate anaerobic bacterium, *Clostridium acetobutylicum*, *J. Bacteriol.* **170**, 817-820.

[264] Speelmans, G., De Vrij, W., Konings, W. N. (1989), Characterization of amino acid transport in membrane vesicles from the thermophilic fermentative bacterium *Clostridium fervidus*, *J. Bacteriol.* **171**, 3788-3795.

[265] Speelmans, G., Poolman, B., Abee, T., Konings, W. N. (1994), The F- or V-type Na^+-ATPase of the thermophilic bacterium *Clostridium fervidus*, *J. Bacteriol.* **176**, 5160-5162.

[266] Höner Zu Bentrup, K., Ubbink-Kok, T., Lolkema, J. S., Konings, W. N. (1997), A Na^+-pumping V_1V_0-ATPase complex in the thermophilic bacterium *Clostridium fervidus*, *J. Bacteriol.* **179**, 1274-1279.

[267] Mead, G. C. (1971), The amino-acid fermenting clostridia, *J. Gen. Microbiol.* **67**, 47-56.

[268] Elsden, S. R., Hilton, M. G. (1979), Amino acid utilization patterns in clostridial taxonomy, *Arch. Microbiol.* **123**, 137-141.

[269] Barker, H. A. (1981), Amino acid degradation by anaerobic bacteria, *Ann. Rev. Biochem.* **50**, 23-40.

[270] Elsden, S. R., Hilton, M. G., Parsley, K. R., Self, R. (1980), The lipid fatty acids of proteolytic clostridia, *J. Gen. Microbiol.* **118**, 115-123.

[271] Dürre, P., Andreesen, J. R. (1982a), Selenium-dependent growth and glycine fermentation by *Clostridium purinolyticum*, *J. Gen. Microbiol.* **128**, 1457-1466.

[272] Stadtman, T. C., Elliott, P., Tiemann, L. (1958), Studies on the enzymic reduction of amino acids. III. Phosphate esterification coupled with glycine reductase, *J. Biol. Chem.* **231**, 961-973.

[273] Arkowitz, R. A., Abeles, R. H. (1989), Identification of acetyl phosphate as the product of clostridial glycine reductase: evidence for an acyl enzyme intermediate, *Biochemistry* **28**, 4639-4644.

[274] Andreesen, J. R. (1994), Glycine metabolism in anaerobes, *Antonie van Leeuwenhoek* **66**, 223-237.

[275] Garcia, G. E., Stadtman, T. C. (1991), Selenoprotein A component of the glycine reductase complex from *Clostridium purinolyticum*: nucleotide sequence of the gene shows that selenocysteine is encoded by UGA, *J. Bacteriol.* **173,** 2093-2098.

[276] Garcia, G. E., Stadtman, T. C. (1992), *Clostridium sticklandii* glycine reductase selenoprotein A: cloning, sequencing, and expression in *Escherichia coli*, *J. Bacteriol.* **174,** 7080-7089.

[277] Kreimer, S., Andreesen, J. R. (1995), Glycine reductase of *Clostridium litorale*. Cloning, sequencing and molecular analysis of the *grdAB* operon that contains two in-frame TGA codons for selenium incorporation, *Eur. J. Biochem.* **234,** 192-199.

[278] Dietrichs, D., Meyer, M., Rieth, M., Andreesen, J. R. (1991), Interaction of selenoprotein P_A and the thioredoxin system, components of the NADPH-dependent reduction of glycine in *Eubacterium acidaminophilum* and *Clostridium litoralis*, *J. Bacteriol.* **173,** 5983-5991.

[279] Lovitt, R. W., Kell, D. B., Morris, J. G. (1986), Proline reduction by *Clostridium sporogenes* is coupled to vectorial proton ejection, *FEMS Microbiol. Lett.* **36,** 269-273.

[280] Kabisch, U. C., Gräntzdörffer, A., Schierhorn, A., Rücknagel, K. P., Andreesen, J. R., Pich, A. (1999), Identification of D-proline reductase from *Clostridium sticklandii* as a selenoenzyme and indications for a catalytically active pyruvoyl group derived from a cysteine residue by cleavage of a proprotein, *J. Biol. Chem.* **274,** 8444-8454.

[281] Möller, B., Hippe, H., Gottschalk, G. (1986), Degradation of various amine compounds by mesophilic clostridia, *Arch. Microbiol.* **145,** 85-90.

[283] Naumann, E., Hippe, H., Gottshcalk, G. (1983), Betaine: new oxidant in the Stickland reaction and methanogenesis from betaine and L-alanine by a *Clostridium sporogenes-Methanosarcina barkeri* coculture, *Appl. Environ. Microbiol.* **45,** 474-483.

[284] Fendrich, C., Hippe, H., Gottschalk, G. (1990), *Clostridium halophilum* sp. nov. and *C. litorale* sp. nov., an obligate halophilic and a marine species degrading betaine in the Stickland reaction, *Arch. Microbiol.* **154,** 127-132.

[285] Dürre, P., Andersch, W., Andreesen, J. R. (1981), Isolation and characterization of an adenine-utilizing, anaerobic sporeformer, *Clostridium purinolyticum* sp. nov., *Int. J. Syst. Bacteriol.* **31,** 184-194.

[286] Wagner, R., Cammack, R., Andreesen, J. R. (1984), Purification and characterization of xanthine dehydrogenase fom *Clostridium acidiurici* grown in the presence of selenium, *Biochim. Biophys. Acta* **791,** 63-74.

[287] Dürre, P., Andreesen, J. R. (1983), Purine and glycine metabolism by purinolytic clostridia, *J. Bacteriol.* **154,** 192-199.

[288] Dürre, P., Andreesen, J. R. (1982b), Anaerobic degradation of uric acid via pyrimidine derivatives by selenium-starved cells of *Clostridium purinolyticum*, *Arch. Microbiol.* **131,** 255-260.

[289] Wachsman, J. T., Barker, H. A. (1954), Characterization of an orotic acid fermenting bacterium, *Zymobacterium oroticum*, nov. gen., nov. spec., *J. Bacteriol.* **68,** 400-404.

[290] Mead, G. C., Adams, B. W., Hilton, M. G., Lord, P. G. (1979), Isolation and characterization of uracil-degrading clostridia from soil, *J. Appl. Bacteriol.* **46,** 465-472.

[291] Hilton, M. G., Mead, G. C., Elsden, S. R. (1975), The metabolism of pyrimidines by proteolytic clostridia, *Arch. Microbiol.* **102,** 145-149.

[292] Stadtman, E. R., Stadtman, T. C., Pastan, I., Smith, L. D. (1972), *Clostridium barkeri* sp. n., *J. Bacteriol.* **110,** 758-760.

[293] Gladyshev, V. N., Khangulov, S. V., Stadtman, T. C. (1996), Properties of the selenium- and molybdenum- containing nicotinic acid hydroxylase from *Clostridium barkeri*, *Biochemistry* **35,** 212-223.

[294] Peters, J. W., Fisher, K., Dean, D. R. (1995), Nitrogenase structure and function; a biochemical-genetic perspective, *Ann. Rev. Microbiol.* **49,** 335-366.

[295] Chen, K. C.-K., Chen, J.-S., Johnson, J. L. (1986), Structural features of multiple *nifH*-like sequences and very biased codon usage in nitrogenase genes of *Clostridium pasteurianum*, *J. Bacteriol.* **166,** 162-167.

[296] Wang, S-Z., Chen, J-S., Johnson, J. L. (1988b), Distinct structural features of the α and β subunits of nitrogenase molybdenum-iron protein of *Clostridium pasteurianum*: an analysis of amino acid sequences, *Biochemistry* **27,** 2800-2810.

[297] Chen, J.-S., Johnson, J. L. (1993), Molecular biology of nitrogen fixation in the clostridia, in: *The Clostridia and Biotechnology* (Woods, D. R., Ed.), pp. 371-392, Stoneham: Butterworth-Heinemann.

[298] Schlessman, J. L., Woo, D., Joshua-Tor, L., Howard, J. B., Rees, D. C. (1998), Conformational variability in structures of the nitrogenase iron proteins from *Azotobacter vinelandii* and *Clostridium pasteurianum*, *J. Mol. Biol.* **280**, 669-685.

[299] Emerich, D. W., Burris, R. H. (1978), Complementary functioning of the component proteins of nitrogenase from several bacteria, *J. Bacteriol.* **134**, 936-943.

[300] Kim, J., Woo, D., Rees, D. C. (1993), X-ray crystal structure of the nitrogenase molybdenum-iron protein from *Clostridium pasteurianum* at 3.0- resolution, *Biochemistry* **32**, 7104-7115.

[301] Golinelli, M-P., Gagnon, J., Meyer, J. (1997), Specific interaction of the [2Fe-2S] ferredoxin from *Clostridium pasteurianum* with the nitrogenase MoFe protein, *Biochemistry* **36**, 11797-11803.

[302] Johnson, J. L., Wang, S-Z., Chen, J-S. (1993), Organization of the nitrogen fixation genes in *Clostridium pasteurianum*, in: *Genetics and Molecular Biology of Anaerobic Bacteria* (Sebald, M., Ed.), pp. 373-381, New York: Springer-Verlag.

[303] Wang, S-Z., Chen, J-S., Johnson, J. L. (1988a), The presence of five *nifH*-like sequences in *Clostridium pasteurianum*: sequence divergence and transcription properties, *Nucleic Acids Res.* **16**, 439-454.

[304] Zinoni, F., Robson, R. M., Robson, R. L. (1993), Organization of potential alternative nitrogenase genes from *Clostridium pasteurianum*, *Biochim. Biophys. Acta* **1174**, 83-86.

[305] Loveless, T. M., Bishop, P. E. (1999), Identification of genes unique to Mo-independent nitrogenase systems in diverse azotrophs, *Can. J. Microbiol.* **45**, 312-317.

[306] Dilworth, M. J., Eady, R. R., Robson, R. L., Miller, R. W. (1987), Ethane formation from acetylene as a potential test for vanadium nitrogenase *in vivo*, *Nature (London)* **327**, 167-168.

[307] Joerger, R. D., Jacobson, M. R., Premakumar, R., Wolfinger, E. D., Bishop, P. E. (1989), Nucleotide sequence and mutational analysis of the structural genes (*anfHDGK*) for the second alternative nitrogenase from *Azotobacter vinelandii*, *J. Bacteriol.* **171**, 1075-1086.

[308] Woods, D. R., Jones, D. T. (1986), Physiological responses of *Bacteroides* and *Clostridium* strains to environmental stress factors, *Adv. Microb. Physiol.* **28**, 1-64.

[309] Labbé, R. G., Shih, N-J. R. (1997), Physiology of sporulation of clostridia, in: *The Clostridia: Molecular Biology and Pathogenesis* (Rood, J. I., McClane, B. A., Songer, J. G., Titball, R. W., Eds.), pp. 21-32, San Diego: Academic Press.

[310] Hsu, E. J., Ordal, Z. J. (1969), Sporulation of *Clostridium thermosaccharolyticum*, *Appl. Microbiol.* **18**, 958-960.

[311] Labbé, R. G., Duncan, C. L. (1975), Influence of carbohydrates on growth and sporulation of *Clostridium perfringens* type A, *Appl. Microbiol.* **29**, 345-351.

[312] Sacks, L. E. (1983), Influence of carbohydrates on growth and sporulation of *Clostridium perfringens* in a defined medium with or without guanosine, *Appl. Environ. Microbiol.* **46**, 1169-1175.

[313] Strasdine, G. A. (1972), The role of intracellular glucan in endogenous fermentation and spore maturation in *Clostridium botulinum* type E, *Can. J. Microbiol.* **18**, 211-217.

[314] Emeruwa, A. C., Hawirko, R. Z. (1973), Poly-β-hydroxybutyrate metabolism during growth and sporulation of *Clostridium botulinum*, *J. Bacteriol.* **116**, 989-993.

[315] Lopez, J. M., Marks, C. L., Freese, E. (1979), The decrease of guanine nucleotides initiates sporulation of *Bacillus subtilis*, *Biochim. Biopys. Acta* **587**, 238-252.

[316] Sacks, L. E., Thompson, P. A. (1977), Increased spore yields of *Clostridium perfringens* in the presence of methylxanthines, *Appl. Environ. Microbiol.* **34**, 189-193.

[317] Labbé, R. G., Nolan, L. L. (1981), Stimulation of *Clostridium perfringens* enterotoxin formation by caffeine and theobromine, *Infect. Immun.* **34**, 50-54.

[318] Craven, S. E., Blankenship, L. C. (1982), Effect of purine derivatives, papaverine hydrochloride, and imidazole on enterotoxin formation by *Clostridium perfringens* type A, *Can. J. Microbiol.* **28**, 851-859.

[319] Setlow, P., Sacks, L. E. (1983), Cyclic AMP is not detectable in *Clostridium perfringens*, *Can. J. Microbiol.* **29**, 1228-1230.

[320] Murrell, W. G. (1967), The biochemistry of the bacterial endospore, *Adv. Microb. Physiol.* **1**, 133-251.

[321] Walker, P. D. (1970), Cytology of spore formation and germination, *J. Appl. Bacteriol.* **33**, 1-12.

[322] Mackey, B. M., Morris, J. G. (1971), Ultrastructural changes during sporulation of *Clostridium pasteurianum*, *J. Gen. Microbiol.* **66**, 1-13.

[323] Cabrera-Martinez, R. M., Setlow, P. (1991), Cloning and nucleotide sequence of three genes coding for small, acid-soluble proteins of *Clostridium perfringens* spores, *FEMS Microbiol. Lett.* **77**, 127-132.

[324] Cabrera-Martinez, R. M., Mason, J. M., Setlow, B., Waites, W. M., Setlow, P. (1989), Purification and amino acid sequence of two small, acid soluble proteins from *Clostridium bifermentans* spores, *FEMS Microbiol. Lett.* **61**, 139-144.

[325] Nicholson, W. L., Setlow, B., Setlow, P. (1991), Ultraviolet irradiation of DNA complexed with α/β-type small, acid-soluble proteins from spores of *Bacillus* or *Clostridium* species makes spore photoproduct but not thymine dimers, *Proc. Natl. Acad. Sci. USA* **88**, 8288-8292.

[326] Hodgkiss, W., Ordal, Z. J. (1966), Morphology of the spore of some strains of *Clostridium botulinum* type E, *J. Bacteriol.* **91**, 2031-2036.

[327] Pope, L., Yolton, D. P., Rode, L. J. (1967), Appendages of *Clostridium bifermentans* spores, *J. Bacteriol.* **94**, 1206-1215.

[328] Yolton, D. P., Pope, L., Williams, M. G., Rode, L. J. (1968), Further electron microscopic characterization of spore appendages of *Clostridium bifermentans*, *J. Bacteriol.* **95**, 231-238.

[329] Pope, L., Rode, L. J. (1969), Spore fine structure in *Clostridium cochlearum*, *J. Bacteriol.* **100**, 994-1001.

[330] Setlow, P., Johnson, E. A. (1997), Spores and their significance, in: *Food Microbiology Fundamentals and Frontiers* (Doyle, M. P., Beuchat, L. R., Montville, T. J., Eds.), pp. 30-65, Washington, DC: ASM Press.

[331] Hyun, H. H., Zeikus, J. G., Longin, R., Millet, J., Ryter, A. (1983), Ultrastructure and extreme heat resistance of spores from thermophilic *Clostridium* species, *J. Bacteriol.* **156**, 1332-1337.

[332] Gould, G. W., Dring, G. J. (1975), Heat resistance of bacterial endospores and the concept of an expanded osmoregulating cortex, *Nature (London)* **258**, 402-405.

[333] Alderton, G., Ito, K. A., Chen, J. K. (1976), Chemical manipulation of the heat resistance of *Clostridium botulinum* spores, *Appl. Environ. Microbiol.* **31**, 492-498.

[334] Ando, Y., Tsuzuki, T. (1983), Mechanism of chemical manipulation of the heat resistance of *Clostridium perfringens* spores, *J. Appl. Bacteriol.* **54**, 197-202.

[335] Bender, G. R., Marquis, R. E. (1985), Spore heat resistance and specific mineralization, *Appl. Environ. Microbiol.* **50**, 1414-1421.

[337] Kihm, D. J., Hutton, M. T., Hanlin, J. H., Johnson, E. A. (1990), Influence of transition metals added during sporulation on heat resistance of *Clostridium botulinum* 113B spores, *Appl. Environ. Microbiol.* **56**, 681-685.

[338] Errington, J. (1993), *Bacillus subtilis* sporulation: regulation of gene expression and control of morphogenesis, *Microbiol. Rev.* **57**, 1-33.

[339] Stragier, P., Losick, R. (1996), Molecular genetics of sporulation in *Bacillus subtilis*, *Ann. Rev. Genet.* **30**, 297-341.

[340] Fitz-James, P. C. (1962), Morphology of spore development in *Clostridium pectinovorum*, *J. Bacteriol.* **84**, 104-114.

[341] Johnston, K., Holland, K. T. (1977), Ultrastructural changes during sporulation of *Clostridium bifermentans*, *J. Gen. Microbiol.* **100**, 217-220.

[342] Long, S., Jones, D. T., Woods, D. R. (1983), Sporulation of *Clostridium acetobutylicum* P262 in a defined medium, *Appl. Environ. Microbiol.* **45**, 1389-1393.

[343] Hoch, J. A. (1993), Regulation of the phosphorelay and the intitiation of sporulation in *Bacillus subtilis*, *Ann. Rev. Microbiol.* **47**, 441-465.

[344] Brown, D. P., Ganova-Raeva, L., Green, B. D., Wilkinson, S. R., Young, M., Youngman, P. (1994), Characterization of *spo0A* homologues in diverse *Bacillus* and *Clostridium* species reveals regions of high conservation within the effector domain, *Mol. Microbiol.* **14**, 411-426.

[345] Jones, D. T., Van Der Westhuizen, A., Long, S., Allcock, E. R., Reid, S. J., Woods, D. R. (1982), Solvent production and morphological changes in *Clostridium acetobutylicum*, *Appl. Environ. Microbiol.* **43**, 1434-1439.

[346] Long, S., Jones, D. T., Woods, D. R. (1984a), The relationship between sporulation and solvent production in *Clostridium acetobutylicum* P262, *Biotechnol. Lett.* **6**, 529-534.

[347] Garnier, T., Cole, S. T. (1988), Studies of UV-inducible promoters from *Clostridium perfringens in vivo* and *in vitro*, *Mol. Microbiol.* **2**, 607-614.

[348] Pich, A., Bahl, H. (1991), Purification and characterization of the DNA-dependent RNA polymerase from *Clostridium acetobutylicum*, *J. Bacteriol.* **173**, 2120-2124.

[349] Sauer, U., Treuner, A., Buchholz, M., Santangelo, J. D., Dürre, P. (1994), Sporulation and primary sigma factor homologous genes in *Clostridium acetobutylicum*, *J. Bacteriol.* **176**, 6572-6582.

[350] Wong, J., Sass, C., Bennett, G. N. (1995), Sequence and arrangement of genes encoding sigma factors in *Clostridium acetobutylicum* ATCC 824, *Gene* **153**, 89-92.

[351] Santangelo, J. D., Kuhn, A., Treuner-Lange, A., Dürre, P. (1998), Sporulation and time course expression of sigma-factor homologous genes in *Clostridium acetobutylicum*, *FEMS Microbiol. Lett.* **161**, 157-164.

[352] Arcuri, E. F., Wiedmann, M., Boor, K. J. (2000), Phylogeny and functional conservation of σ^E in endospore-forming bacteria, *Microbiology* **146**, 1593-1603.

[353] Foster, S. J., Johnstone, K. (1990), Pulling the trigger: the mechanism of bacterial spore germination, *Mol. Microbiol.* **4**, 137-141.

[354] Gibbs, P. A. (1967), The activation of spores of *Clostridium bifermentans*, *J. Gen. Microbiol.* **46**, 285-291.

[355] Craven, S. E., Blankenship, L. C. (1985), Activation and injury of *Clostridium perfringens* spores by alcohols, *Appl. Environ. Microbiol.* **50**, 249-254.

[356] Gould, G. W. (1970), Germination and the problem of dormancy, *J. Appl. Bacteriol.* **33**, 34-49.

[357] Douglas, F., Hambleton, R., Rigby, G. J. (1973), An investigation of the oxidation-reduction potential and of the effect of oxygen on the germination and outgrowth of *Clostridium butyricum* spores, using platinum electrodes, *J. Appl. Bacteriol.* **36**, 625-633.

[358] Takagi, A., Kawata, T., Yamamoto, S. (1960), Electron microscope studies on ultrathin sections of spores of the *Clostridium* group, with special reference to the sporulation and germination process, *J. Bacteriol.* **80**, 37-46.

[359] Hoeniger, J. F. M., Headley, C. L. (1968), Cytology of spore germination in *Clostridium pectinovorum*, *J. Bacteriol.* **96**, 1835-1847.

[360] Hoeniger, J. F. M., Headley, C. L. (1969), Ultrastructural aspects of spore germination and outgrowth in *Clostridium sporogenes*, *Can. J. Microbiol.* **15**, 1061-1066.

[361] Samsonoff, W. A., Hashimoto, T., Conti, S. F. (1970), Ultrastructural changes associated with germination and outgrowth of an appendage-bearing clostridial spore, *J. Bacteriol.* **101**, 1038-1045.

[362] Miyata, S., Moriyama, R., Sugimoto, K., Makino, S. (1995b), Purification and partial characterization of a spore cortex-lytic enzyme of *Clostridium perfringens* S40 spores, *Biosci. Biotech. Biochem.* **59**, 514-515.

[363] Chen, Y., Miyata, S., Makino, S., Moriyama, R. (1997), Molecular characterization of a germination-specific muramidase from *Clostridium perfringens* S40 spores and nucleotide sequence of the corresponding gene, *J. Bacteriol.* **179**, 3181-3187.

[364] Miyata, S., Kozuka, S., Yasuda, Y., Chen, Y., Moriyama, R., Tochikubo, K., Makino, S. (1997), Localization of germination-specific spore-lytic enzymes in *Clostridium perfringens* S40 spores detected by immunoelectron microscopy, *FEMS Microbiol. Lett.* **152**, 243-247.

[365] Urakami, K., Miyata, S., Moriyama, R., Sugimoto, K., Makino, S. (1999), Germination-specific cortex-lytic enzymes from *Clostridium perfringens* S40 spores: time of synthesis, precursor structure and regulation of enzyme activity, *FEMS Microbiol. Lett.* **173**, 467-473.

[366] Miyata, S., Moriyama, R., Miyahara, N., Makino, S. (1995a), A gene (*sleC*) encoding a spore-cortex-lytic enzyme from *Clostridium perfringens* S40 spores; cloning, sequence analysis and molecular characterization, *Microbiology* **141**, 2643-2650.

[367] Shih, N-J., Labbé, R. G. (1996), Characterization and distribution of amylases during vegetative cell growth and sporulation of *Clostridium perfringens*, *Can. J. Microbiol.* **42**, 628-633.

[368] Allison, C., Macfarlane, G. T. (1990), Regulation of protease production in *Clostridium sporogenes*, *Appl. Environ. Microbiol.* **56**, 3485-3490.

[369] Gehin, A., Gelhaye, E., Raval, G., Petitdemange, H. (1995), *Clostridium cellulolyticum* viability and sporulation under cellobiose starvation conditions, *Appl. Environ. Microbiol.* **61**, 868-871.

[370] Gerischer, U., Dürre, P. (1992), mRNA analysis of the *adc* gene region of *Clostridium acetobutylicum* during the shift to solventogenesis, *J. Bacteriol.* **174**, 426-433.

[371] Walter, K. A., Bennett, G. N., Papoutsakis, E. T. (1992), Molecular characterization of two *Clostridium acetobutylicum* ATCC 824 butanol dehydrogenase isozyme genes, *J. Bacteriol.* **174**, 7149-7158.

[372] Fischer, R. J., Helms, J., Dürre, P. (1993), Cloning, sequencing and molecular analysis of the *sol* operon of *Clostridium acetobutylicum*, a chromosomal locus involved in solventogenesis, *J. Bacteriol.* **175**, 6959-6969.

[373] Sauer, U., Dürre, P. (1995), Differential induction of genes related to solvent formation during the shift from acidogenesis to solventogenesis in continuous culture of *Clostridium acetobutylicum*, *FEMS Microbiol. Lett.* **125**, 115-120.

[374] Bahl, H. (1993), Heat shock response and onset of solvent formation in *Clostridium acetobutylicum*, in: *The Clostridia and Biotechnology* (Woods, D. R., Ed.), pp. 247-259, Stoneham: Butterworth-Heinemann.

[375] Tummala, S. B., Welker, N. E., Papoutsakis, E. T. (1999), Development and characterization of a gene expression reporter system for *Clostridium acetobutylicum* ATCC 824, *Appl. Environ. Microbiol.* **65**, 3793-3799.

[376] Bertram, J., Kuhn, A., Dürre, P. (1990), Tn*916*-induced mutants of *Clostridium acetobutylicum* defective in regulation of solvent formation, *Arch. Microbiol.* **153**, 373-377.

[377] Mattsson, D. M., Rogers, P. (1994), Analysis of Tn*916*-induced mutants of *Clostridium acetobutylicum* altered in solventogenesis and sporulation, *J. Ind. Microbiol.* **13**, 258-268.

[378] Wilkinson, S. R., Young, D. I., Morris, J. G., Young, M. (1995), Molecular genetics and the initiation of solventogenesis in *Clostridium beijerinckii* (formerly *Clostridium acetobutylicum*) NCIMB 8052, *FEMS Microb. Rev.* **17**, 275-285.

[379] Sauer, U., Dürre, P. (1992), Possible function of $tRNA^{Thr}{}_{ACG}$ in regulation of solvent formation in *Clostridium acetobutylicum*, *FEMS Microbiol. Lett.* **100**, 147-154.

[380] Nair, R. V., Green. E. M., Watson, D. E., Bennett, G. N., Papoutsakis, E. T. (1999), Regulation of the *sol* locus genes for butanol and acetone formation in *Clostridium acetobutylicum* ATCC 824 by a putative transcriptional repressor, *J. Bacteriol.* **181**, 319-330.

[381] Ravagnani, A., Jennert, K. C. B., Steiner, E., Grünberg, R., Jefferies, J. R., Wilkinson, S. R., Young. D. I., Tidswell, E. C., Brown, D. P., Youngman, P., Morris, J. G., Young, M. (2000), Spo0A directly controls the switch from acid to solvent production in solvent-forming clostridia, *Mol. Microbiol.* **37**, 1172-1185.

[382] Kashket, E. R., Cao, Z-Y. (1995), Clostridial strain degeneration, *FEMS Microbiol. Rev.* **17**, 307-315.

[383] Gottschal, J. C., Morris, J. G. (1982), Continuous production of acetone and butanol by *Clostridium acetobutylicum* growing in turbidostat culture, *Biotechnol. Lett.* **4**, 477-482.

[384] Meinicke, B., Bahl, H., Gottschalk, G. (1984), Selection of an asporogenous strain of *Clostridium acetobutylicum* in continuous culture under phosphate limitation, *Appl. Environ. Microbiol.* **48**, 1064-1065.

[385] Woolley, R. C., Morris, J. G. (1990), Stability of solvent production by *Clostridium acetobutylicum* in continuous culture: strain differences, *J. Appl. Bacteriol.* **69**, 718-728.

[386] Ross, R., Delia, J., Mooney, R., Chesbro, W. (1990), Nutrient limitation of two saccharolytic clostridia: secretion, sporulation, and solventogenesis, *FEMS Microbiol. Ecol.* **74**, 153-164.

[387] Long, S., Jones, D. T., Woods, D. R. (1984b), Initiation of solvent production, clostridial stage and endospore formation in *Clostridium acetobutylicum* P262, *Appl. Microbiol. Biotechnol.* **20**, 256-261.

[388] Rogers, P., Palosaari, N. (1987), *Clostridium acetobutylicum* mutants that produce butyraldehyde and altered quantities of solvents, *Appl. Environ. Microbiol.* **53**, 2761-2766.

[389] Dürre, P., Kuhn, A., Gottschalk, G. (1986), Treatment with allyl alcohol selects specifically for mutants of *Clostridium acetobutylicum* defective in butanol synthesis, *FEMS Microbiol. Lett.* **36**, 77-81.

[390] Janati-Idrissi, R., Junelles, A. M., El Kanouni, A., Petitdemange, H., Gay, R. (1987), Selection of mutants of *Clostridium acetobu-*

tylicum defective in acetone synthesis, *Ann. Inst. Past. Microbiol.* **138**, 313-323.

[391] Landuyt. S. L., Hsu, E. J. (1992), Preparation of refractile spores of *Clostridium thermosaccharolyticum* involves a solventogenic phase, *Appl. Environ. Microbiol.* **58**, 1797-1800.

[392] Sebald, M. (1993), Mutations, in: *Genetics and Molecular Biology of Anaerobic Bacteria* (Sebald, M., Ed.), pp. 64-97, New York: Springer-Verlag.

[393] Granum, P. E., Stewart, G. S. A. B.(1993), Molecular biology of *Clostridium perfringens* enterotoxin, in: *Genetics and Molecular Biology of Anaerobic Bacteria* (Sebald, M., Ed.), pp. 235-247, New York: Springer-Verlag.

[394] Kokai-Kun, J. F., McClane, B. A. (1997), The *Clostridium perfringens* enterotoxin, in: *The Clostridia: Molecular Biology and Pathogenesis* (Rood, J. I., McClane, B. A., Songer, J. G., Titball, R. W., Eds.), pp. 326-357, San Diego: Academic Press.

[395] McClane, B. A. (1998), New insights into the genetics and regulation of expression of *Clostridium perfringens* enterotoxin, *Curr. Topics Microbiol. Immunol.* **225**, 37-55.

[396] Duncan, C. L. (1973), Time of enterotoxin formation and release during sporulation of *Clostridium perfringens* type A, *J. Bacteriol.* **113**, 932-936.

[397] Czeczulin, J. R., Hanna, P. C., McClane, B. A. (1993), Cloning, nucleotide sequencing, and expression of the *Clostridium perfringens* enterotoxin gene in *Escherichia coli, Infect. Immun.* **61**, 3429-3439.

[397a] McClane, B. A. (1997), *Clostridium perfringens*, in: *Food Microbiology Fundamentals and Frontiers* (Doyle, M. P., Beuchat, L. R., Montville, T. J., Eds.), pp. 305-326, Washington, DC: ASM Press.

[398] Duncan, C. L., Strong, D. H., Sebald, M. (1972), Sporulation and enterotoxin production by mutants of *Clostridium perfringens, J. Bacteriol.* **110**, 378-391.

[399] Melville, S. B., Labbe, R., Sonenshein, A. L. (1994), Expression from the *Clostridium perfringens cpe* promoter in *C. perfringens* and *Bacillus subtilis, Infect. Immun.* **62**, 5550-5558.

[400] Czeczulin, J. R., Collie, R. E., McClane, B. A. (1996), Regulated expression of *Clostridium perfringens* enterotoxin in naturally *cpe*-negative type A, B, and C isolates of *C. perfringens, Infect. Immun.* **64**, 3301-3309.

[401] Brynestad, S., Iwanejko, L. A., Stewart, G. S. A. B., Granum, P. E. (1994), A complex array of Hpr consensus DNA recognition sequences proximal to the enterotoxin gene in *Clostridium perfringens* type A, *Microbiology* **140**, 97-104.

[402] Strauch, M. A., Hoch, J. A. (1993), Transition-state regulators: sentinels of *Bacillus subtilis* post-exponential gene expression, *Mol. Microbiol.* **7**, 337-342.

[403] Zhao, Y., Melville, S. B. (1998), Identification and characterization of sporulation-dependent promoters upstream of the enterotoxin gene (*cpe*) of *Clostridium perfringens, J. Bacteriol.* **180**, 136-142.

[404] Hoch, J. A., Silhavy, T. J. (Eds.) (1995), *Two-Component Signal Transduction*, Washington, DC: ASM Press.

[405] Rood, J. I. (1998), Virulence genes of *Clostridium perfringens, Ann. Rev. Microbiol.* **52**, 333-360.

[406] Shimizu, T., Okabe, A., Rood, J. I. (1997), Regulation of toxin production in *Clostridium perfringens*, in: *The Clostridia: Molecular Biology and Pathogenesis* (Rood, J. I., McClane, B. A., Songer, J. G., Titball, R. W., Eds.), pp. 451-470, San Diego: Academic Press.

[407] Hundsberger, T., Braun, V., Weidmann, M., Leukel, P., Sauerborn, M., Eichel-Streiber, C. V. (1997), Transcription analysis of the genes *tcdA-E* of the pathogenicity locus of *Clostridium difficile, Eur. J. Biochem.* **244**, 735-742.

[408] Dupuy, B., Sonenshein, A. L. (1998), Regulated transcription of *Clostridium difficile* toxin genes, *Mol. Microbiol.* **27**, 107-120.

[409] Ketley, J. M., Mitchell, T. J., Haslam, S. C., Stephen, J., Candy, D. C. A., Burdon, D. W. (1986), Sporogenesis and toxin A production by *Clostridium difficile, J. Med. Microbiol.* **22**, 33-38.

[410] Kamiya, S., Ogura, H., Meng, X. Q., Nakamura, S. (1992), Correlation between cytotoxin production and sporulation in *Clostridium difficile, J. Med. Microbiol.* **37**, 206-210.

[411] Haslam, S. C., Ketley, J. M., Mitchell, T. J., Stephen, J., Burdon, D. W., Candy, D. C. A. (1986), Growth of *Clostridium difficile* and production of toxins A and B in complex and defined media, *J. Med. Microbiol.* **21**, 293-297.

[412] Osgood, D. P., Wood, N. P., Sperry, J. F. (1993), Nutritional aspects of cytotoxin production by *Clostridium difficile, Appl. Environ. Microbiol.* **59**, 3985-3988.

[413] Yamakawa, K., Kamiya, S., Meng, X. Q., Karasawa, T., Nakamura, S. (1994), Toxin production by *Clostridium difficile* in a defined medium with limited amino acids, *J. Med. Microbiol.* **41**, 319-323.

[414] Hammond, G. A., Johnson, J. L. (1995), The toxigenic element of *Clostridium difficile* strain VPI 10463, *Mic. Pathog.* **19**, 203-213.

[415] Hammond, G. A., Lyerly, D. M., Johnson, J. L. (1997), Transcriptional analysis of the toxigenic element of *Clostridium difficile, Microb. Pathog.* **22**, 143-154.

[416] Moncrief, J. S., Barroso, L. A., Wilkins, T. D. (1997), Positive regulation of *Clostridium difficile* toxins, *Infect. Immun.* **65**, 1105-1108.

[417] Missiakis, D., Raina, S. (1998), The extracytoplasmic function sigma factors: role and regulation, *Mol. Microbiol.* **28**, 1059-1066.

[418] Onderdonk, A. B., Lowe, B. R., Bartlett, J. G. (1979), Effect of environmental stress on *Clostridium difficile* toxin levels during continuous cultivation, *Appl. Environ. Microbiol.* **38**, 637-641.

[420] Yamakawa, K., Karasawa, T., Ikoma, S., Nakamura, S. (1996), Enhancement of *Clostridium difficile* toxin production in biotin-limited conditions, *J. Med. Microbiol.* **44**, 111-114.

[421] Minton, N. P. (1995), Molecular genetics of clostridial neurotoxins, *Curr. Topics Microbiol. Immunol.* **195**, 161-194.

[422] Henderson, I., Davis, T., Elmore, M., Minton, N. P. (1997), The genetic basis of toxin production in *Clostridium botulinum* and *Clostridium tetani*, in: *The Clostridia: Molecular Biology and Pathogenesis* (Rood, J. I., McClane, B. A., Songer, J. G., Titball, R. W., Eds.), pp. 261-294, San Diego: Academic Press.

[423] Collins, M. D., East, A. K. (1998), Phylogeny and taxonomy of the food-borne pathogen *Clostridium botulinum* and its neurotoxins, *J. Appl. Microbiol.* **84**, 5-17.

[424] Kubota, T., Yonekura, N., Hariya, Y., Isogai, E., Isogai, H., Amano, K., Fujii, N. (1998), Gene arrangement in the upstream region of *Clostridium botulinum* type E and *Clostridium butyricum* BL6340 progenitor toxin genes is different from that of other types, *FEMS Microbiol. Lett.* **158**, 215-221.

[425] Marvaud, J-C., Gibert, M., Inoue, K., Fujinaga, Y., Oguma, K., Popoff, M. R. (1998b), *botR/A* is a positive regulator of botulinum neurotoxin and associated non-toxin protein genes in *Clostridium botulinum* A, *Mol. Microbiol.* **29**, 1009-1018.

[426] Patterson-Curtis, S. I., Johnson, E. A. (1989), Regulation of neurotoxin and protease formation in *Clostridium botulinum* OkraB and HallA by arginine, *Appl. Environ. Microbiol.* **55**, 1544-1548.

[427] Leyer, G. J., Johnson, E. A. (1990), Repression of toxin production by tryptophan in *Clostridium botulinum* type E, *Arch. Microbiol.* **154**, 443-447.

[428] Davis, T. O., Henderson, I., Brehm, J. K., Minton, N. P. (2000), Development of a transformation system for group II, non-proteolytic *Clostridium botulinum* type B strains, *J. Mol. Microbiol. Biotechnol.* **2**, 59-69.

[429] Schantz, E. J., Johnson, E. A. (1992), Properties and use of botulinum toxin and other microbial neurotoxins in medicine, *Microbiol. Rev.* **56**, 80-89.

[430] Nakamura, S., Serikawa, T., Yamakawa, K., Nishida, S., Kozaki, S., Sakaguchi, G. (1978), Sporulation and C_2 toxin production by *Clostridium botulinum* type C strains producing no C_1 toxin, *Microbiol. Immunol.* **22**, 591-596.

[431] Mueller, J. H., Miller, P. A. (1956), Essential role of histidine peptides in tetanus toxin production, *J. Biol. Chem.* **223**, 185-194.

[432] Tsunashima, I., Sato, K., Shoji, K., Yoneda, M., Amono, T. (1964), Excess supplementation of certain amino acids to medium and its inhibitory effect on toxin production by *Clostridium tetani, Biken. J.* **7**, 161-163.

[433] Mellanby, J. (1968), The effect of glutamate on toxin production by *Clostridium tetani, J. Gen. Microbiol.* **54**, 77-82.

[434] Marvaud, J-C., Eisel, U., Binz, T., Niemann, H., Popoff, M. R. (1998a), TetR is a positive regulator of the tetanus toxin gene in *Clostridium tetani* and is homologous to BotR, *Infect. Immun.* **66**, 5698-5702.

[435] Charles, J.-F., Nicolas, L., Sebald, M., Debarjac, H. (1990), *Clostridium bifermentans* serovar *malaysia*: sporulation, biogenesis of inclusion bodies and larvicidal effect on mosquito, *Res. Microbiol.* **141**, 721-733.

[436] Nicolas, L., Hamon, S., Frachon, E., Sebald, M., Debarjac, H. (1990), Partial inactivation of the mosquitocidal activity of *Clostridium bifermentans* serovar *malaysia* by extracellular proteinases, *Appl. Microbiol. Biotechnol.* **34**, 36-41.

[437] Nicolas, L., Charles, J-F., Debarjac, H. (1993), *Clostridium bifermentans* serovar *malaysia*: Characterization of putative mosquito larvicidal proteins, *FEMS Microbiol. Lett.* **113**, 23-28.

[438] Terracciano, J. S., Rapaport, E., Kashket, E. R. (1988), Stress- and growth phase-associated proteins of *Clostridium acetobutylicum*, *Appl. Environ. Microbiol.* **54**, 1989-1995.

[439] Pich, A., Narberhaus, F., Bahl, H. (1990), Induction of heat shock proteins during initiation of solvent formation in *Clostridium acetobutylicum*, *Appl. Microbiol. Biotechnol.* **33**, 697-704.

[440] Narberhaus, F., Giebeler, K., Bahl, H. (1992), Molecular characterization of the *dnaK* gene region of *Clostridium acetobutylicum*, including *grpE*, *dnaJ*, and a new heat shock gene, *J. Bacteriol.* **174**, 3290-3299.

[441] Rüngeling, E., Laufen, T., Bahl, H. (1999), Functional characterisation of the chaperones DnaK, DnaJ and GrpE from *Clostridium acetobutylicum*, *FEMS Microbiol. Lett.* **170**, 119-123.

[442] Narberhaus, F., Bahl, H. (1992), Cloning, sequencing and molecular analysis of the *groESL* operon of *Clostridium acetobutylicum*, *J. Bacteriol.* **174**, 3282-3289.

[443] Sauer, U., Dürre, P. (1993), Sequence and molecular characterization of a DNA region encoding a small heat shock protein of *Clostridium acetobutylicum*, *J. Bacteriol.* **175**, 3394-3400.

[444] Ciruela, A., Cross, S., Freedman, R. B., Hazlewood, G. P. (1997), Sequence and transcriptional analysis of *groES* and *groEL* genes from the thermophilic bacterium *Clostridium thermocellum*, *Gene* **186**, 143-147.

[445] Cross, S. J., Ciruela, A., Poomputsa, K., Romaniec, M. P. M., Freedman, R. B. (1996), Thermostable chaperonin from *Clostridium thermocellum*, *Biochem. J.* **316**, 615-622.

[446] Bahl, H., Müller, H., Behrens, S., Joseph, H., Narberhaus, F. (1995), Expression of heat shock genes in *Clostridium acetobutylicum*, *FEMS Microbiol. Rev.* **17**, 341-348.

[447] Woods, D. R. (Ed.) (1993), *The Clostridia and Biotechnology*, Stoneham: Butterworth-Heinemann.

[448] Mauchline, M. L., Davis, T. O., Minton, N. P. (1999), Clostridia, in: *Manual of Industrial Microbiology and Biotechnology 2nd. Ed.* (Demain, A. L., Davies, J. E., Eds.), pp. 475-490, Washington, DC: ASM Press.

[449] Phillips-Jones, M. K. (1993), Bioluminescence (*lux*) expression in the anaerobe *Clostridium perfringens*, *FEMS Microbiol. Lett.* **106**, 265-270.

[450] Quixley, K. W. M., Reid, S. J. (2000), Construction of a reporter gene vector for *Clostridium beijerinckii* using a *Clostridium* endoglucanase gene, *J. Mol. Microbiol. Biotechnol.* **2**, 53-57.

[451] Rogers, P., Gottschalk, G. (1993), Biochemistry and regulation of acid and solvent production in clostridia, in: *The Clostridia and Biotechnology* (Woods, D. R., Ed.), pp. 25-50, Stoneham: Butterworth-Heinemann.

[452] Schuster, K. C., Mertens, F., Gapes, J. R. (1999), FTIR spectroscopy applied to bacterial cells as a novel method for monitoring complex biotechnological processes, *Vibr. Spectrosc.* **19**, 467-477.

4
Genetic Tools for Solventogenic Clostridia

Seshu B. Tummala, Christopher Tomas, Latonia M. Harris, Neil E. Welker, Frederick B. Rudolph, George N. Bennett, Eleftherios T. Papoutsakis

4.1
Introduction

Metabolic engineering (ME) is the directed modification of cellular metabolism and/or properties through the introduction, deletion and/or modification of genes using recombinant DNA methods. ME can be employed to generate strains with desirable cellular traits either for industrial/environmental applications, or for genetic studies [1, 2] While the concept of ME has been used for strain improvement studies and process development for many years, the advent of recombinant DNA technology has enabled cellular modifications using targeted genetic modifications instead of and/or in addition to the traditional mutagenesis and selection approach.

The industrial history and future potential of solventogenic clostridia have been extensively reviewed [3-5]. The past 10 years have witnessed substantial developments in the genetics of solventogenic clostridia, and the completed genome sequence of *Clostridium acetobutylicum* ATCC 824 will most certainly accelerate progress in this arena. Although several genetic tools had been available since the early 1990s (e. g., reviewed in [4, 6]), the genetic and metabolic engineering toolbox for solventogenic clostridia remains incomplete. In this chapter, we review and discuss several key genetic tools with emphasis on recently developed ones, including the development of antisense RNA technology, gene inactivation by recombination, and gene-expression reporter systems. These tools are crucial for metabolic engineering studies to generate strains for genetic studies and industrial or environmental applications.

4.2
Genetic manipulation of C. *acetobutylicum*: overview

In order to alter the genetic characteristics of *C. acetobutylicum*, several technologies need to be available. These include:

- an assortment of vectors with suitable replication origins, antibiotic resistance markers, and restriction sites for cloning. A variety of vectors allows a choice among host strains (e. g., hosts that are genetically modified and already contain a resistance element) and for various purposes (e. g., to examine regulation of gene expression by a reporter system or to overexpress a protein in a particular circumstance).
- A method for routine introduction of plasmid constructs into the host.
- Effective recombination of introduced DNA with the chromosome so that defined gene alterations can be made. Recombination procedures require an efficient means for introducing DNA into the cells, either by conjugation, electroporation, or natural competence, coupled with appropriate selections, counter selections, or indicators for screening of products. Consequently, recombination technologies require knowledge and development of the other techniques.

4.3
Methods for introducing DNA into *Clostridium acetobutylicum*

The first method used for introducing plasmids into *C. acetobutylicum* was conjugation with heterologous donor strains bearing plasmids such as pAMß1 (reviewed by Young [7]). This method has been extended and widely used in transposon mutagenesis and was aided by expansion to include a wider variety of donors, even *E. coli* [8]. Chemically induced transformation of *C. acetobutylicum* and related strains has been reviewed by Reysett et al. [9] and Minton et al. [10]. The chemical transformation procedure made efforts to limit autolysin action by using an autolysin mutant of strain N1-4 or autolysin inhibitors [11, 12] strain N1-4. Other work by Soutschek-Bauer et al. [13] on *C. thermohydrosulfuricum* and by Yoshino et al. [14] on solventogenic *Clostridium* strains used high concentrations of PEG to allow transformation.

Upon the development of electroporation in the 1980s and its adaptation to bacterial species, the technique was applied to *C. acetobutylicum* and related Clostridia [9, 10]. Work with *C. perfringens* showed that various antibiotics could be used for selection [15-20]. Strains N1-4 [11], *C. beijerinckii* 8052 [21], *C. beijerinckii* NRRL B-592 [22] and *C. acetobutylicum* ATCC 824 [23] were transformed by electroporation during this period of interest. Some other strains, however, were not easily transformed.

Strain-specific protocols have been developed for the transformation of *C. acetobutylicum*. These have been reviewed [5, 6, 24, 25]. Several replicons from Gram-positive bacteria have been used in conjunction with a macrolide, lincosamide, and streptogramin B resistance element (MLS^r), tetracycline resistance, or chloramphenicol resistance for selection of transformation in *C. acetobutylicum* [5, 6, 24, 26]. In strain ATCC 824, we have identified a restriction system (*Cac*824I) that frequently cuts DNA originating from *E. coli* or other G+C-rich DNA [23] and prevents efficient transformation. We have developed a novel *in vivo* system employing the cloned Φ3T3 methylase system for methylating DNA (in *E. coli*) specifically at *Cac*824I recognition sites, and thereby protecting it from restriction [27].

4.4
Strategies for improvement of electroporation efficiencies

In the current *C. acetobutylicum* electroporation protocol [23], the plasmid DNA is first methylated by passage through the *E. coli* strain, ER2275, which bears pAN1. This plasmid has an origin compatible with the *E. coli*–clostridial shuttle vectors and other cloning or recombination vectors and bears the Φ3T3 methylase gene [27]. Other constructs bearing the Φ3T3 methylase gene have been made which have a different origin of plasmid replication [28]. A number of articles reviewing the technique of electroporation and the application to related microorganisms have appeared [29-33]. Electroporation of clostridia has been reviewed and discussed [9,10] and various factors have been reported that improve efficiency in various clostridial strains. While a number of studies have led to conditions providing good efficiency, some aspects of the technique may not have been fully evaluated and optimized. Such factors include: cell growth conditions; electroporation solutions; electroporation electrical parameters; size, quality and purity of DNA; recovery and selection conditions. Particulars of these factors are discussed below.

- Growth stage: The preparation of cells from the optimal stage of growth has been shown to be important for a variety of organisms [11, 16, 21].
- Growth conditions: The growth of Gram-positive cells on glycine to weaken the cell wall has been shown to improve certain electroporations [34-36], and growth of the cells at higher temperature also aided efficiency [37].
- Cell pretreatment: This can affect the efficiency of introduction of DNA into cells. The use of alkaline-tris buffer [13, 14] and lengthy preincubation at 4 °C have increased DNA uptake in some circumstances [38]. The addition of autolysin inhibitors such as choline [12] and the use of lysostaphin predigestion [20] have been employed to improve efficiency.
- Electroporation conditions: Effects of electroporation temperature have also been found important with certain strains [39]. The electroporation solutions have been adjusted through the use of melibiose rather than sucrose as osmotic stabilizer [10], and adjustment of cell density in the electroporation cuvette can also have an effect [16].
- DNA conditions: Purity of the DNA and presence or removal of substances used in the DNA isolations can affect efficiency [16]. The addition of inert polymers to make a two-phase system has also increased efficiency in some cell lines [40].

- Electrical parameters: Optimization of voltage gradients and time constants for size and type of DNA is an important consideration [41, 42]. The use of a radio frequency (RF) modulated pulse for better control of the detrimental effects of the pulse and improved survival of electroporated cells has led to higher efficiency [43, 44].
- Recovery treatment: The recovery of transformed cells after electroporation is also important [44]. Time, temperature and culture broth are factors affecting recovery. A low pH (6.4) recovery medium was important in the case of *C. perfringens* [16, 20]. Differences in recovery efficiency with various antibiotic selections have been observed [45], and such differences may be overcome by the addition of chelating agents.

4.5
Chromosomal genetic recombination

4.5.1
Overview

A method for introducing DNA into the *C. acetobutylicum* chromosome is critical for long-term applications with metabolically engineered cells. Chromosomal incorporation would promote culture stability and eliminate the need to use antibiotics for plasmid maintenance. Likewise, the ability to specifically inactivate chromosomal genes is critical both from the standpoint of creating production strains and as a method for determining *in vivo* functions of cloned genes.

While homologous recombination is a frequently used technique to generate strains with defined genotypes in many bacteria, it has not been much exploited in the solventogenic clostridia. Reasons for this are: the inability to do crosses by using strains with integrated mobilization elements which stimulate high-frequency recombination in conjugation processes; the lack of development of transducing phages that are so useful with *E. coli*; and the low frequency of transformation and the apparent lack of a competence system as exists in *B. subtilis*. While efforts have been made to achieve chromosomal recombination and reports have been published with related clostridial species [46, 47], the techniques have not been widely used [48]. The methods of DNA transfer through mating have been restricted essentially to transfer of plasmids [10] or conjugative transposons [7, 49], which have special mechanisms for these processes and these have not been used for movement of chromosomal markers. Thus, these important routes to defined strains with multiple mutations, so important in the development of genetics of other organisms, have been unavailable in *C. acetobutylicum*, necessitating an almost exclusive reliance on molecular biology techniques.

In order to construct mutant strains deficient in a defined function, several requirements are needed:

- A copy of the gene or portion of the gene to be inactivated. This can be obtained from restriction digest of cloned DNA or by PCR amplification of a segment from the chromosome using primers designed from the DNA sequence of the region.
- A vector which allows for preparation and characterization of the construct. The vector needs a selectable marker to allow for isolation of the desired construct upon recombination with the host chromosome.
- An efficient means of introducing the construct into the cell. This can be done by using either a nonreplicative construct and a means of efficient transformation such as electroporation or a plasmid with a conditional replication so that cells bearing the recombined construct can be isolated from those carrying the plasmid construct. The plasmid carrying the DNA to be recombined can also be lost and the host recombinants screened by replica means, but this is a more time-consuming process.
- Efficient recombination between the host chromosome and the introduced construct. In many bacteria this is naturally efficient enough to carry out construction of useful strains. However, in clostridia the general recombination system seems rather weak from empirical observations. The level of recombination is determined by the action of recombination proteins on the DNA and the amount of recombination proteins in the cell (e. g., recA) is important in altering the recombination frequency.

4.5.2
Use of non-replicative plasmids

The first gene inactivations or "gene knock-out mutations" by defined recombination were reported by Wilkinson and Young [50]. These were made by recombination between a segment of a gene cloned on a non-replicative plasmid [51] and the homologous chromosomal gene via a Campbell type integration event which results in disruption of the chromosomal gene [50, 52]. Chromosomal segments as small as 336 bp allowed integration. Excision and loss of plasmid could occur indicating the possibility of replacing chromosomal alleles with designed mutations. Also the amplification of genes was observed [50]. Using this approach mutations in *gutD* and *spo0A* have been made and analyzed [52], indicating a role of Spo0A in regulating solventogenesis [53]. In these experiments with *C. beijerinckii* 8052 the plasmid was introduced by conjugation methods.

We have used a similar integrative strategy to integrate genes in *C. acetobutylicum* ATCC 824. In our case, the non-replicative plasmid DNA, a derivative of plasmid pJC4, was introduced by electroporation. We have applied the approach to inactivation of genes of metabolic interest [54, 55]. Briefly, the method used by Green [54, 55] involves cloning an internal gene segment into the Em^r vector, pJC4, a vector which cannot replicate in *C. acetobutylicum*. The pJC4 derivative is then methylated by growing it in a host bearing pAN1 [27] and the DNA mixture is isolated. Since pAN1 is a lower copy plasmid, most of the preparation is the desired plasmid for recombination. The DNA is then electroporated into *C. acetobutylicum* using conditions as described [27] and the cells plated for selection of Em^r colonies. The cells for electroporation are prepared fresh from active cultures [27, 54, 55]. The method has not given recombinants with a number of other genes attempted and unexpected variants have also been observed [56]. However, characterized mutant strains have exhibited interesting fermentation properties.

A recombinant strain with an inactivated butyrate kinase gene [55] exhibited low butyrate kinase (BK) activity and had considerably decreased butyrate (>75 %) and increased butanol (>40 %) production compared to the wild-type strain [57]. Analysis of three recombinant strains with the inactivated AAD (*aad*) gene [54] showed no acetone (since the CoA transferase genes needed for acetone formation follow the *aad* gene in an operon arrangement) and only very low butanol formation. The latter result confirmed that AAD [58] is necessary for butanol formation [59]. An artificially created acetone operon (*ace*) was recombined into the chromosome [4]. The product strain (which now should have a duplication of the acetone pathway genes) exhibited increased acetone and butanol production. These studies demonstrate that with further development, gene inactivation and chromosomal integration in *C. acetobutylicum* are feasible technologies, with the potential to generate stable high producing strains for industrial applications.

4.5.3
Use of replicative plasmids

One method for obtaining high numbers of cells containing the desired construct so that recombination can take place is to use cells which have the plasmid construct already in them, and then eliminate the cells in which chromosomal recombination has not taken place. This can be done by selecting for the antibiotic resistance marker associated with the gene cloned on the plasmid (usually Em^r) under conditions where the plasmid cannot replicate. Although efforts have been made to find temperature-sensitive plasmids which work well in *C. acetobutylicum*, reports of their successful use in the formation of mutants through recombination have not appeared. Although plasmid loss under non-selecting conditions can be used, this is a more time-consuming approach. Nevertheless, such an approach has been successfully applied in the case of generating a *spo0A* mutation of *C. acetobutylicum* ATCC 824 [60]. This approach involves the use of pLHKO, a replicative chloramphenicol resistant vector containing a multiple cloning site. First, an internal fragment from the targeted gene is cloned into the multiple cloning site. The MLSr marker is then cloned into the middle of the internal gene fragment resulting in a replicative vector capable of double crossover integration into the chromosome at the target site. The pLHKO derivative is methylated and introduced into *C. acetobutylicum* using electroporation. The strain carrying the pLHKO derivative is grown on plates without antibiotic selection and replica plated daily onto fresh plates. After five days, the culture is replica plated onto plates separately containing erythromycin and chloramphenicol. Bacteria which are capable of growth on erythromycin but not on chloramphenicol are isolated. These are bacteria which may have incorporated the MLSr gene into the target area of the genome via double crossover integration. The precise nature of the genetic alteration is determined using PCR amplification of the chromosomal region followed by sequencing. The use of this method for targeted double crossover integration into the chromosome shows great promise as a means of producing stable recombinant strains with gene replacements as well as gene inactivations.

4.5.4 Future directions

Conditions for transformation with plasmids may be different from those ideal for isolating products of chromosomal recombination with introduced DNA. Some strategies to improve chromosomal recombination in other bacterial cells include the use of large fragments to obtain recombination with the chromosome [61, 62]. Recent work has shown that alkaline denaturation of the DNA introduced into *Streptomyces coelicolor* greatly increased homologous recombination [63].

Improvement of the recombination frequency through increasing the level of recombination proteins is also a reasonable strategy. In studies of recombination, a limiting factor is the assembly of a RecA filament [64, 65]. An important prerequisite for effective formation of the RecA coated filament that is active in recombination, is the processing of the substrate DNA to a form with a 3′ single-stranded end. The activation of the DNA is accomplished by a variety of processing nucleases essential for recombination [65, 66]. Subsequent to these interactions, a variety of redundant recombination proteins can carry out the reactions leading to completion of homologous recombination. Efforts to enhance recombination via elevation of RecA levels and to study recombination genes and proteins in clostridia may lead to improved recombination procedures in the future.

4.6 Plasmids to complement recombinant erythromycin resistant (MLS^r) strains

4.6.1 Plasmids with a thiamphenicol/tetracycline resistance marker

To date, replicative and non-replicative plasmids used for insertional gene inactivation have carried genes conferring resistance to erythromycin (MLS^r; [67, 68]. This made overexpression of genes from plasmids carrying an erythromycin resistance marker unfeasible, limiting the number of possible genetic manipulations possible in a single strain. Alternative replicative plasmids have been developed which contain either a chloramphenicol acetyl transferase (*cat*) gene (pIMPTH; [67]) or a *tet*M gene (pTLH1; [57]) which encode resistance to thiamphenicol and tetracycline, respectively.

The *aad* gene, encoding an aldehyde/alcohol dehydrogenase (AAD), and the butyrate operon (*ptb* and *buk*) have been cloned into pIMPTH to create plasmids pTHAAD and pTHBUT, respectively. These plasmids were used to complement *aad* and *buk* gene inactivation mutants. Complementation of the *aad* mutant

(PJC4AAD) with pTHAAD resulted in restoration of AAD activity and restored butanol production to wild type levels, while complementation of the *buk* mutant (PJC4BK) with pTHBUT restored BK activity and butyrate production [67]. The *aad* gene has also been cloned into pTLH1 to create plasmid pTAAD. The overexpression of *aad* from pTADD in the *buk* mutant strain PJC4BK resulted in increased ethanol production, but failed to increase butanol levels. This result suggests that *aad* is not a limiting factor in butanol production in pH≥5.0 fermentations [57]. pTAAD was also used to overexpress *aad* in a *solR* mutant (*solR* is a putative regulatory gene of the *sol* locus). SolRH(pTAAD) produced significantly higher levels of butanol, acetone, and ethanol (21 %, 45 %, and 62 %, respectively) than the SolRH strain [69]. The use of plasmids carrying antibiotic resistance genes other than erythromycin to complement gene inactivation mutants is a powerful tool for studying gene function.

4.6.2
Tetracycline but not clarithromycin inhibits solvent but not acid formation [69]

The use of two antibiotics, namely clarithromycin (a pH stable derivative of erythromycin) and tetracycline, to select for the inactivation mutant SolRH and pTAAD, respectively, was found to severely inhibit growth and solvent production. Use of both antibiotics resulted in final butyrate levels up to 1000 % higher than SolRH grown with neither antibiotic. Optical densities in fermentations which used both antibiotics for selection were approximately one-half those in fermentations without antibiotic selection (A600: 5.5 versus 11). Final solvent levels were decreased by about 80 %. Use of tetracycline alone resulted in the same increase in final butyrate levels, decrease in optical density, and resulted in an over 90 % reduction of solvent levels. Use of only clarithromycin in SolRH(pTAAD) fermentations was shown to result in no alteration of growth rate or product formation relative to SolRH(pTAAD) fermentation without antibiotic selection.

To avoid the negative effects of tetracycline on solvent production, SolRH(pTAAD) and SolRH(pTLH1) fermentations were performed without tetracycline. The presence of pTAAD and pTLH1 were confirmed by plasmid DNA isolation over the life of the fermentation. Comparative plating studies on tetracycline selective and non-selective plates also showed that both pTAAD and pTLH1 were stable over the course of a fermentation (ca. 3 days).

4.7
Gene expression reporter systems in *C. acetobutylicum*

A gene-expression reporter system for *C. acetobutylicum* and other solventogenic clostridia is an indispensable genetic tool. A reporter system would allow one to study the expression of both autologous and heterologous promoters in *C. acetobutylicum* and understand their regulation. An understanding of promoter strength and regulation could lead to more effective clostridial expression vectors. An increased knowledge of promoter strength and regulation can be coupled with current gene inactivation and antisense RNA strategies to develop more complex metabolic engineering strategies for increasing solvent production.

Until recently, no effective gene-expression reporter systems for *C. acetobutylicum* were available. Therefore, characterization of clostridial promoters had not been reported. Since *E. coli* genes are poorly expressed in *C. acetobutylicum,* the use of traditional *E. coli* reporter genes such as *lacZ* had not reported for use in *C. acetobutylicum.* Green Fluorescent Protein (GFP) has also been used as a reporter gene in many bacterial systems. However, GFP did not appear to be a good candidate for use as a reporter gene for *C. acetobutylicum* because of its requirement for oxygen which is necessary for the development of the chromophore responsible for fluorescence [70]. In 1993, a gene-expression reporter system for *Clostridium perfringens* was developed which utilized the *luxA* and *luxB* genes from *Vibrio fischeri* as the reporter gene [71]. Similar to GFP, the use of these bioluminescent genes as reporter genes for *C. acetobutylicum* are not feasible due to the need for oxygen in the reaction that produces the bioluminescence. A chloramphenicol acetyltransferase gene, *catP,* from *C. perfringens* was also considered for potential use as a reporter gene in a reporter system for *C. acetobutylicum.* Although adequate as a reporter gene for *C. perfringens* [72, 73], the traditional chloramphenicol acetyltransferase (Cat) assay was not well suited for use with *C. acetobutylicum,* because this species contains high levels of non-specific CoA transferases which would interfere with the Cat assay. Thus, none of the well-established reporter genes in bacterial systems seemed to be good candidates for use in a reporter system to effectively monitor gene expression in *C. acetobutylicum.*

In 1991, Burchhardt and Bahl [74] cloned and analyzed the *lacZ* gene from *Clostridium thermosulfurogenes* EM1 (later classified as *Thermoanaerobacterium thermosulfurogenes* EM1 [8]). They proposed that this gene would be an excellent candidate as a reporter gene for use in *C. acetobutylicum* [74]. Since *C. acetobutylicum* ATCC 824 lacks endogenous β-galactosidase activity [75], a reporter system with this *lacZ* as a reporter gene would be more sensitive to promoter activity.

Tummala et al. [76] demonstrated the development of a gene-expression reporter system (pHT3) with the *lacZ* gene from *T. thermosulfurogenes* EM1 as the reporter gene for use in *C. acetobutylicum* ATCC 824 (Figure 1). In this study, several experiments were performed to characterize the reporter system and assess its potential use for promoter characterization studies. The detection of β-galactosidase activity in time course studies with the reporter system and the phosphotransbutyrylase *(ptb),* thiolase *(thl),* and acetoacetate decarboxylase *(adc)* promoters cloned upstream of the *lacZ* gene from *T. thermosulfurogenes* EM1 demonstrated that the reporter gene produced a functional β-galactosidase in *C. acetobutylicum.* In addition, time course studies revealed significant differences in β-galactosidase specific activity profiles between strains containing the different promoters thus suggesting that the reporter system developed in this study is able to effectively distinguish between different promoters. The stability of the β-galactosidase produced by the reporter gene was also examined in the strains containing the reporter system and the *ptb* or *thl* promoters using chloramphenicol treatment to inhibit protein syn-

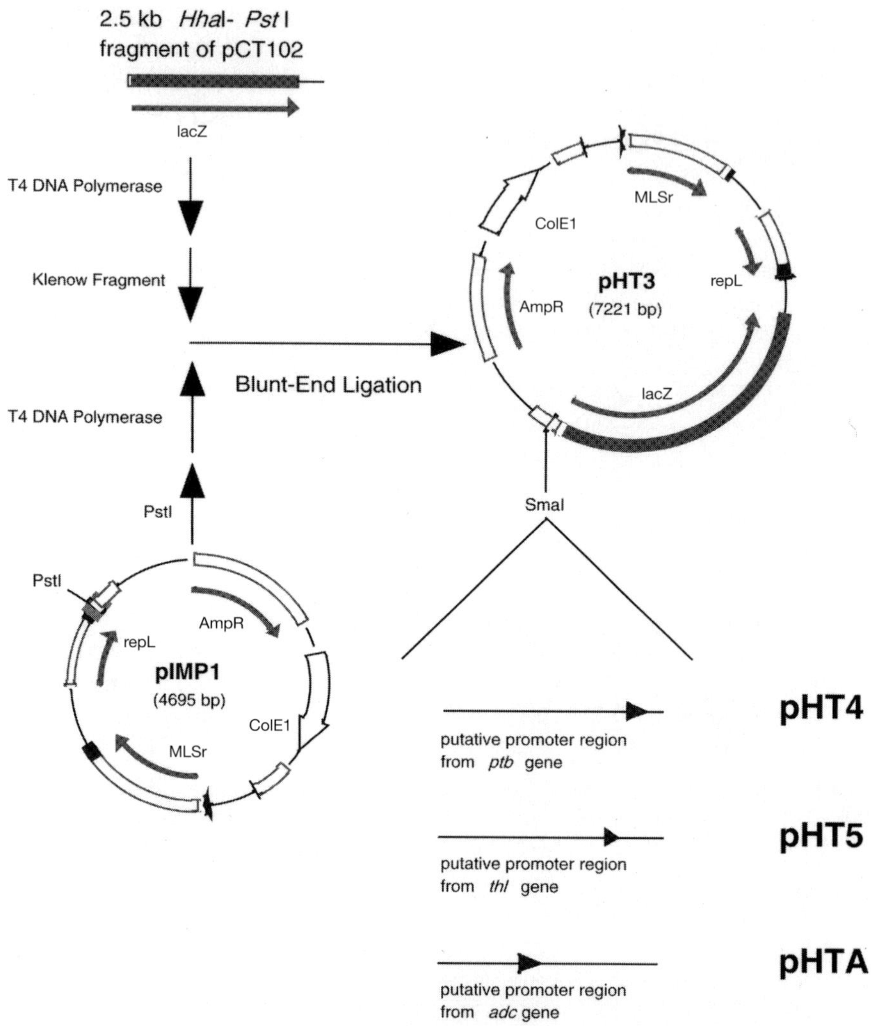

Figure 1. Construction of the reporter system, pHT3, and three test vectors, pHT4, pHT5, and pHTA. For each plasmid, the location and direction of transcription for relevant genes are indicated (arrows). Relevant restriction sites are shown. Abbreviations: lacZ, *lacZ* gene from *T. thermosulfurogenes* EM1; AmpR, ampicillin resistance genes; ColE1, Col E1 origin of replication; MLSr, macrolide-lincosamide-streptogramin B resistance gene; repL, the OriII origin of replication from pIMP1.

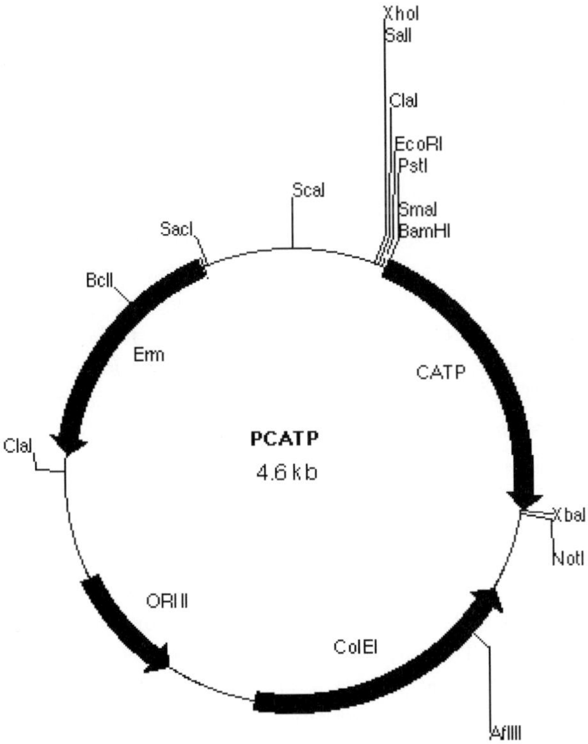

Figure 2. The reporter plasmid pCatP. The plasmid contains the MLS resistance element specifying erythromycin resistance, the OriII origin of replication from pIMP1, a ColE1 origin of replication for replication in *E. coli*, and a *catP* gene encoding chloramphenicol acetyltransferase used as a reporter. The arrows show the orientation of each element shown. The promoterless *catP* gene can be expressed if a promoter fragment is cloned in the appropriate orientation in the multiple cloning site upstream from the gene. In most constructs fragments were introduced between the EcoRI and BamHI sites.

thesis. A semi-quantitative interpretation of these data using a model of protein production indicated that the β-galactosidase produced by the *lacZ* gene from *T. thermosulfurogenes* EM1 is stable in the exponential phase of growth. Also, in pH-controlled fermentations of the 824 strain containing the reporter system and the *ptb* promoter, the kinetics of β-galactosidase formation from the *ptb* promoter and phosphotransbutyrylase formation from its own autologous promoter were found to be similar.

Recently, other gene-expression reporter systems have been developed for clostridia. One such reporter system was developed by Soucaille and co-workers (Toulouse, France). In this reporter system, the *gusA* gene is used as the reporter gene in the plasmid pEGusA-P. The *gusA* gene codes for a β-glucuronidase which can be detected using the fluorimetric GUS assay [77, 78]. Ravagnani et al. [53] used a modified pEGusA-P vector called pRGus to measure the *in vivo* effects of modifying the OA boxes of the *ptb* and *adc* genes on gene expression in *Clostridium beijerinckii*.

In another study by Bennett and co-workers (Houston, TX), a promoterless version of the *catP* gene [17] was amplified and subcloned into a newly constructed shuttle vector, pSA12, to form the reporter plasmid pCatP (Figure 2). The *catP* gene product in this study was detected using a spectrophotometric assay described by Shaw [79]. To ensure that this reporter system functioned properly, the *ptb, thl, adc, solR* and 15 deletion promoters of the *aad* p1 and p2 promoters of *C. acetobutylicum* were cloned into pCatP and were all found to function in *E. coli*. Expression studies indicate differences among the 15 deletion constructs that should be useful in defining the essential parts of *aad* promoters and are consistent with the primer extension studies of *aad* expression performed by Nair et al. [58].

4.8 Antisense RNA in *C. acetobutylicum*

Many naturally occurring antisense RNA (asRNA) systems in prokaryotes have been well-studied and characterized.The different asRNA systems display a wide variety of mechanisms of action, binding pathways, and kinetics [80]. Generally, naturally occurring asRNAs in prokaryotes are small untranslated transcripts that pair to target mRNAs via complimentary stem-loop interactions between them and the target mRNA. The binding of asRNA to target mRNA can either prevent target mRNA translation by hindering target mRNA ribosome binding site interactions with the translational machinery of the cell (i. e., ribosomes) or by altering the structure of the target mRNA such that ribonucleases can then degrade the target mRNA (Figure 3).

Many studies have been performed to determine the association rate for naturally occurring and mutated asRNA binding with target mRNA in prokaryotes. These studies show that by increasing the association rate constant between asRNA and target mRNA the inhibition of the target gene expression can also be increased [81-83]. Thus, it appears that the most important consideration in the design of asRNA is the ability of the asRNA to bind the target mRNA in a fast, efficient manner.

One method of designing asRNA with higher annealing rates to a specific target mRNA is to incorporate stem-loop structures into the artificial asRNA that are complementary to stem-loop structures in the target mRNA. The incorporation of complementary loop structures into the design of asRNA was initially reported but with limited success [84]. However, more recent results suggest that if the stem-loop of the target RNA contains a YUNR (Y-pyrimidine and R-purine) motif at the beginning of the stem-loop, the secondary structure of the stem-loop is more conducive to binding with asRNA which will enhance both the asRNA-target RNA association rate and *in vivo* downregulation of the target protein [88].

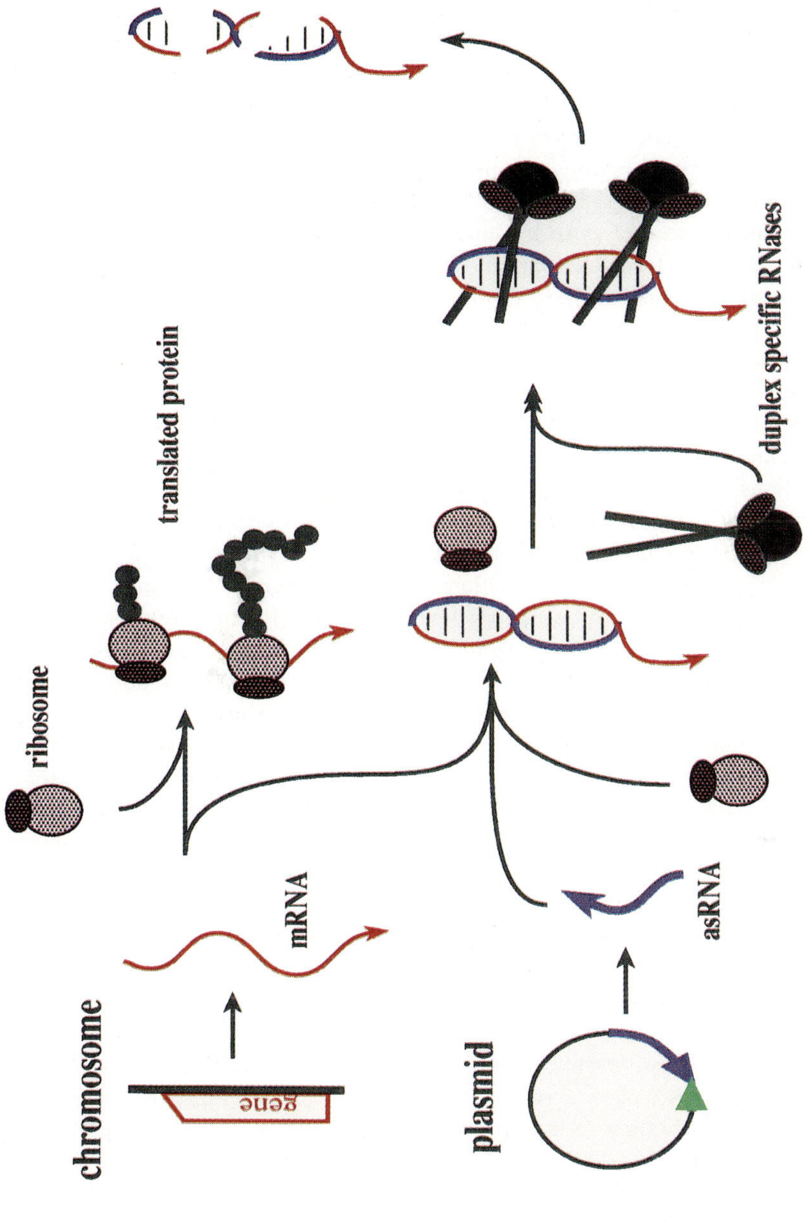

Figure 3. A representation of the likely process by which antisense RNA alters prokaryotic gene expression.

Another technique used in the design of effective asRNA is to search the target mRNA for regions with little or no secondary structure and then develop asRNA with high complementarity to these regions. Regions with little or no secondary structure on the target mRNA provide excellent sites for the design of complementary asRNA, because these regions tend to be less stable and will look to stabilize themselves by binding to complementary asRNA [86]. One popular, experimental method that exploits the local secondary structure of a given RNA sequence for the development of asRNA is *in vitro* selection of asRNA [87]. This technique involves searching for asRNA that bind with target mRNA with the highest association rate constants as determined from non-denaturing polyacrylamide gel electrophoresis [83, 88]. Those asRNA that have the highest association rate constants are thought to bind to regions of the target mRNA that have little or no secondary structure.

Theoretical analyses and the use of computer algorithms can also be performed on any known RNA sequence to determine local secondary structure. These types of analyses can provide information on the extent of local folding, on the energy difference of defined sequence stretches in the single-stranded vs. the double-stranded state, and on structural elements that could be important for RNA duplex formation [86]. Using this information, asRNA can then be developed to be complementary to regions of predicted low secondary structure [89]. Nonetheless, when using this method, one should be careful of interpreting predicted RNA structures, because the science of RNA folding is still emerging and prediction of RNA structures has not been perfected.

Most of the studies discussed above have not been performed with antisense RNA in clostridia. However, an antisense RNA molecule has been reported to occur naturally in solventogenic clostridia. AsRNA has been shown to be involved in the regulation of glutamine synthetase (GS) in a solventogenic *Clostridium* species (formerly *C. acetobutylicum* P262). In 1992, Fierro-Monti et al. showed that expression of a 43-base asRNA occurred in cells containing the cloned glutamine synthetase (*glnA*) gene. They also showed that cells grown in conditions conducive to GS synthesis (namely, nitrogen-limiting conditions) contained low levels of the asRNA compared to the amount of *glnA* transcripts, while those conditions which were unfavorable to GS synthesis (i.e., nitrogen-rich conditions) contained high levels of asRNA transcript compared to *glnA* transcript levels [90]. This suggests the 43-base asRNA plays a role in the regulation of glutamine synthetase [91]. This study was the first to indicate that asRNA could control gene expression in *C. acetobutylicum*.

The first (and so far only) study involving synthetic antisense RNA and clostridia was performed by Desai and Papoutsakis. They examined the effectiveness of asRNA strategies for the metabolic engineering of *C. acetobutylicum* [92]. In this study, Desai and Papoutsakis developed two different asRNA molecules against two genes involved in the primary metabolic network of *C. acetobutylicum* ATCC 824. One molecule was directed towards the mRNA of the butyrate kinase (*buk*) gene, while the other one was directed towards the mRNA of the phosphotransbutyrylase (*ptb*) gene. The *buk*-asRNA was 102 nucleotides long and had 87% complementarity to *buk*. The *ptb*-asRNA was 698 nucleotides long and had 96 % complementarity to *ptb*.

Strains containing the *buk* and *ptb* asRNA showed remarkably different enzyme levels and solvent production profiles when compared to controls. A strain of *C. acetobutylicum* ATCC 824 containing the *buk*-asRNA exhibited 85 % lower butyrate kinase (BK) activity than the control strain. This strain also resulted in 50 and 35 % higher final concentrations of acetone and butanol, respectively, than the control strain. In addition, a strain of *C. acetobutylicum* ATCC 824 containing the *ptb*-asRNA molecule exhibited 70 % lower phosphotransbutyrylase activity and 96 and 75 % lower final acetone and butanol concentrations, respectively, than the control. Based on these results, the potential for the manipulation of *C. acetobutylicum*'s primary metabolic network using asRNA strategies was established.

Conclusions

Large progress has been achieved over the last few years in the development of tools for genetic and metabolic engineering of solventogenic clostridia. These include an improved electrotransformation protocol, gene inactivation by recombination, antisense RNA, functional vectors with two antibiotic resistances, and gene-expression reporter systems. Such tools have proven invaluable in metabolic engineering studies and generation of superproducing strains, and form the foundation for functional genomic studies in view of the completion of the genome sequence of *Clostridium acetobutylicum* ATCC 824. Nevertheless, the development of more efficient and simpler versions of these genetic tools will simplify genetic analysis and metabolic engineering undertakings. Furthermore, more specialized tools (such as those needed to study DNA/protein or protein/protein interactions, or to efficiently identify and clone differentially expressed genes) are urgently needed. Finally, the development of new high throughput tools (such as DNA arrays, protein arrays, and hybrid arrays) for genome-wide studies together with easy to use bioinformatics tools are now a high priority goal in order to bring solventogenic clostridia research into the 21st century.

Acknowledgements

This work was supported in part by a National Science Foundation (USA) grants (BES-9905669, BES-0001288) and a National Institutes of Health (USA) Pre-doctoral Biotechnology Training Grant (GM 08449).

References

[1] Bailey, J. E., Introduction, in: *Metabolic Engineering* (Lee, S., Papoutsakis , E. T., Eds.), New York, Marcel Dekker, **1999**, pp. XI-XVI.

[2] Lee, S. Y., Papoutsakis, E.T., *The Challenges and Promise of Metabolic Engineering in Metabolic Engineering* (Lee, S. Y., Papoutsakis, E. T., Eds.), New York, Marcel Dekker, **1999**, pp. 1-12.

[3] Jones, D. T., Woods, D. R., Acetone-butanol fermentation revisited, *Microbiol. Rev.* **1986**, 50, 484-524.

[4] Papoutsakis, E. T., Bennett, G. N., Molecular Regulation and Metabolic Engineering of Solvent Production by *Clostridium acetobutylicum*, in: *Metabolic Engineering* (Lee, S. Y., Papoutsakis, E. T., Eds.), New York, Marcel Dekker, **1999**, pp. 253-279.

[5] Woods, D. R., The genetic engineering of microbial solvent production, *Trends Biotechnol.* **1995**, 13, 259-264.

[6] Papoutsakis, E. T., Bennett, G. N., Cloning, structure, and expression of acid and solvent pathway genes of *Clostridium acetobutylicum*, in: *The Clostridia and Biotechnology* (Woods, D. R., Ed.), Stoneham, MA, Butterworth-Heinemann, **1993**, pp. 157-199.

[7] Young, M., Development and exploitation of conjugative gene transfer in Clostridia, in: *The Clostridia and Biotechnology* (Woods, D. R., Ed.), Stoneham, MA, Butterworth-Heinemann, **1993**, pp. 99-117.

[8] Duerre, P., Transposons in Clostridia, in: *The Clostridia and Biotechnology* (Woods, D. R., Ed.), Stoneham, MA, Butterworth-Heinemann, **1993**, pp. 227-246.

[9] Reysett, G., Sebald, M., Transformation/electrotransformation of Clostridia, in: *The Clostridia and Biotechnology* (Woods, D. R., Ed.), Stoneham, MA, Butterworth-Heinemann, **1993**, pp. 151-156.

[10] Minton, N. P., Brehm, J. K., Swinfield, T-J., Whelan, S. M., Mauchline, M. L., Bodsworth, N., Oultram, J. D., Clostridial Cloning Vectors, in: *The Clostridia and Biotechnology* (Woods, D. R., Ed.), Stoneham, MA, Butterworth-Heinemann, **1993**, pp. 119-150.

[11] Reysett, G., Transformation and electrotransformation in Clostridia, in: *Genetics and Molecular Biology of Anaerobic Bacteria* (Sebald, M., Ed.), New York, Springer-Verlag, **1988**, pp. 111-119.

[12] Reysett, G., Hubert, J., Podvin, L., Sebald, M., Transfection and transformation of *Clostridium acetobutylicum* strain N1-4081 protoplasts, *Biotechnol. Tech.* **1988**, 2, 199-204.

[13] Soutschek-Bauer, E., Hartl, L., Staudenbauer, W. L., Transformation of *Clostridium thermohydrosulfuricum* DSM568 with plasmid DNA, *Biotechnol. Lett.* **1985**, 7, 705-710.

[14] Yoshino, S., Yoshino, T., Hara, S., Ogata, S., Hayashida, S., Construction of shuttle vector plasmid between *Clostridium acetobutylicum* and *Escherichia coli*, *Agric Biol Chem.* **1990**, 54, 437-441

[15] Allen, S. P., Blaschek H. P., Electroporation-induced transformation of intact cells of *Clostridium perfringens*, *Appl. Environ. Microbiol.* **1988**, 54, 2322-2324.

[16] Allen, S. P., Blaschek H. P., Factors involved in the electroporation-induced transformation of *Clostridium perfringens*, *FEMS Microbiol. Lett.* **1990**, 58, 217-220.

[17] Kim, A. Y., Blaschek, H. P., Construction of an *Escherichia coli–Clostridium perfringens* shuttle vector and plasmid transformation of *Clostridium perfringens*, *Appl. Environ. Microbiol.* **1989**, 55, 360-365.

[18] Phillips-Jones, M. K., Plasmid transformations of *Clostridium perfringens* by electroporation methods, *FEMS Microbiol. Lett.* **1990**, 54, 221-226.

[19] Phillips-Jones, M. K., Introduction of recombinant DNA into *Clostridium* spp., *Methods Mol. Biol.* **1995**, 47227-235.

[20] Scott, P. T., Rood, J. I., Electroporation-mediated transformation of lysostaphin-treated *Clostridium perfringens*, *Gene* **1989**, 82, 327-333

[21] Oultram, J. D., Loughlin, M., Swinfield, T-J., Brehm, J. K., Thompson, D. E., Minton, J. P., Introduction of plasmids into whole cells of *Clostridium acetobutylicum* by electroporation, *FEMS Microbiol. Lett.* **1988**, 56, 83-88

[22] Birrer, G. A., Chesbro, W. R., Zsigray, R. M., Electro-transformation of *Clostridium beijerinckii* NRR B-592 with shuttle plasmid pHR106 and recombinant derivatives, *Appl. Microbiol. Biotechnol.* **1994**, 41, 32-38.

[23] Mermelstein, L. D., Welker, N. E., Bennett, G. N., Papoutsakis, E. T., Expression of cloned homologous fermentative genes in *Clostridium acetobutylicum* ATCC 824, *Bio/Technology* **1992**, 10, 190-195.

[24] Woods, D. R., *The Clostridia and Biotechnology*, Stoneham, MA, Butterworth-Heinemann, **1993**.

[25] Young, M., Minton, N. P., Staudenbauer, W. L., Recent advances in the genetics of the clostridia, *FEMS Microbiol. Rev.* **1989**, 5, 301-325.

[26] Blaschek, H. P., White, B. A., Genetic systems development in the clostridia, *FEMS Microbiol. Rev.* **1995**, 17, 349-356.

[27] Mermelstein, L. D., Papoutsakis, E. T., *In vivo* methylation in *Escherichia coli* by the *Bacillus subtilis* phage phi 3T I methyltransferase to protect plasmids from restriction upon transformation of *Clostridium acetobutylicum* ATCC 824, *Appl. Environ. Microbiol.* **1993**, 59, 1077-1081.

[28] Wong, J., Genetic regulation in *Clostridium acetobutylicum* ATCC 824, *Thesis*, Rice University, Houston, TX, **1995**.

[29] Drury, L., Transformation of bacteria by electroporation, *Methods Mol. Biol.* **1994**, 31, 1-8.

[30] Dunny, G. M., Lee, L. N., LeBlanc, D. J., Improved electroporation and cloning vector system for gram-positive bacteria, *Appl. Environ. Microbiol.* **1991**, 57, 1194-1201.

[31] Kubicka, P., Kramaric, R., Electroporation of cosmid DNA into bacterial cells, *Trends Genet.* **1994**, 10, 5-5.

[32] Luchansky, J. B., Muriana, P. M., Klaenhammer, T. R., Application of electroporation for transfer of plasmid DNA to *Lactobacillus, Lactococcus, Leuconostoc, Listeria, Pediococcus, Bacillus, Staphylococcus, Enterococcus,* and *Propionibacterium*, *Mol. Microbiol.* **1988**, 2, 637-646.

[33] Miller, J. F., Bacterial transformation by electroporation, *Methods Enzymol.* **1994**, 235, 375-385.

[34] McIntyre, D. A., Harlander, S. K., Improved electroporation efficiency of intact *Lactococcus lactis* subsp. lactis cells grown in defined media, *Appl. Environ. Microbiol.* **1989**, 55, 2621-2626.

[35] Na, S., Shen, T., Xiao, W., Jia, P., The factors affecting transformation efficiency of coryneform bacteria by electroporation, *J. Biotechnol.* **1995**, 11, 193-198.

[36] Wells, J. M., Wilson, P. W., Le Page, R. W., Improved cloning vectors and transformation procedure for *Lactococcus lactis*, *J. Appl. Bacteriol.* **1993**, 74, 629-636.

[37] Glenn, A. W., Roberto, F. F., Ward, T. E., Transformation of *Acidiphilium* by electroporation and conjugation, *Can. J. Microbiol.* **1992**. 38, 387-393.

[38] Argnani, A., Leer, R. J., van Luijk, N., Pouweis, P. H., A convenient and reproducible method to genetically transform bacterial of the genus *Bifidobacterium*, *Microbiology.* **1996**, 142, 109-114.

[39] Wards, B. J., Collins, D. W., Electroporation at elevated temperatures substantially improves transformation efficiency of slow-growing mycobacteria, *FEMS Microbiol. Lett.* **1996**, 145, 101-105.

[40] Hui, S. W. Stoicheva, N., Zhao, Y. L., High-efficiency loading, transfection, and fusion of cells by electrporation in two-phase polymer systems, *Biophys. J.* **1996**, 71, 1123-1130.

[41] Hui, S. W., Effects of pulse length and strength on electroporation efficiency, *Methods Mol. Biol.* **1995**, 55, 29-40.

[42] Sheng, Y., Mancino, V., Birren, B., Transformation of *Escherichia coli* with large DNA molecules by electroporation, *Nucleic Acids Res.* **1995**, 23, 1990-1996

[43] Chang, D. C., Hunt, J. R., Zheng, W., Gao, P.-Q., Application of radio-frequency electric field in electroporation and electrofusion, in: *Guide to Electrotransformation and Electrofusion* (Chang, D. C., Chassy, B. M., Saunders, J. A., Sowers, A. E., Eds.), New York, Academic Press, **1992**, pp. 303-326.

[44] Tyurin, M., Padda, R., Huang, K.-X., Wardwell, S., Caprette, D., Bennett, G. N., Electrotransformation of *Clostridium acetobutylicum* ATCC 824 using high-voltage radio-frequency modulated square pulses, *J. Appl. Microbiol.* **2000**, 88, 220-227.

[45] Steele, C., Zhang, S., Shilltoe, E. J., Effect of different antibiotics on efficiency of transformation of bacteria by electroporation, *Biotechniques* **1994**, 17, 360-365

[46] Jones, D. T., Jones, W. A., Woods, D. R., Production of recombinants after protoplast fusion in *Clostridium acetobutylicum*, *J. Gen. Microbiol.* **1985**, 131, 1213-1216.

[47] Jones, D. T., Woods, D. R., Gene transfer recombination and gene cloning in *Clostridium acetobutylicum*, *Microbiol. Sci.* **1986**, 3, 19-22.

[48] Jones, D.T, Mutagenesis and its application in biotechnology, in: *The Clostridia and Biotechnology* (Woods, D. R., Ed.) Stoneham, MA, Butterworth-Heinemann, **1993**, pp. 77-97.

[49] Duerre, P., Transposons in Clostridia, in: *The Clostridia and Biotechnology* (Woods, D. R., Ed.), Stoneham, MA, Butterworth-Heinemann, **1993**, pp. 227-246.

[50] Wilkinson, S. R., Young, M., Targeted integration of genes into the *Clostridium acetobutylicum* chromosome, *Microbiology* **1994**, 140, 89-95

[51] Williams, D. R., Young, D. I., Young, M., Conjugative plasmid transfer from *Escherichia coli* to *Clostridium acetobutylicum, J. Gen. Microbiol.* **1990**, 136, 819-826

[52] Wilkinson, S. R., Young, D. I., Morris, J. G., Young, M., Molecular genetics and the initiation of solventogenesis in *Clostridium beijerinckii* (formerly *Clostridium acetobutylicum*) NCIMB 8052, *FEMS Microbiol. Rev.* **1995**, 17, 275-285.

[53] Ravagnani, A., Jennert, K. C., Steiner, E., Grunberg, R., Jefferies, J. R., Wilkinson, S. R., Young, D. I., Tidswell, E. C., Brown, D. P., Youngman, P., Morris, J. G., Young, M., Spo0A directly controls the switch from acid solvent production in solvent-forming clostridia, *Mol. Microbiol.* **2000**, 37, 1172-1185

[54] Green, E. M., Bennett, G. N., Inactivation of an aldehyde/alcohol dehydrogenase gene from *Clostridium acetobutylicum* ATCC 824, *Appl. Biochem. Biotechnol.* **1996**, 57-58, 213-221.

[55] Green, E. M., Boynton, Z. L., Harris, L. M., Rudolph, F. B., Papoutsakis, E. T., Bennett, G. N., Genetic manipulation of acid formation pathways by gene inactivation in *Clostridium acetobutylicum* ATCC 824, *Microbiology* **1996**, 142, 2079-2086.

[56] Wong, J., Bennett, G. N., Recombination-induced variants of *Clostridium acetobutylicum* ATCC 824 with increased solvent production, *Curr. Microbiol.* **1996**, 32, 349-356.

[57] Harris, L. M., Desai, R. P., Welker, N. E., Papoutsakis, E. T., Characterization of recombinant strains of the *Clostridium acetobutylicum* butyrate kinase inactivation mutant: need for new phenomenological models for solventogenesis and butanol inhibition? *Biotechnol Bioeng.* **2000**,67, 1-11.

[58] Nair, R. V., Bennett, G. N., Papoutsakis, E. T., Molecular characterization of an aldehyde/alcohol dehydrogenase gene from *Clostridium acetobutylicum* ATCC 824, *J. Bacteriol.* **1994**, 176, 871-885

[59] Nair, R. V., Papoutsakis, E. T., Expression of plasmid-encoded and in *Clostridium acetobutylicum* M5 restores vigorous butanol production, *J. Bacteriol.* **1994**, 176, 5843-5846.

[60] Harris, L. M., The regulatory role of SpoOA in *C. acetobutylicum* solvent production, Ph.D Thesis, Northwestern University, Evanston, IL, **2001**.

[61] Balasubramanian, V., Pavelka, M. S., Bardorov, S. S., Martin, J., Weisbrod, T. R., McAdam, R. A., Bloom, B. R., Jacobs, W. R., Allelic exchange in *Mycobacterium tuberculosis* with long linear recombination substrates, *J. Bacteriol.* **1996**,178, 273-279.

[62] Charles, T. C., Doty, S. L., Nester, E. W., Construction of *Agrobacterium* strains by electroporation of genomic DNA and its utility in analysis of chromosomal virulence mutations, *Appl. Environ. Microbiol.* **1994**, 60, 4192-4194.

[63] Oh, S. H., Chater, K. F., Denaturation of circular or linear DNA facilitates targeted integrative transformation of *Streptomyces coelicolor* A3(2): Possible relevance to other organisms, *J. Bacteriol.* **1997**, 179, 122-127.

[64] Lloyd, R. G., Low, B., Some genetic consequences of changes in the level of recA gene function in *Escherichia coli* K-12, *Genetics* **1976**, 84, 675-695.

[65] Lloyd, R. G., Low, K. B., Homologous recombination in *Escherichia coli and Salmonella* (F. C. Neidhardt, Ed.), Washington, DC, Am. Soc. Microbiol. Press, **1996.**

[66] Smith, G. R., Amundsen, S. K., Dabert, P., Taylor, A. F., The initiation and control of homologous recombination in *Escherichia coli, Philos. Trans. R. Soc. Lond. B Biol. Sci.* **1995**, 347, 13-20.

[67] Green, E. M., Bennett, G. N., Genetic manipulation of acid and solvent formation in *Clostridium acetobutylicum* ATCC 824, *Biotechnol. Bioeng.* **1998**, 58, 217-221.

[68] Mermelstein, L. D., Papoutsakis, E. T., Evaluation of macrolide and lincosamide antibiotics for plasmid maintenance in low pH *Clostridium acetobutylicum* ATCC 824 fermentations, *FEMS Microbiol. Lett.* **1993**, 113, 71-76.

[69] Harris, L. M., Desai, R. D., Blank, L., Welker, N. E., Papoutsakis, E. T., Fermentation characterization and flux analysis of recombinant strains of *Clostridium acetobutyli-*

cum with an inactivated regulatory (solR) gene, *J. Ind. Microbiol. Biotechnol.*, submitted.

[70] Heim, R., Prasher, D. C., Tsien, R. Y., Wavelength mutations and posttranslational autoxidation of green fluorescent protein, *Proc. Natl. Acad. Sci. USA* **2000**.

[71] Phillips-Jones, M. K., Bioluminescense (lux) expression in the anaerobe *Clostridium perfringens*, *FEMS Microbiol. Lett.* **1993**, 106, 265-270.

[72] Bullifent, H. L., Moir, A., Titball, R. W., The construction of a reporter system and use for the investigation of *Clostridium perfringens* gene expression, *FEMS Microbiol. Lett.* **1995**, 131, 99-105.

[73] Matsushita, C., Matsushita, O., Koyama, M., Okabe, A., A *Clostridium perfringens* vector for the selection of promoters, *Plasmid* **1994**, 31, 317-319.

[74] Burchhardt, G., Bahl, H., Cloning and analysis of the beta-galactosidase-encoding gene from *Clostridium thermosulfurogenes* EM1, *Gene* **1991**, 106, 13-19.

[75] Yu, P.-L., Smart, J. B., Ennis, B. M., Differential induction of beta-galactosidase and phospho-beta-galactosidase activities in the fermentation of whey permeate by *Clostridium acetobutylicum*, *Appl. Microbiol. Biotechnol.* **1987**, 26, 254-257.

[76] Tummala, S. B., Welker, N. E., Papoutsakis, E. T., Development and characterization of a gene expression reporter system for *Clostridium acetobutylicum* ATCC 824, *Appl. Environ. Microbiol.* **1999**, 65, 3793-3799.

[77] Jefferson, R. A., The GUS reporter gene system, *Nature* **1989**, 342, 837-838.

[78] Jefferson, R. A., Kavanagh, T. A., Bevan, M. W., GUS fusions: beta-glucuronidase as a sensitive and versatile gene fusion marker in higher plants, *EMBO J.* **1987**, 6, 3901-3907.

[79] Shaw, W. V., Chloramphenicol acetyltransferase from chloramphenicol-resistant bacteria, *Methods Enzymol.* **1975**, 43, 737-755.

[80] Wagner, E. G., Simons, R. W., Antisense RNA control in bacteria, phages, and plasmids, *Annu. Rev. Microbiol.* **1994**, 48, 713-742

[81] Hjalt, T., Wagner, E. G., The effect of loop size in antisense and target RNAs on the efficiency of antisense RNA control, *Nucleic Acids Res.* **1992**, 20, 6723-6732.

[82] Hjalt, T. A., Wagner, E. G., Bulged-out nucleotides protect an antisense RNA from RNase III cleavage, *Nucleic Acids Res.* **1995**, 23, 571-579.

[83] Persson, C., Wagner, E. G., Nordstrom, K., Control of replication of plasmid R1: kinetics of *in vitro* interaction between the antisense RNA, CopA, and its target, CopT, *EMBO J.* **1988**, 7,3279-3288.

[84] Engdahl, H. M., Hjalt, T. A., Wagner, H. G., A two unit antisense RNA cassette test system for silencing of target genes, *Nucleic Acids Res.* **1997**, 25, 3218-3227.

[85] Franch, T., Petersen, M., Wagner, E. G., Jacobsen, J. P., Gerdes, K., Antisense RNA regulation in prokaryotes: rapid RNA/RNA interaction facilitated by a general U-turn loop structure, *J. Mol. Biol.* **1999**, 294, 1115-1125.

[86] Nellen, W., Sczakiel, G., *In vitro* and *in vivo* action of antisense RNA, *Mol. Biotechnol.* **1996**, 6, 7-15.

[87] Sczakiel, G., The design of antisense RNA, *Antisense Nucleic Acid Drug Dev.* **1997**, 7, 439-444

[88] Rittner, K., Burmeister, C., Sczakiel, G., *In vitro* selection of fast-hybridizing and effective antisense RNAs directed against the human immunodeficiency virus type 1, *Nucleic Acids Res.* **1993**, 21, 1381-1387

[89] Patzel, V., Sczakiel G., Theoretical design of antisense RNA structures substantially improves annealing kinetics and efficacy in human cells, *Nature Biotechnol.* **1998** 16, 64-68.

[90] Fierro-Monti, I. P., Reid, S. J., Woods, D. R., Differential expression of a *Clostridium acetobutylicum* antisense RNA: implications for regulation of glutamine synthetase, *J. Bacteriol.* **1992**, 174, 7642-7647.

[91] Woods, D. R., Reid, S. J., Regulation of nitrogen metabolism, starch utilization, and the *ß-hbd-adh1* gene cluster in *Clostridium acetobutylicum*, *FEMS Microbiol. Rev.* **1995**, 17, 299-306.

[92] Desai, R. P., Papoutsakis, E. T., Antisense RNA strategies for the metabolic engineering of *Clostridium acetobutylicum*, *Appl. Environ. Microbiol.* **1999**, 65, 936-945.

5
Applied Acetone–Butanol Fermentation

David T. Jones

5.1
Background to the applied acetone-butanol fermentation

5.1.1
Introduction

The ability of many clostridial species to degrade a wide range of plant polysaccharides to produce organic acids, alcohols, or other neutral solvents could provide a practical route for the conversion of renewable biomass and agricultural wastes to bulk chemicals and fuels [1-4]. Thermophilic species enabling the fermentation process to be operated at high temperatures, and species capable of degrading hemicellulose and crystalline cellulose, have been identified as having potential for applications in biotechnology [3, 4]. However, due to the low concentrations and yields of the end products produced, none of these approaches have reached a stage where commercialization is viable.

The only successful industrial fermentation process of this type utilizing *Clostridium* species has been the acetone-butanol (AB) fermentation. The AB fermentation was used as the major route for the commercial production of these solvents during the first part of the twentieth century [5], and along with the ethanol fermentation, it played a key role in the development of the chemical industry in the USA and Britain [6]. Until the early 1960s, the fermentation route competed successfully with synthetic processes, but it was eventually superseded by the cheaper petrochemical-based process. However, the AB fermentation process has remained of interest due to its potential application in biotechnology.

5.1.2
Development of the applied AB fermentation

The origins and development of the applied AB fermentation during the First World War, for the production of acetone as a strategic war material for munitions, has been described in some detail previously [5, 7]. The applied AB fermentation process was initially developed in Britain but a shortage of fermentation substrate required the fermentation process to be transferred to Canada. A second AB fermentation plant was established at Terre Haute in the USA following the entry of the USA into the war. At the end of the war, the US plant was auctioned off to private enterprise and became operational in 1920 as the Commercial Solvents Corporation (CSC). As part of the purchase agreement, CSC also acquired worldwide patent rights to both the Weizmann process and the strain, which was subsequently named *Clostridium acetobutylicum* [8]. This effectively excluded all competitors from operating the applied AB process until the end of 1935. An extensive research facility was established at CSC that undertook development of new strains for use on maize mash, which was used as the fermentation substrate at that time. Although a number of new strains were isolated from various sources, none of these proved to be superior to the original Weizmann strain [6].

By the beginning of the 1930s, the abundant supply of cheap molasses provided an attractive alternative to maize mash as a substrate for the AB fermentation process. The availability of this less expensive and more convenient sugar-based fermentation substrate provided a strong incentive to switch to this raw material. However, although extensive trials were undertaken using existing starch-fermenting strains of *C. acetobutylicum*, an economically viable molasses fermentation process was never achieved.

A search for more effective isolates of solvent-producing clostridia for use with molasses was begun, and by 1932 CSC had established a commercially viable AB fermentation process using this new raw material. During the period 1936 to 1940 CSC filed a number of US patents covering the industrial production of acetone and butanol from molasses, utilizing a succession of new saccharolytic strains [7]. The first successful industrial strain was named *Clostridium saccharo-acetobutylicum* and was described in a series of patents filed in 1936, 1937, and 1938. This new industrial strain was capable of fermenting molasses without the requirement for the inversion of the sucrose, at sugar concentrations of around 6 %, giving solvent yields approaching 30 %. A culture of the prototype strain used by CSC was lodged with the Northern Regional Research Laboratory (NRRL) culture collection in 1945 (NRRL B-591). This industrial strain appears to have been in use from around 1935 as the main production strain for the fermentation of molasses by both CSC in the USA and at a newly opened fermentation plant in Britain named Commercial Solvents, Great Britain (CS-GB). This strain was superseded a few years later by a superior industrial strain. Two variants of this new strain were patented by CSC in 1938 under the name of *Clostridium saccharo-butyl-acetonicum-liquefaciens*. This new industrial strain allowed shorter fermentation times, produced solvent yields of 30-33 %, gave improved solvent ratios, and enabled sugar concentrations of around 6.5 % to be routinely used. Following its introduction, it was utilized as the main production strain by both CSC and CS-GB. An example of this industrial strain was lodged with NRRL in 1946 (NRRL B-643). In 1940 another new industrial strain was patented by CSC under the name of *Clostridium granulobacter acetobutylicum*. The performance of this strain was similar to the *C. saccharo-butyl-acetonicum-liquefaciens* strain, but produced slightly higher solvent concentrations and enabled even higher sugar concentrations to be fermented [9].

Following the expiry of the Weizmann patent in 1936, a number of new commercial AB fermentation processes were established in the USA and Puerto Rico in competition with CSC [5]. The AB fermentation process also began operating in South Africa and Japan around this time. Prior to the expiry of the patent, CSC decided to establish a fermentation plant at Bromborough, in Britain, designed for both ethanol and butanol production. This was apparently undertaken as a joint venture with the Distillers Corporation Ltd., and the new CS-GB plant became operational at the end of 1935.

From this time on the commercial AB fermentation continued to operate on a world-wide basis and reports indicate that at various times the process functioned in Egypt, the Soviet Union, Formosa, mainland China, and Brazil [5]. During the Second World War the AB fermentation again made a major contribution to the war effort. After the war a few more solvent-producing strains were patented [7]. Amongst the most notable of these was a new molasses-fermenting industrial strain utilized by the Sanraku Distiller's Company in Japan for the industrial production of solvents. This new high butanol-producing industrial strain was patented in 1960 under the name of *C. saccharoperbutylacetonicum* [10]. Two variants of this strain were lodged with the American Type Culture Collection (ATCC 27021 and 27022).

Although molasses-based fermentations improved the economic viability of the commercial AB fermentation, increasing competition from the petrochemical industry after the Second World War led to the closure of most of the AB fermentation plants in the western world by the beginning of the 1960s. The process continued to operate in South Africa until the beginning of the early 1980s and has continued to operate in China up until recent times.

5.1.3 Taxonomic status of the solvent-producing clostridia

Many of the original industrial strains of solvent-producing clostridia patented by the various companies appear to have been lost. Fortunately, examples of a number of the key industrial strains have survived in various culture collections around the world. During the thirty years following the Second World War little research was undertaken on the AB fermentation process and the various names utilized for patent purposes fell into disuse. It was only towards the end of the 1970s, when the potential of the solvent-producing clostridia for application in biotechnology began to be re-evaluated, that there was a significant renewed interest in this group of organisms. By this time the majority of the surviving industrial strains had become referred to as *C. acetobutylicum*, with some strains with sufficient distinguishable features being classified as *Clostridium beijerinckii*. However, as new scientific knowledge accumulated during the 1980s and early 1990s it became increasingly apparent that many of the strains in common use differed quite significantly in both their physiological and genetic characteristics. This suggested that the various strains that had come to be classified as *C. acetobutylicum* did not constitute a closely related homologous group. This has subsequently been born out by a number of molecular taxonomy studies utilizing biotyping, genomic DNA fingerprinting,

16S rRNA gene sequencing, and DNA-DNA re-association [11-13]. These studies have established that various industrial and culture collection strains of solvent-producing clostridia can be separated into four species.

A group of closely related industrial strains isolated and developed for their ability to produce acetone and butanol from industrial starch-based substrates, originally classified as *C. acetobutylicum*, have been confirmed as constituting a single species. Industrial strains belonging to the *C. acetobutylicum* species have been found to be only distantly related to the various industrial strains that were isolated and developed later for solvent production from sugar-based substrates. The various saccharolytic industrial strains have now been shown to belong to three closely related species [11, 13].

The majority of these saccharolytic strains has been confirmed as belonging to the *C. beijerinckii* species. The industrial strains isolated and developed in Japan in the early 1960s, that were named *C. saccharoperbutylacetonicum*, have been confirmed to form another distinct species. A further group of industrial strains, isolated and developed in the USA by CSC and patented under the name *C. saccharobutyl-acetonicum-liquefaciens*, which were used extensively for commercial solvent production in the USA, Britain, and South Africa, have been shown to constitute a third saccharolytic species. The name *Clostridium saccharobutylicum* has been proposed for this newly identified species [9]. Many of these saccharolytic industrial strains are still classified as *C. acetobutylicum* and require reclassification [14].

5.1.4
Documentation of the applied AB fermentation

A large amount of published material exists relating to the commercial applied AB fermentation process that operated in the western world from the beginning of the 1920s to the beginning of the 1980s (refer to the review by Jones and Woods [5]). However, although these publications contain a wealth of valuable information, the majority were written from the perspective of providing a general overview of the fermentation. Much of the specific detailed information regarding the actual performance of the fermentation process operated by the individual commercial companies was therefore omitted. This is understandable as this type of specific detailed process information would have been commercially sensitive at the time of writing. Unfortunately, most of this information has been lost as there was little incentive to make it available once the commercial fermentation process ceased to operate.

The author was privileged to have enjoyed a close working relationship with the National Chemical Products (NCP) company in South Africa that operated a commercial AB fermentation process from 1935 until 1982. This batch fermentation process appears to have been typical of the applied AB fermentation processes operated by a number of companies world-wide. The reason the process survived in South Africa was due to the plentiful supply of competitively-priced and locally produced cane molasses and coal, coupled with the lack of competition from a local petroleum industry [15]. The association with the company as a consultant during the late 1970s and early 1980s, provided both hands-on experience with the applied

AB process, as well as access to the confidential company reports, operating protocols, and production data. After the fermentation ceased operation, the company agreed to the information being released and access to the company archives and strain collection was provided. This classified material, along with information from the publications by Spivey [15] and Robson and Jones [16], has been utilized to provide a detailed account of the history and operation of the commercial batch AB fermentation run by NCP, much of which has not been published before.

5.2 The applied batch fermentation operated by NCP

5.2.1 The history of the commercial AB fermentation in South Africa

The origins of the National Chemical Products (NCP) company in South Africa dates back to 1935, when the National Maize Products company was established in Germiston, South Africa, for the purpose of producing ethanol by fermentation of maize, for use as a motor fuel as well as for the manufacture of methylated spirits and solvents. The new factory constructed at this site, situated in the maize-growing area in South Africa, consisted of an ethanol fermentation plant, an AB fermentation plant, a yeast plant, a vinegar installation, and a dry-ice plant along with a laboratory. The plant was designed and erected by a French engineering company Ets. Barbet, and incorporated licensed technology developed by Usines de Melle in France. The batch ethanol fermentation process utilized the Barbet technology, employing an amylo process involving the treatment of pre-cooked maize with a mucor mould, to convert starch to fermentable sugars. The batch AB fermentation process also used maize as a substrate and utilized the Melle strain.

The AB fermentation plant was commissioned in 1936 following fulfillment of guarantees by the contractors. The original batch AB fermentation, using maize as the substrate, took between 40 to 50 h to reach completion. The distillation systems that were installed in the new plant, combined the advantages of technology developed by both the Barbet Company and the Usines de Melle, and were said to have provided the company with one of the most up-to-date fermentation and distillation facilities in the world at that time. From the start, the AB fermentation process was reported to have run smoothly and by 1937 the sales of secondary products produced by the company exceeded the sales of fuel alcohol. The increase in demand for acetone and butanol led to a number of additions and refinements to the ancillary plant and facilities, resulting in a substantial reduction in production costs. During 1939 the rising price of maize forced the company to switch to molasses as the substrate for the ethanol fermentation. As a result, the name of the company was changed to NCP in 1940. However, the company continued to use maize as the substrate for the AB fermentation. In 1939 the fledgling company struck major problems regarding the disposal of fermentation effluents when it was barred from discharging untreated effluent directly into the municipal waterways. As a result, the entire fermentation operation at Germiston was threatened with closure. These difficulties were overcome by the installation of a costly in-house process for effluent disposal.

The collapse of France during the early stages of the Second World War seriously affected the sales of acetone and butanol. Company records indicate that at this time a substantial amount of solvents were being exported to Europe. Consequently, the factory was threatened with closure in 1940. However, company fortunes changed when at the end of that year NCP was contracted to produce war materials for the South African Government. The demand for acetone to make cordite for use in the manufacture of munitions escalated rapidly, requiring that the plant be operated at double its original capacity to supply needs. At this time the company was able to run around 56 fermentations per month. As part of the war effort, a synthetic acetone plant was also established at the Germiston factory, to augment the output of the AB fermentation. This synthetic plant was based on a catalytic process for converting acetylene to acetone developed by a subsidiary of the Distillers Company Limited in Britain. This synthetic plant operated from 1942 to 1944.

In December 1943, the factory experienced problems with a suspected bacteriophage infection of the AB fermentation and technical advise was obtained from the CSC in the USA. Continuing problems with the maize fermentation, coupled with the increasing importance of maize as a strategic war material, led to the decision to convert the fermentation process to the use of blackstrap molasses. At the beginning of 1944, trials were begun with molasses using the Melle culture, but these gave poor results. In mid 1944, a number of industrial strains were supplied by CSC and factory trials were undertaken at NCP. By the beginning of 1945, the AB fermentation had been fully converted to molasses utilizing the CSC strains. After some initial teething problems, the substitution was successfully achieved, resulting in considerable cost savings, enabling the company to reduce prices.

NCP became a public company in 1944, and in 1946 the Distillers Company of London took a substantial shareholding. As a result, the Distillers Company, along with its organic chemical subsidiary CS-GB, made available a wealth of technical experience and know-how relating to the production of organic chemicals. In 1947, serious difficulties were experienced with the AB fermentation resulting in a substantial reduction in the yield and output of the AB fermentations. These were attributed to bacteriological culture problems, associated with the quality of molasses adversely affected by a prolonged drought. These problems resulted in NCP switching over to the use of British production strains supplied by CS-GB. These strains remained in use in the factory until late the following year, while comparative investigations were undertaken by NCP and the Distillers Company. By the end of 1948, the problems experienced with the AB fermentation had been overcome and from this time on, until the late 1970s, the batch AB fermentation process operated at NCP ran with remarkable efficiency and reliability, with few problems being encountered. This reliability is well illustrated by the performance figures documented in Tables 1 and 2. In 1965, the Distillers Company transferred its shares to a wholly owned subsidiary, Distillers Chemical and Plastics Limited. This company subsequently became BP Chemicals (UK). In 1967, a South

Table 1. NCP industrial fermentation data for the period of 1961-1971

Years	*Cultures*	*Number of runs*	*Percentage of set sugar (%)*	*Period of fermentation (h)*	*Total solvents concentration (g/l)*	*Acetone ratio*	*Yield (%)*
1961	P172	1839	5.89	31	17.8	34.8	30.3
1962	P172, P188	1812	5.89	33	17.7	34.3	30.0
1963	P188	1896	5.90	31	17.7	35.5	29.9
1964	P188, P192	2206	6.11	30	18.7	36.1	30.5
1965	P192, P193	2083	6.31	31	18.8	36.3	29.8
1966	P186, P199	2124	5.80	30	18.4	36.3	29.1
1967	P194, P201, P213	2148	6.22	30	18.2	35.4	29.0
1968	P201, P214, P218, P220	2202	6.37	31	18.5	34.7	29.0
1969	P213, P218, P230	2103	6.43	29	18.3	35.3	28.5
1970	P239, P243	2101	6.44	29	18.3	34.8	28.4
1971	P239	2104	6.57	31	17.5	33.8	26.6

Table 2. Comparison of NCP industrial AB fermentation data for 1961 and 1971

Months in 1961	*Cultures*	*Number of runs*	*Percentage of set sugar (%)*	*Period of fermentation (h)*	*Total solvents concentration (g/l)*	*Acetone ratio*	*Yield (%)*
January	P172 (116), other strains(39)	155	5.92	30	17.8	38.1	30.1
February	P172 (75), E29(65)	155	5.84	30	17.5	35.0	30.0
March	P172 (119), other strains (21)	140	5.81	30	17.5	34.8	30.1
April	P172 (136), other strains (19)	155	5.90	31	17.6	34.6	29.8
May	P172 (135), other strains (15)	150	5.89	30	17.8	33.6	30.2
June	P172 (144), other strains (16)	160	5.94	31	18.1	34.2	30.5
July	P172 (159)	159	5.98	31	18.2	34.8	30.4
August	P172 (145), other strains (10)	155	5.96	29	18.1	35.4	30.4
September	P172 (144), other strains (11)	155	5.81	30	17.6	34.6	30.3
October	P188 (116), other strains (34)	150	5.87	31	17.9	34.2	30.5
November	P188 (88), P172 (57), others (10)	155	5.88	31	18.0	34.3	30.6
December	P172 (142), other strains (8)	150	5.92	32	18.0	34.4	30.4

Table 2. (continued)

Months in 1971	*Cultures*	*Number of runs*	*Percentage of set sugar (%)*	*Period of fermentation (h)*	*Total solvents concentration (g/l)*	*Acetone ratio*	*Yield (%)*
January	239 (185)	185	6.40	30	17.9	33.6	28.0
February	239 (181)	181	6.58	30	18.3	33.2	27.8
March	239 (167)	167	6.67	30	18.2	34.5	27.3
April	293 (159), other strains (23)	182	6.68	30	17.7	32.9	26.5
May	239 (167)	180	6.75	29	18.5	34.1	27.4
June	239 (190)	190	6.70	31	17.5	33.3	26.2
July	239 (150)	150	6.35	33	16.2	33.4	25.5
August	P239 (163), other strains (4)	167	6.36	31	16.1	31.9	25.3
September	P239 (104), other strains (79)	183	6.60	32	17.3	33.0	26.2
October	P239 (171)	171	6.44	33	17.4	34.4	27.0
November	P239 (180)	180	6.66	33	17.4	35.7	26.1
December	P239A (168)	168	6.66	34	17.4	36.2	26.1

African holding company, Sentrachem, acquired all BP Chemicals shares in NCP to ensure that control remained in South African hands and the NCP Company became one of the founder companies of the Sentrachem Group.

At the end of July 1980, a go-slow schedule for the AB fermentation was introduced due to a shortage of molasses caused by a serious drought. Towards the end of the year, production ceased completely and during the whole of 1981, operation of the AB fermentation was suspended due to problems encountered in obtaining molasses. At the beginning of 1982, an economic feasibility study was undertaken resulting in a decision to recommission the AB fermentation plant. Production in the refurbished plant commenced again around mid year and the process ran for a few months. However, the solvent concentration and yields obtained were erratic and ongoing problems were encountered with bacterial contamination. Later that year, the decision was made to cease operations and mothball the plant indefinitely. The plant has subsequently been demolished.

5.2.2
Origins and history of NCP industrial strains

The original Melle production strain was supplied by the French company and was used by NCP for solvent production from maize, from 1936 until 1944. Archival material indicates that this strain was a direct descendant of the original Weizmann *C. acetobutylicum* strain used in France during the First World War for the Production of acetone [5]. However, this is unconfirmed as no example of this strain has survived.

In 1945, the escalating cost of maize and its importance as a strategic war material resulted in a decision to convert the AB fermentation to molasses. To facilitate the conversion, NCP acquired several industrial production strains from CSC in the USA. These sub-cultures were supplied in the form of spores dried on sterile soil sealed in glass vials. One of the original glass vials supplied by CSC labeled BAS/B3 and dated 1944 was retained unopened in the NCP strain collection. This vial was opened in 1994 and a viable culture was recovered. In 1946, CSC also lodged an example of their main production strain with the NRRL culture collection (NRRL B-643).

At least two batches of the CSC strains were supplied to NCP. The first of these industrial cultures are listed in the records as BAS/2 and BAS/3 and were trialed in factory production fermentations from May to July of 1944. Initially, the main culture used was BAS/2 but due to the poor results obtained this was substituted by the BAS/3 culture. In August 1944, four new cultures of BAS were received from CSC designated BAS/A, A1, A2, and B. The relationship between the BAS/A and B strains and the BAS/2 and 3 strains is not known.

Towards the end of 1945, NCP began to produce its own cultures from the American BAS/B3 strains. The NCP company kept detailed records of all of the strains developed and their records indicated that all the earliest NCP production strains were derived from a single BAS strain. Of the five strains recorded which were prepared by NCP at this time, three still exist as viable spores. These are BAS/B/136

NCP 1946, BAS/B/SW/136 NCP 1946, and BAS/B3/SW/336 NCP 1946. The company continued to prepare its own cultures and each time they prepared a new strain from a previous culture it was given a new serial number. The numbered codes assigned to the main production strains exceeded 280 strains and from 1948 were preceded by the letter P to designate a production strain. However, although all of these production strains appear to have been trialed in full-scale fermentations, not all of these were utilized for extended production runs. Individual production strains were often used continuously until the spore stock became depleted, as long as no deterioration in performance occurred. In some cases, a single production strain was utilized for more than two years to set over 4,000 individual fermenters. In addition to production strains, the company also propagated a considerable number of strains for experimental purposes that were assigned other codes. From the data available in the NCP archives, it has been possible to construct a detailed chronological family tree of all the strains generated by NCP during the period 1945 to 1982 (data not shown).

Towards the end of 1946, serious problems were experienced with the AB fermentation process at NCP that were attributed to a drought affecting molasses quality. In an attempt to overcome these difficulties, factory trials were undertaken at NCP during the first three months of 1947 using the Bromborough 34/11 culture supplied by CS-GB. In March, the Bromborough 34/11 strain was adopted as the main production strain and continued in use until the end of September 1948. During this time, 188 fermentation runs were carried out with an average solvent concentration of 14.7 g/l. During this period, both NCP and the Distillers Company Research Facility at Epsom undertook comparative laboratory studies with the NCP and Bromborough cultures, as well as optimization of cane molasses as the fermentation substrate. High solvent yields were achieved with both strains and it was concluded that neither of the production strains conferred discernible advantage. As a result of these studies, the factory reverted to the use of NCP production strains during the later part of 1947, and although solvent concentrations remained somewhat erratic, many of the fermentations produced between 18-19 g/l of solvents. By the end of the following year, the factory was consistently achieving solvent concentrations of 17-18 g/l with occasional concentrations as high as 21.5 g/l being recorded. At this time, the monthly supply of cultures from the Bromborough factory were discontinued. Unfortunately, no examples of these early Bromborough strains survived. One example of a later phage-immunized Bromborough strain, designated 37/3 supplied in 1950 has, however, survived along with a subculture of the strain prepared by NCP at this time (Table 3). None of these later Bromborough cultures were utilized as standard production strains by NCP.

5.2.3 Analysis and characterization of surviving NCP strains

Of the approximately 300 production strains propagated by NCP, only 31 of the original spore stocks are known to have survived. Samples of all of these strains were transferred to the Department of Microbiology, University of Cape Town, South

Africa around the time the fermentation process was closed. The NCP strain collection was later transferred to the Department of Microbiology, at the University of Otago, in New Zealand. With one exception, all of the original NCP spore stocks tested yielded viable cultures, even though some of this material was over fifty years old (unpublished results). Several different times and temperatures were used to heat-activate the spores and the activated spores were germinated in both please define abbreviation (CBM) and TYA culture media [11]. No one set of conditions was found to be optimal for achieving germination. In some instances, germination was protracted and erratic, suggesting that probably only a small proportion of the spores that were originally present remained viable. However, all of the strains recovered resulted in very actively-fermenting cultures that generated solvent yields and concentrations equivalent to those produced by freshly prepared spore stocks. All of the vegetative cultures recovered, subsequently sporulated well on supplemented CBM agar, and new spore stocks were prepared and maintained in sterile distilled water at 4 °C. The surviving NCP strains, including their source, date of preparation, and origin are listed in Table 3.

A study carried out by Keis [14] revealed that the industrial strains utilized by NCP consisted of two different species. One group of strains were classified as *C. beijer-*

Table 3. Surviving CSC, CS-GB, and NCP industrial strains (see also note added in proof)

***C. saccharobutylicum* Group 1 strains**

Source	*Code*	*Date*	*Origin*	*Mixed*
CSC	BAS/B3	1944	CSC	–
CS-GB	37/3 (Imm)	1950	CS-GB	–
NCP	P108	1955	P104	–
NCP	P172	1958	P118	+
NCP	P195	1965	P157	–
NCP	P199	1966	P157	–
NCP	P200	1966	P118	+
NCP	P202	1966	P201	+
NCP	P220	1968	P201	–
NCP	P249	1972	P157	–
NCP	P254	1974	P217	+
NCP	P262	1974	P200	–
NCP	P265	1974	P201	+
NCP	P268	1974	P266	–

Table 3. (continued)

***C. saccharobutylicum* Group 2 strains**

Source	*Code*	*Date*	*Origin*	*Mixed*
NRRL	B643	1946	CSC	–
NCP	BAS/B3/SW/336	1946	CSC BAS/B3	+
NCP	P162	1957	CS-GB 37/3	–
NCP	P258	1974	P252	–
NCP	P272	1977	?	+

***C. beijerinckii* strains**

Source	*Code*	*Date*	*Origin*	*Mixed*
NCP	BAS/B/136	1946	CSC BAS/B	–
NCP	BAS/B/SW/136	1946	CSC BAS/B	–
NCP	BAS/B3/SW/336	1946	CSC BAS/B3 .	+
NCP	P106	1954	P104	–
NCP	P172	1958	P118	+
NCP	P193	1964	P157	–
NCP	P200	1966	P118	+
NCP	P202	1966	P118	+
NCP	P254	1974	P217	+
NCP	P259	1974	P251	–
NCP	P260	1974	P217	–
NCP	P261	1974	P251	–
NCP	P263	1974	P244	–
NCP	P264	1974	P199	–
NCP	P265	1974	P201	+
NCP	P270	1976	P264	–
NCP	P271	1976	?	–
NCP	P272	1977	?	+
NCP	P280	1980	P271	–

inckii and the other group belonged to the newly named *C. saccharobutylicum* species. Strains belonging to this species could be further divided into two subgroups on the basis of differences in their genomic DNA fingerprints. The two species can be readily distinguished on the basis of rifampicin and phage sensitivity and gelatin hydrolysis [11, 17]. The NCP strains belonging to the *C. beijerinckii* species were resistant to rifampicin, sensitive to the phage CA1, and do not hydrolyze gelatin.

Application of these biotyping methods enabled an initial analysis of all the surviving NCP culture collection strains to be undertaken [17]. The clear zones resulting from the hydrolysis of gelatin that formed around the colonies of strains belonging to *C. saccharobutylicum,* coupled with differences in colony morphology, made it possible to distinguish between the two species. Based on the biotyping results, seven of the 31 soil spore stocks were found to contain a mixture of the two species, including one of the original spore stocks produced by NCP in 1946 (Table 3). These findings indicate that examples of both species were provided as part of the original batch of cultures obtained from CSC in 1946, and both variants of the two *C. saccharobutylicum* genomic subgroups were also represented. Both taxonomic and fermentation studies support the conclusion that the NCP strains identified as *C. saccharobutylicum* species are descendants of various strains (BAS/B3, NRRL B-643, and 37/3) patented by CSC in 1938 under the name of *C. saccharo-butyl-acetonicum-liquefaciens,* and the NCP strains identified as *C. beijerinckii* are descendants of strains patented by CSC in 1940 under the name of *C. granulobacter acetobutylicum* [9].

These findings indicate that, unbeknown to the company, from the very start of the molasses fermentation at NCP, the company was utilizing some inocula that consisted of mixtures of these two species. To confirm the results obtained by biotyping, pure cultures of the seven mixed spore stocks were obtained. Genomic DNA fingerprint patterns of the 38 pure strains were then determined using the infrequent-cutting restriction endonucleases *Apa*I, *Sma*I, and *Eag*I and pulsed-field gel electrophoresis [11]. The result of this analysis revealed that all of the 19 strains belonging to the *C. beijerinckii* species exhibited the same genomic DNA fingerprints. The 19 strains belonging to the *C. saccharobutylicum* species exhibited similar but not identical DNA fingerprints. Fourteen of the strains exhibiting a common DNA fingerprint pattern were assigned to subgroup 1, represented by the BAS/B3 strain provided by CSC in 1945. The other five strains were assigned to subgroup 2 represented by the NRRL B643 strain lodged by CSC in 1946. The NRRL strain did, however, differ from the four strains from the NCP collection in that it exhibited an additional band in the DNA profile when cut with *Apa*I. It is interesting to note that no perceptible variation was observed in genomic DNA fingerprint profiles of any the strains belonging to either species during 35 years of propagation. This suggests that the macro-structure of the genomes of these two species of solvent-producing clostridia are stably maintained.

From the detailed records of the NCP production strains in the NCP archives, it was possible to construct a complete family tree pinpointing the origins and relationships of the 38 extant strains. However, it was surprising to find that the documented relationships did not exhibit a good correlation with the molecular taxonomy findings. In several instances strains that were cited as being direct ancestors were found to be different species or subgroups. An example is the 37/3 phage-immunized strain supplied by CS-GB which belonged to subgroup 1 of *C. saccharobutylicum*, while the P162 strain derived from it was found to have a subgroup 2 DNA fingerprint type. This suggests that even where a particular spore stock appeared to be pure, some tubes within the batch may have contained low levels of mixed spores. In addition, it is likely that cross-contamination could have occurred. The reason that the company remained unaware that they were using a mixture of species was that spore stocks were always prepared from bulk liquid cultures.

The fact that both species are capable of producing high concentrations and yield of solvents, and that many of the original spore stocks contained mixtures of the two species, indicates that both species were utilized routinely in the industrial AB fermentation process at NCP. It is known that both subgroups of *C. saccharobutylicum* produce potent bacteriocin-like substances that are capable of inhibiting the growth of strains belonging to all of the other subgroups of the *C. beijerinckii* species, with the exception of strains belonging to the DNA fingerprint subgroup used by the NCP factory [11].

A preliminary study was undertaken on the effect of using mixed cultures in molasses fermentation (unpublished results). In these experiments, mixtures of vegetative cultures or mixtures of spore inocula were prepared in both equal and mixed proportions. In both cases, the fermentation containing mixed cultures gave similar results to pure control cultures. In some instances, more or less equal proportions of the two species were recovered at the end of mixed fermentations. In other cases, one or the other of the two species were observed to predominate, largely or entirely. When the original mixture contained a predominance of one species, in most cases this was the only species recovered at the end of the fermentation. These findings support the conclusions that the two species used by NCP both produced very similar results and were able to co-exist in mixed fermentations without any significant deterioration in fermentation performance occurring. These observations raise many intriguing questions regarding the relationship between the two solvent-producing species.

5.2.4
The NCP batch AB fermentation process

Strain propagation and culture maintenance

The standard method used for the maintenance of stock cultures was to preserve each strain in the form of endospores dried in sterile soil. In order to prepare a new production strain from an existing high solvent-yielding production culture, a small sample of soil spore culture was heat-shocked in saline at 70 °C for 90 to 120 s, followed by rapid chilling. The heat-activated spore suspension of the selected production strain was then used to inoculate a flask of liquid potato-glucose culture medium. After several days growth at 35 °C a suitable crop of spores developed. A small volume of the broth was heat-activated and used to inoculate a fresh flask of potato-glucose medium and allowed to grow at 34 °C for 48 h. Several cycles of heat activation with subsequent growth and sporulation in potato-glucose medium were sometimes undertaken until large microscopically "phase bright" spores, associated with vigorous, motile cultures resulted. The final culture was allowed to stand at 34 °C for 3 to 4 days to ensure adequate sporulation, whereupon aliquots of the liquid culture were transferred to plugged glass tubes containing sterile sand or soil and allowed to dry. Before adoption of a new batch of spores as a production strain, cultures were checked for fermentation performance and for bacterial and bacteriophage contamination in the laboratory. High solvent-yielding cultures were then trialed in full-scale production fermentations interspersed over a period of several weeks. If these full-scale production runs proved satisfactory, the new production strain was introduced when required. Although it was considered that freshly prepared spores gave a higher percentage outgrowth and resulted in more vigorous fermentations, spore preparations dried on sterile sand remained viable for long periods of time without any significant loss of viability or fermentation performance (see section 5.2.3).

For almost the entire period that NCP operated the AB fermentation process, the mode of selection for high-yielding cultures was an indirect process employing mixed bulk liquid cultures. From time to time, attempts were made to isolate pure colonies from plates but these were frequently non-motile or weakly motile and resulted in poor fermentations. During the last few years that the fermentation process operated, a technique for isolating motile high-yielding colonies from plates was developed, and pure cultures maintained as spores, stored in sterile distilled water at 4 °C were produced. These pure cultures were considered to show less variability from fermentation to fermentation.

Inoculum preparation

Each fermenter inoculum was started from a fresh suspension of heat-activated spores dried in sterile soil. The seed inoculum of vegetative cells was then built-up in the culture laboratory, by employing a number of progressively larger cultures designated as the A1, A2, B, and C stages (Table 4) as illustrated in Figure 1.

Before transfer to the A1 stage, a small amount of the dried soil containing the spores was heat-shocked at 70 °C for 90 s in saline, and then rapidly cooled in an

Table 4. The seed stages used for fermenter inoculum development

Seed stage	*Culture medium*	*Volume (l)*	*Normal duration (h)*
A1	Potato/Glucose	0.15	10
A2	Molasses	0.5	6
B	Molasses	3.5	6
C	Molasses	9.0 (x 2)	9
Prefermenter	Molasses	3500	9

Figure 1. (A)

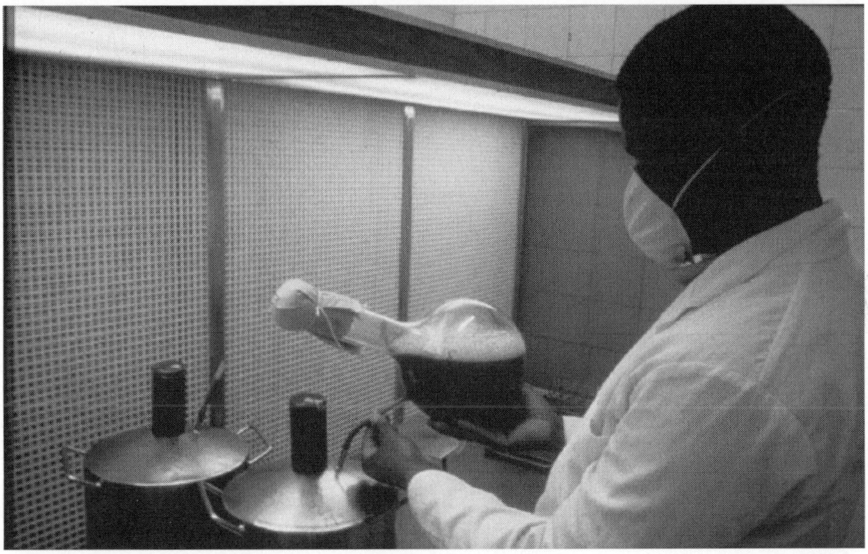

Figure 1. (B)

Figure 1. The NCP acetone-butanol fermentation plant and process.
(A) Seed inoculum stages used for the routine inoculation of the commercial AB fermentation at NCP; (B) the process utilized for the routine transfer of the seed stage inoculum under asepetic conditions in a laminar flow hood; (C) view of the 3,500 liter prefermenters used for the inoculation of the commercial stage fermenters; (D) a view of the tops of the commercial stage fermenters from the upper access platform; (E) a side view of one of the commercial stage fermenters with part of the fermenter house wall removed; (F) view of the evaporator plant used for processing of the waste cell biomass.

Figure 1. (C)

Figure 1. (D)

Figure 1. (E)

Figure 1. (F)

ice bath. The cooled, heat-activated spore suspension was then allowed to stand for several hours before the entire spore suspension was transferred to the A1 stage.

The culture flasks were equipped with side tubes so that aseptic transfer from flask to flask was easily carried out. Other precautions taken to minimize the risk of contamination during inoculation, included the use of laminar flow hoods and face masks (Figure 1). The culture laboratory was maintained under positive air pressure with filtered air, and the incubator rooms were equipped with UV lamps.

The state of the culture at the time of transfer of each seed stage was considered to be critical. Transfer took place when the culture reached the late log-phase growth, before it reached the pH breakpoint. It was important that there was no evidence of granulose being laid down, or of the development of spindle-shaped cells, as once cells had begun differentiation into this morphological state, poor growth occurred in subsequent stages. Vigorous motility normally developed after a few hours in the A1 stage and it was observed that there was a strong correlation between motility and good solvent production. The absence of vigorous motility after heat-shocking, or indeed at any time in the subsequent seed stages, was viewed as an indication that poor solvent yields would be obtained in the final fermentation, even though the cell population might appear normal in other respects.

Transfer times from seed stage to seed stage, and from the final stage to the plant fermenter, were considered to be vital for the proper development of the inoculum and were carefully adhered to. The exact time of transfer at each stage tended to vary according to the strain used and the composition of the culture medium, and was determined by careful observation (Table 4).

For much of the period that the batch AB fermentation process was run at NCP, no prefermenter stage was utilized. Standard practice was to inoculate each 90,000-litre plant fermenter with three C stage cultures. This provided an inoculum of approximately 27 l giving an inoculation ratio of 1/3000. Due to the small size of the inoculum, extreme care was taken with the plant operation to ensure asepsis since the AB fermentation was highly susceptible to contamination. The stainless steel cans for the C stage were equipped with a beveled spout designed to fit a steam-sterilized rubber tube off the side of the prefermenter. In order to reduce the chances of contamination, antiseptic aerosol was applied around the port whilst inoculation took place.

During the 1970s, the deteriorating quality of molasses as a fermentation substrate resulted in an increasing number of failed fermentations. This made the use of a larger inoculum desirable and a decision was taken to install a prefermenter stage. Two 9-litre C stage cultures were used to inoculate the 3,500-litre prefermenter, which in turn became the inoculum used for the 90,000 l production fermentation, giving a final inoculation ratio of 1/26 (Figure 1). Several important advantages were gained by the introduction of a prefermenter stage. These included shorter fermentation times, increased yields, and reduced risk of bacterial contamination. There was some indication, that the use of a much larger inoculum also allowed the maximum set concentration of sugars that could be fermented in batch culture, to be increased.

The AB fermentation plant

The production plant consisted of twelve commercial stage fermenters each with a working volume of 90,000 l. The cylindrical fermenters had hemispherical end sections to withstand the positive pressure maintained during sterilization and fermentation (Figure 1). They were constructed of mild steel, clad with stainless steel and were not equipped with any type of mechanical agitator. The production vessels were all interlinked by a manifold of pipes and all feed and exhaust lines were steamed whilst not in use to maintain sterility.

Mash formulation

Blackstrap molasses produced as a by-product from the local cane-sugar industry was used as the fermentation substrate. This material provided the entire carbohydrate source and also supplied nitrogen, phosphorous trace elements, and buffering capacity. Ammonium sulphate was added to supplement the organic nitrogen present in the molasses, and at the time the fermentation reached the pH breakpoint, a 25% solution of ammonia liquor was fed into the fermentation vessel. This acted as an additional nitrogen source and was used as a means of pH control. Calcium carbonate was also added to the molasses mash to enhance the buffering capacity. This was a necessary additive, particularly when ammonium salts were employed, because of its pH regulatory effect.

For much of the period that the fermentation operated corn-steep liquor, produced as a by-product from a neighboring glucose factory, it was added at a concentration of about 1.5% by volume. This provided an additional source of organic nitrogen and vitamins, and no phosphate addition was required when the corn-steep liquor was employed. It was found that although the fermentation did proceed without the presence of corn-steep liquor, ammonia on its own was not adequate to achieve good solvent yields. The amino acids, polypeptides, and growth stimulants present in corn-steep liquor had a beneficial effect on solvent production.

On arrival at the factory, the molasses obtained from the various sugar mills was pooled in an attempt to obtain as much consistency as possible, and stored in large storage tanks. The extensive variations in the molasses required the adjustment of nutrient additions in order to optimize the fermentations. The molasses to be used in the AB fermentation was pumped to a mixing vessel where the other components of the fermentation medium were added and mixed. Recycled water from the evaporators was adjusted to pH 11 with calcium hydroxide, and used to dilute molasses that was weighed in from a molasses head tank, to give a fermentable sugar concentration between 5, 8, and 6.2%.

Mash sterilization and vessel preparation

After formulation and mixing, the molasses mash was sterilized continuously by direct steam injection, using a combination of a plate heat-exchanger and holding tank to give the required retention time of 4 min at 128 °C. The sterilized mash was cooled to 34 °C in the heat-exchanger by cold incoming mash with the end section being chilled by cooling water. Additional calcium hydroxide was used to adjust the pH to 6.0-6.3 after mash sterilization and transferred aseptically to the appropriate sterile fermenter.

Prior to filling, each fermenter was steam-sterilized and flushed with sterile carbon dioxide gas, so that the formation of a vacuum during cooling was prevented by the creation of a positive pressure that prevented non-sterile air being drawn into the vessel.

During the filling process, the mash was flushed for two h with sterile carbon dioxide in order to thoroughly mix the components and strip out any remaining dissolved oxygen, to reduce the redox potential to a level that ensured completely anaerobic conditions. Finally, if required, the pH was adjusted to 5.8-6.0 with ammonia. Once filling of the fermenter was completed, carbon dioxide was used to continue to agitate the mash for 20 min before and after inoculation to ensure good mixing.

Fermenter inoculation and fermenter pressure

The original inoculation process involved the aseptic connection of the final seed stage stainless steel "C" cans to the inoculation port on the side of the fermenter. The later inoculation process was achieved by pumping in the 3,500 l inoculum from the prefermenter. After inoculation, the fermenter pressure was adjusted to 35 kPa with carbon dioxide and held at that value for about five h, when pressure build-up through gas production in the fermentation had begun. At this point, the pressure was reduced to between 15-18 kPa to allow for gas pressure build-up over the fermentation period. Once gas evolution by fermentation pushed the pressure back up to 35 kPa, the gas recovery line was opened. The maintenance of a positive pressure in the fermenters at all times reduced the likelihood of contamination.

The fermentation cycle during batch production

Under normal operating conditions, the individual fermenters were prepared and inoculated at approximately three-hour staggered intervals, except during periods allocated to maintenance and cleaning. The data presented in Table 1 represents the averaged results obtained during the ten-year period from the end of 1960 to the end of 1970. The data presented in Table 2 give a comparison of the averaged results obtained for the years 1961 and 1971. From this data, it can be seen that the average fermentation time remained very constant over this period, ranging between about 29 and 33 h per month with an average for the ten-year period being 30.5 h per fermentation. With around 30 h for the fermentation and 18 h to empty steam and refill the fermenter, the total cycle of cleaning, sterilization, filling, fermentation, and emptying of the vessels took around 48 h. Theoretically, this would have allowed eight fermenters to be set within each 24-h period. In practice during normal operation over this ten-year period, the plant averaged just under six fermenters being set each day with an average of around 2,078 fermentations being performed per year. When not in use, the fermenter vessels and all feed lines were steam-sterilized and the pipe work associated with each fermenter was left under steam.

Monitoring and process control of the fermentation

Throughout the entire fermentation process the culture laboratory personnel were involved in monitoring the progress of fermentation. Before inoculation the pH and the SG of the mash were checked. From the sixth hour of fermentation, samples were removed every hour and the pH and SG were checked. Every second sample was examined microscopically to monitor cell morphology, motility, cell numbers, and evidence of contamination, and the gas evolution rate was measured. The series of typical physiological and morphological changes that take place as a normal fermentation proceeds, enabled an experienced observer to pinpoint the stage of a typical fermentation to within an hour or two. The typical profile obtained for the batch fermentation has been documented previously [15, 18] and has not been included.

Physiology of the batch fermentation

The solvent-producing clostridia are obligate anaerobes that require a redox potential of −250 mV or less, for adequate growth and solvent production. The morphology exhibited by these organisms can vary quite considerably, depending on the strain of organism used and the formulation of the culture medium. However, the two species of solvent-producing clostridia used by NCP as their production strains, show remarkably similar morphology and behavior in the molasses fermentation. This distinct and relatively constant variation in morphology with time, can be used by an experienced observer to assess the age and progress of a fermentation with considerable accuracy. The development of vigorous motility was used as an important diagnostic characteristic but only occurred during the initial stage of the fermentation. The typical industrial batch fermentation in molasses exhibited three, more or less, distinct phases. The first phase was characterized by a slow decrease in pH, together with a corresponding increase in the total titratable acidity. This phase that continued for approximately 18 h corresponded with a continuing decrease in the fermentable sugars in the molasses mash. At around the 18 h, the fermentation reached a “pH breakpoint” after which the pH began to increase again, corresponding with a decrease in the concentration of titratable acid, as acetate and butyrate were converted to solvents. Gas production began almost immediately after inoculation, but only became readily detectable after several hours had elapsed. Hydrogen and carbon dioxide were produced throughout the fermentation in approximately equal volumes until at about 30 h, when the fermentable sugar concentration dropped to around zero, whereupon gas evolution abruptly ceased. During the final phase, solvent production leveled off and the pH curve flattened out at a final value of around pH 5.8 and in typical fermentations, gas evolution ceased quite abruptly between 29 and 33 h, which was taken to indicate the cessation of fermentation.

Mass balance, solvent yields, and conversion efficiencies

Each 90,000-l fermenter contained approximately 5,850 kg of fermentable sugars of which approximately 59 % was lost as gaseous carbon dioxide (2900 kg) and hydrogen (117 kg) and 32 % was converted to solvents (butanol, 1053 kg; acetone, 526 kg;

and ethanol, 175 kg) while 6 % was converted to biomass, and 3 % remained as residual acid and other metabolites.

The fermenters were set with a relatively low concentration of fermentable sugars from 5.8-6.6 %. The reason for this, was that solvents become toxic over the 2 % level and higher sugar concentrations could not be fermented without waste. Utilizing invert sugars of 6.2 %, it was possible to achieve average yields of around 18 g/l total solvents at a conversion efficiency of 29-30 %. However, a small percentage of the invert sugar was unfermentable. If this fraction was subtracted, conversion efficiencies ranged from 31 to 32 % (Tables 1 and 2). Solvents were produced in the approximate ratio of 6:3:1 with proportional limits of 59-67 % for butanol, and 30-38 % for acetone, with ethanol production being constant at around 3 %.

Problems with deterioration of molasses quality

Over the period that the AB fermentation was operated, the improvements in efficiency of sugar extraction made by the local sugar industry, resulted in molasses of poorer and poorer quality as a fermentation substrate. In mid-season, total invert sugar concentrations of less than 45 % became common, resulting in molasses of a low sugar to high ash ratio. The deteriorating molasses quality affected fermentation performance and the standard yields and concentrations were not always attained.

To maintain the same levels of solvents yields and concentrations it was necessary to increase the set fermentable sugar concentrations (Table 1). However, the high ash loading resulted in inhibitory effects. It was found that with quality poor molasses it was essential to utilize lower sugar concentrations in the mash so that conversion efficiency was not sacrificed. This type of substrate inhibition of the fermentation became an important limiting factor and led to the search for more osmo-tolerant cultures, as well as exploring the possibility of the adoption of a fed-batch rather than the traditional batch process. However, these improvements were never implemented.

Recovery of solvents

At the end of the fermentation, the fermented molasses mash was pumped to a beer well that fed two continuous beer distillation columns run in parallel. The fermented mash was fed at a constant rate to the top of the beer still containing 30 perforated plates. A stream containing approximately 50 % water and 50 % mixed solvents (ranging from 40-60 %) was removed overhead, separating it from the stillage slops (a mix of water and cell biomass). The mixed solvents (high wine) that were recovered, were separated by batch fractionation in a high wine batch still, connected to a condenser and decanter unit. Three streams were obtained containing separate fractions. One of these contained acetone, the other butanol, and the third contained a mixture of ethanol and isopropanol. These three fractions were transferred to the crude butanol receiver, the crude acetone receiver, and a middle fraction receiver. These were then purified using a butanol batch still, an acetone batch still, and A middle fraction batch still linked to a condenser and decanter. The butanol was dried by removal of the distillate through a decanter. A high boiling fraction containing higher alcohols, esters, and organic acids was also obtained. Butanol and acetone were distilled to a high purity. A large proportion of the butanol was used in-house by NCP as a feedstock for other chemical syntheses. The ethanol recovered was of a low grade and the mixed ethanol distillate was used to make methylated spirits. The distillation of the three streams continued to be run as a batch process, as the existing equipment achieved satisfactory results. Although the distillation process could have been converted to an entirely continuous process, NCP never undertook the modernization.

Recovery of by-product gases

Carbon dioxide was produced in large quantities, accounting for more than 50 % of the carbon in the fermentable sugar in the molasses, and was recovered as a by-product. The gas generated during the AB fermentation process was first passed through a scrubber to remove traces of solvents. Carbon dioxide was then recovered by a process of absorption and desorption from a potassium carbonate solution. The recovered carbon dioxide was then combined with that generated by the ethanol fermentation plant, compressed, and then purified by passage though activated carbon and silica gel units. Finally, it was liquefied by cooling and sold as bulk gas, bottled gas, or converted to dry ice. The exit gases from the carbon dioxide recovery plant contained approximately 80 % hydrogen and 20 % carbon dioxide by volume, and was vented to the atmosphere. For a period, the exit gas was used as a fuel and as a hydrogenation medium.

Utilization of fermentation wastes

The AB and ethanol fermentation processes operated by NCP generated significant volumes of effluent and initially the disposal of these high BOD (biologocal oxygen demand) wastes caused a major problem. The NCP distillery at Germiston was land-locked with no major water course available for effluent disposal. This required the development of effective *in situ* methods of effluent treatment. Several methods of waste disposal were used at NCP during the 48-year period that the AB fermentation operated. Initially, the stillage was incinerated to produce a high potash fertilizer, but the cost of this method of waste disposal proved greater than the value of the fertilizer. The AB fermentation was known to produce a beer containing riboflavin, B complex vitamins, and other growth factors in relatively large concentrations. Advantage of this was taken to market the dried stillage as a vitamin supplement for animal feeds. However, the price of vitamins dropped to a point where this process also became uncompetitive.

The final method developed at NCP for the handling of fermentation wastes involved combining the stillage from the AB fermentation with the stillage from the ethanol fermentation, to produce Dried Molasses Distillers Solubles (DMDS). This product was then incorporated into a wide range of animal feed formulations. To produce the DMDS, the combined distillation slops were first evaporated under vacuum in multiple effect evaporators, to produce a thick concentrate containing 50 % solids, that was then spray-dried (Figure 1). This process resulted in approximately 97 % of the fermentation waste leaving the factory in bags as a dry product. The highly successful "Rumevite" system developed by a sister company, used

DMDS as the basis for the production of various balance feed supplements for ruminants supplied as concentrates or in block form. This was marketed both nationally and internationally resulting in the stillage becoming an economically valuable by-product of the AB fermentation process. The DMDS product was shipped around the world, with the international distribution handled by Rumumco Ltd. of Burton-on-trent in the UK. Most of the water evaporated from the stillage, was condensed and recycled as the make-up water and added to the fermentation substrate. Any excess was disposed of in the municipal sewerage system.

Problems with bacterial contamination

From time to time, problems with bacterial contamination of the AB fermentation were experienced at NCP. In most cases contamination was caused by lactic acid bacteria of various types, the majority of which were never fully identified. However, by paying strict attention to cleanliness and sterility in the plant, bacterial infections occurred very infrequently and never posed a major problem. The major causes of bacterial contamination appear to have occurred through errors in operating procedures resulting in unsteamed pockets of residual mash or beer being left in the complex pipe work. A typical symptoms of fermentations contaminated by lactic acid bacteria was a continuously decreasing pH after the time the breakpoint would normally have occurred, resulting in a final pH of 4.5 or less after 30 h. Gas evolution was normally also poor and the resulting sluggish fermentation yielded very little solvents. No practical procedures existed to recover these contaminated fermentations.

Problems with bacteriophage contamination

The AB fermentation run by NCP also suffered a number of phage infections. As these have been reviewed in some detail in a recent publication [19] only a brief summary is given here. During the period that NCP operated the AB fermentation process, at least four confirmed phage infections were reported as well as another two cases of suspected but unconfirmed phage contamination. The first reported phage infection occurred during late 1943 and early 1944, while the company was still operating a maize fermentation process using the Melle strain. Erratic fermentations with substantially reduced solvent yields were again reported in 1947 that were suspected to have been caused by a phage infection but this was not confirmed. As a result of these difficulties, the company embarked on trials using phage-immunized cultures provided from the UK that extended through until the mid-1950s. Another phage infection occurred in 1960, when one of the factory fermenters exploded. Shortly after the explosion, sluggish and erratic fermentations were encountered in most of the production fermenters. The presence of a phage was confirmed by electron microscopy and the entire plant was closed for cleaning and disinfection. Phage contamination also occurred in 1976, following a switch from the use of carbon dioxide to unfiltered nitrogen as the gas utilized in the fermenter filling operation. A number of the fermenters in the plant were affected and exhibited sluggish fermentations with poor solvent yields and problems were encountered in starting-up fermentations. The presence of two different phages was confirmed by electron microscopy. As a result of this phage infection, the entire plant was closed for ten days to allow cleaning and disinfection. A further suspected infection occurred in 1977. The symptoms encountered included reduced solvent yields and slightly prolonged fermentation times. Evidence suggested that the production strain in use at the time had become lysogenic, but this has never been confirmed. The problems with erratic fermentations continued well into 1978. Similar problems recurred later in 1979 and early 1980a and this time the wide-spread occurrence of a pseudo-lysogenic phage was confirmed by electron microscopy and phage propagation. The symptoms produced by this infection were reduced solvent yields and extended fermentation times. As the symptoms of the infection were less severe than those encountered with previous virulent infections, a decision was made to continue to operate the fermentation process while steps were taken to eradicate the infection. Infected cultures were replaced by a phage-free production strain, fermenters and lines were steam cleaned, and improved culture practices and plant hygiene was introduced. As a consequence, the fermentation process returned to normal over a period of a few weeks.

5.3 Recent advances and developments

5.3.1 Scientific advances over the last 25 years

The abrupt rise in petroleum prices during the 1970s provided a strong incentive to re-examine the feasibility of producing liquid fuels and chemical feedstocks by microbial conversion of renewable biomass. These developments were in part responsible for the renewed interest in the applied AB fermentation that has continued for the last 25 years.

Research activities initiated in the late 1970s intensified during the 1980s, and led to a much better understanding of the basic biochemistry and physiology of the solvent-producing clostridia [1, 5, 20]. Much of this research was directed towards investigating the various factors which influence and regulate the outcome of the fermentation, with the aim of trying to optimize conditions for solvent production. These studies led to an increased understanding of the various mechanisms involved in the regulation of solvent production and the nature of solvent toxicity and tolerances. They also drew attention to the role of cell differentiation and cell degeneration in the fermentation process. In some instances the outcomes on the studies suggested practical ways of controlling and manipulating the environmental parameters during the fermentation, to produce a shift in metabolism, to achieve increases in the yield and concentration of the end products. Basic research on the genetics of solvent-producing clostridia was also initiated during the 1980s that included the development of mutagenic techniques, gene transfer systems, and the use of plasmids for the construction of cloning vectors [21]. The cloning and sequencing of various clostridial genes and their expression in *Escherichia coli* was also successfully accomplished [22]. Studies of a more practical nature investigated the use of alternative fermentation substrates [4, 5]. In a number of cases, these were linked with developments aimed at the processing of lignocellulose to produce fermentable substrates at a competitive cost.

During the 1990s, basic research using molecular genetic techniques intensified, making a substantial contribution to our understanding of control mechanisms governing cell metabolism, including nutrient uptake and solventogenesis [4, 23-25]. These types of approaches have been enhanced by the development of various new tools such as transposon mutagenesis and other gene knockout systems. The application of molecular genetic techniques also provided a much clearer understanding of the taxonomic and phylogenetic relationships of the industrial solvent-producing clostridia [7]. However, the discovery that researchers were working with four separate species of bacteria has added a further level of complexity and diversity to the field. The physical and genetic mapping of three of the four species has been undertaken [26, 27] and the sequencing of the *C. acetobutylicum* genome has been completed by Genome Therapeutics Inc. with the support of the US DOE [25]. Although as yet, genetic engineering has not contributed to any major practical advances for the AB fermentation, it has opened up approaches that could be utilized to improve strains to overcome some of the limitations of the fermentation process. Such approaches include the enhancement of solvent tolerance [28] and strain stability [29].

5.3.2
Advances in process technology over the last 25 years

The advances in our knowledge of the biochemistry, physiology, genetics, and taxonomy that have occurred over the last 25 years, have been matched by developments in process technology and engineering. Major advances in large-scale cell cultivation techniques and continuous culture have occurred [30-32]. Various cell immobilization and cell recycle techniques for biomass retention and enhanced solvent productivity have also been developed. A variety of alternative product removal and recovery techniques have also been tested [30, 32]. Advances have also been made in combining continuous with alternative methods for the online recovery of solvents from the fermentation broth. Progress has also been made in the development of technologies for the processing of lignocellulose materials for the production of fermentable substrates that could enable economically viable processes to be developed [4].

Advances in both scientific knowledge and the process technology have together provided the potential to improve the original applied batch AB fermentation process. These developments now provide a much more comprehensive framework for the development of industrial-scale processes for the production, extraction, and processing of solvents by the fermentation route. However, until now most of these advances have been mainly of academic interest. There have been very few practical attempts to scale-up and assess the commercial viability of the applied AB fermentation.

5.3.3
Intrinsic limitations affecting the applied AB fermentation

Despite the advances that have been made, the applied AB fermentation suffers from a number of intrinsic limitations that mitigate against its re-establishment as a viable commercial fermentation process. These are summarized below:

- Acetone and butanol are low cost, bulk chemicals with a commercial value only marginally higher than the cost of substrates used to produce them. To be economically viable, the applied fermentation has to produce large volumes using very reliable process technology. Significant improvements in economic viability can probably only by achieved if newer, cheaper substrates become available.
- The biochemical pathways used by the solvent-producing clostridia limit the achievable yield of solvents to around 33 %. The theoretical limits governing substrate yields are already being approached with current technology.
- The AB fermentation is a complex process involving physiological, morphological, and developmental changes in the bacterium and the production of solvents only occurs late in the batch fermentation cycle. The long fermentation times and need for large fermenter capacity coupled with a significant level of fermentation failure, all contribute to the low productivity of the batch fermentation. Continuous fermentation can enhance productivity, but the complexity of the process detracts from applying this approach commercially.
- As currently operated, the applied AB fermentation has an absolute requirement for sterility, as contamination causes major problems and reduces productivity. The need to sterilize the fermentation substrate and operate a sterile process adds a considerable cost burden.
- Butanol is highly toxic and limits the final solvent concentrations achievable in the fermentation. The low concentration of solvents produced requires energy intensive recovery processes, generating high recovery costs.
- The fermentation process produces significant volumes of effluent which require specific processes for handling, treatment, and processing. These can contribute to the operating costs of the fermentation.

The combination of intrinsic limitations, associated with the use of the fermentation route for the production of solvents, places serious limitations on its commercially viability. Unless approaches can be developed to counter some of these limitations, it is unlikely that the fermentation route will become a viable proposition in the foreseeable future.

5.3.4
Substrate and product markets

The quantities of acetone and butanol produced world-wide are extremely large and currently virtually all production is via petrochemical synthesis. As a result, the price of these products are to a large extent dependent on the price of crude oil and prices can fluctuate considerably on the open market from year to year [33].

Acetone, butanol, and isopropanol are currently used as bulk chemicals as well as chemical feedstocks for the synthesis of a wide range of other products. In addition, these solvents also have potential for use as liquid fuel, as mixtures in diesel and petrol, or as diesel supplements. They can also be used to increase stability and the temperature and water tolerance of mixtures of methanol or ethanol and conventional fuels, to improve the octane value and to help improve carbon dioxide emission values. However, for the fermentation route to become a viable option for fuel production, these products would need to be produced in large volumes at competitive prices. Unless the cost of fermentation substrates and the energy requirements for sterilization and product separation could be greatly reduced, the production of liquid fuels is not likely to be feasible for most countries. An exception to this could occur if the fermentation process were to be established for strategic reasons.

One of the major intrinsic limitations of the applied AB fermentation is the small differential that exists between substrate and product prices. In the traditional batch fermentation process, the substrate costs contributed over 60 % of total production costs [34]. The current prices commanded by purpose-grown crops in most developed countries makes it unlikely that the fermentation route using these types of substrates could compete economically with oil-derived products at the present price of crude oil [33]. The re-establishment of the AB fermentation process will therefore only be possible if suitable low-cost, high-volume substrates can be found.

From the 1920s, through to the present, the scientific literature contains numerous reports of AB fermentation trials conducted with alternative low-cost substrates [4, 5]. In addition, internal research reports produced by NCP from 1945 through to 1982 include numerous reports on optimization studies and trials undertaken to assess alternative fermentation substrates. None of these research findings were published and comments in the published literature indicate that many of the companies operating the industrial AB fermentation undertook similar unpublished studies. Virtually every feasible crop and agricultural waste appears to have been investigated at one time or other with varying levels of success [4, 5]. The solvent-producing clostridia exhibit a remarkable ability to degrade and utilize most plant polysaccharides and sugars with the exception of lignocellulose. However, although it has proved possible to produce solvents from most naturally-occurring agricultural crops and wastes in laboratory trials, problems relating to volume, supply, transport, and seasonality have tended to create difficulties for the potential commercialization of these substrates.

Recently, an AB fermentation research program financed by the Commission of the European Union, with support from the Austrian Ministry of Science and Transport, has led to the investigation of potential niche markets involving the conversion of cheap, low-grade substrates to solvents [35]. This concept involves the use of relatively small rural-based AB fermentation plants able to use a variety of surplus and low-grade agricultural substrates such as frozen potato, corn contaminated by mycotoxins, and sugar beet surpluses [36]. Recent assessments have indicated that small AB fermentation plants of this type could meet the specific needs of local rural regions in a similar way to small ethanol distilleries. In Europe, these bulk-chemical fermentations were successfully operated for many decades [35].

This type of approach is supported by the conclusions reached in earlier studies by Marlatt and Datta [37], who concluded that the AB fermentation process probably had higher capital costs but lower production costs than acetone and butanol produced by petrochemical processes.

The key to the use of the fermentation route for the production of bulk chemicals probably lies in the development of effective economic processes for the production of lignocellulose hydrolysates as fermentation substrates [38]. Currently, these technologies remain experimental and unproven, but further research and development might well enhance the economic feasibility of this process.

Numerous studies have been undertaken to evaluate and optimize the utilization of cellulose and hemicellulose hydrolysates as substrates for the AB fermentation [4, 5]. The use of co-cultures of cellulolytic and solventogenic species for lignocellulose utilization have also been investigated [39, 40]. An alternative strategy is the genetic manipulation of solvent-producing strains to improve cellulose and hemicellulose utilization. The discovery of apparently unexpressed genes encoding for endogenous cellulase enzymes, as a result of the recently completed genome sequencing of *C. acetobutylicum*, raises intriguing possibilities. If these genes could be reactivated it might be possible to modify these bacteria to metabolize cellulose directly.

To date, the most comprehensive investigation of the use of lignocellulose as a substrate for the applied AB fermentation, was the program undertaken in France during the 1980s. This program was established as a result of a strategic decision by the French Government for the replacement of 10 % of the nations liquid fuel needs by gasoline substitutes. The Institut Français du Pétrole (IFP) and its engineering subsidiary Technip undertook an investigation to produce butanol and isopropanol as blending agents to enable methanol to be added to petrol as a fuel extender. As part of this program the feasibility of producing solvents using Jerusalem artichokes and sugar beet was investigated [41]. A new strain of *C. beijerinckii* (formally *C. acetobutylicum*) IFP 903 (ATCC 39057) was isolated by enrichment culture from Jerusalem artichokes at the Institute National Agronomique in Paris and a butanol-resistant mutant of this strain IFP 904 (ATCC 39058) was developed at the IFP.

The project included the construction of a pilot-plant for the conversion of biomass to acetone and butanol [42]. The pre-industrial-scale plant at Soustons was initially designed to use mainly cereal straw and corn stover as the substrate for the AB fermentation process [43]. The aim was to produce one ton of solvents from six to seven tons of this raw material. The biomass was pre-treated by steam explosion and then hydrolyzed using a cellulase complex produced in a separate process in the plant. The hydrolysates were then fermented to acetone and butanol by solvent-producing clostridia and the solvents were separated by distillation. The AB fermentation was operated as a batch process on both a pilot-scale and in the demonstration-scale and yields of up to one ton solvents from 7.7 tons of corn cobs were reported. The stillage was concentrated, anaerobically digested, or recycled [41]. An economic evaluation of the process concluded that lignocellulose as a substrate had the highest production costs but offered the greatest potential [44]. The viability of using lignocellulose as a substrate was also strongly dependent on the market value of lignin produced as a by-product [45].

5.3.5
Solvent yields and by-product recovery

During the course of the AB fermentation, most of the energy contained in the fermentation substrate is retained in the high energy end products that include organic acids, solvents, and gaseous hydrogen. Only around 5 % of the available energy is diverted to the creation of new cell biomass. However, due to the complex nature of the catabolic pathways utilized by the solvent-producing clostridia, the yield of solvents is limited to a practical maximum of around 33 %. This intrinsic limitation is perceived to be one of the drawbacks of the AB fermentation. Although it has been shown to be possible to obtain slight reductions in the yield of gaseous products with a corresponding increase in solvent yield by sparging with carbon monoxide and other techniques [46], it is unlikely that significantly higher solvent yields can be achieved.

It appears that it was possible to operate a commercially viable AB fermentation with yields of around 28 %. If average yields above 33 % could be sustained, then the process economics would improve slightly. However, even if yields approached a theoretical maximum, this alone would not make the process economical. The costs of the substrate and the costs associated with solvent recovery have a much greater impact on commercial viability. Solvents yields of 26-28 % would make commercial feasibility marginal.

The economic viability of the traditional industrial batch AB fermentation plant of the type operated by NCP, was heavily dependent on the recovery and sale of other by-products. The marginal economics of the AB fermentation required that as much of the other fermentation by-products as possible be utilized, and it was possible to achieve a high degree of operating efficiency in such integrated plants.

The potential role of by-products was considered in a recent analysis of the feasibility of establishing relatively small rural-based AB fermentation plants based on niche markets, using a variety of low-grade or surplus agricultural substrate [33]. Although the sale of stillage as animal feed provided an important income for the previous world-wide AB fermentation operated last century, the animal feed market in Europe was not considered to be reliable enough to include this as a source of income. A more feasible option was considered to be the production of methane from the stillage. The feasibility calculations for the niche fermentation were based on the assumption that the hydrogen, and possibly methane, would be used to generate energy for the plant. The sale of carbon dioxide was also discounted on the basis that the carbon dioxide market in Europe is currently less than the installed production capacity [33]. It was considered that profits would be too small to make a major impact on the process economics and the gas would therefore be vented to the atmosphere. Possible income from the sale of by-products were therefore ignored in this economic analysis. The conversion of stillage from the AB fermentation into bacterial cells accumulating thermoplastic polyesters of polyhydroxyalkanoate, has also been investigated [47].

5.3.6
Productivity and continuous culture

The batch AB fermentation is a complex process involving physiological, morphological, and developmental changes in the bacterium during the course of the fermentation. The precise triggers controlling these changes are still not fully understood [25, 48]. The traditional applied batch AB fermentation process involved reasonably long fermentation times to reach completion. Early commercial fermentations took from 50 to 60 h [9]. This was reduced to around 30 h by the introduction of new industrial strains and reducing the ratio of the starting inoculum. To increase productivity during the Second World War, CS-GB resorted to a semi-continuous operation involving a complete fermenter of late exponential-stage cells grown under sterile conditions, and being used as the inoculum for three fermenters filled to volume with unsterilized molasses. Although the solvent yields obtained were slightly lower it enabled the normal fermentation cycle of 90 h to be reduced to between 30 to 32 h [6].

The relatively long fermentation time for the batch process required a large fermenter capacity which resulted in a relatively low volumetric productivity in these plants. The use of continuous fermentation provides scope for improving the economic viability of the applied AB fermentation through achieving a significant improvement in volumetric productivity [30]. Continuous culture has also provided a valuable tool for experimental studies. Single-stage continuous culture techniques have been used by many workers to investigate fundamental aspects of the AB fermentation, including the evaluation of the effects of different nutrients, metabolites, and physical factors [30, 31]. These studies have also provided useful information about the regulatory mechanisms and kinetics of solvent production. The major limitation associated with the uses of single-stage continuous culture, is the generation of different populations of cells exhibiting metabolism and growth rates that result in the periodic oscillation in cell populations and end products [49]. This prevents the establishment of a true steady-state and limits solvent concentrations and yields obtained. These disadvantages have limited the use of single-stage continuous culture to an experimental role and has precluded its adoption for the applied fermentation.

To overcome some of these limitations, two-stage or multistage continuous fermentation systems have been investigated for both experimental applications and commercial use. In these systems, solvent production normally occurs during the second stage of the fermentation when little or no increase in cell biomass occurs. The results obtained with such systems indicate that multi-stage continuous flow systems, designed to retain high concentrations of non-growing, solvent-producing cells, are greatly superior to conventional continuous culture systems. Two main experimental approaches for the retention and re-use of cell biomass in flow-through systems have been utilized. These are cell immobilization and cell recycling. Many methods are available for immobilizing cells, including entrapment [32].

Probably the most successful method used for the solvent-producing clostridia has been the immobilization of cells by adsorption onto bonechar and used in packed bed or fluidized bed reactors for continuous solvent production. Such systems have been successfully operated with integrated systems for continuous product removal by adsorption, gas stripping, pervaporation, and liquid-liquid extraction [50]. Cell recycling using ultrafiltration was also demonstrated as a successful method for retaining biomass and increasing productivity in the AB fermentation [30, 31]. This approach has been used in conjunction with *in situ* recovery of solvents from the effluent stream.

Information in the literature relating to attempts to scale-up continuous culture processes for the applied AB fermentation is quite limited. Patent applications indicate that at least some of the companies that operated the industrial batch AB fermentation process did pursue the feasibility of converting to continuous fermentation. However, commercial sensitivity has tended to keep the results achieved confidential.

A patent on continuous fermentation was lodged by CSC [51] and it can be concluded from the patent that the test trials were conducted, but no detailed description of this work was published. The patent states that the continuous process could be continued for 30 days or longer and that a significant reduction in the residence time and increase in the production capacity of the given plant could be achieved, resulting in considerable savings in labor, steam, electric power, and other expenses. However, it appears that the process described was never applied in large-scale [52].

The most extensive reports of pilot-scale multistage fermentation systems for the continuous production of solvents relate to a process developed in the USSR after the Second World War [53-56]. A pilot unit was built at the Dokshukino plant in 1958 that included a battery of eleven fermenters [54]. The first fermenter was equipped with a stirrer and it served as an inoculator and activator. Based on the results of the pilot-plant, the continuous AB fermentation was introduced in 1960 into the Dokshukino production plant employing three batteries of seven to eight fermenters with volumes of 220 and 270 m^3. The introduction of the continuous cascade system was reported to have resulted in a 20% increase in productivity saving 64.4 kg of starch per ton of solvents produced [56]. A more recent report on a patented design for the production of solvents from maize using a cascade fermentation system utilizing four stages was published by Marlatt and Datta [37].

The most recent information on the scale-up of a continuous AB fermentation process has come from a research project financed by the Commission of the European Union with support from the Austrian Ministry of Science and Transport. This project has involved the building and testing of a pilot-plant for the fermentation of agricultural starch products to acetone and butanol. A group in Vienna led by Richard Gapes have had the responsibility for designing, building, and running the pilot-plant and for undertaking support work in the laboratory [57, 58]. The pilot-plant was built and run at a commercial agricultural distillery in Lower Austria during 1997 and 1998 [35]. The pilot-plant was designed to be able to run in fully continuous operation, but also capable of batch-wise processing. A two-stage process was selected to overcome the problems of oscillations and to permit greater experimental freedom with the second stage system. The first stage had a working volume of 50 l and the second stage approximately 150 l. The substrates used have included both semi-synthetic media containing xylose, glucose, and sucrose, along with a variety of agricultural products such as whey, potato, corn, silage, and wheat [58]. The initial intention was to focus on the organic fraction of municipal solid waste, with the emphasis later switched to the use of low-cost starch produce, as was traditionally used by the ethanol fermentation industry in Europe. This change of emphasis came about through interest expressed by a commercial ethanol fermentation facility run by a farmer's co-operative, and the initial pilot-scale operation was redesigned and constructed in the commercial ethanol plant. This provided a number of advantages as most of the plant infrastructure, substrate storage, and substrate preparation equipment were provided by the production facility. However, the location of the plant in a commercial distillery also proved to be a disadvantage as substrate delivery and the availability of services, such as water and electricity, were governed by commercial production requirements [35].

Pilot-plant experiments were intended to test the performance and reliability of fermenting commercially available agricultural substrates in an industrial environment, and to provide scale-up experience. The AB fermentation plant was run in parallel to the production of crude ethanol from different starchy raw materials, including potato, wheat, rye, and maize and in total the equivalent of over four years of continuous fermentation were performed [58]. It was reported that, in general, the results obtained in the pilot-plant matched, or were superior to, laboratory results and demonstrated the ability of the process to reproducibly perform stable fermentations for extended periods.

Significant improvements in laboratory-scale continuous AB fermentation, including increased productivity, long-term stability, and lack of degeneration changes have been demonstrated by a number of authors. Experience gained with the scale-up in the Dokshukino plant in the former USSR [56], and from the Austrian pilot-plant have indicated that stability can be achieved with commercial-scale continuous culture, and the problems of contamination and degeneration can be minimized. In spite of the complex nature of the AB fermentation, pilot-scale trials and full-scale continuous operation in the Dokshukino plant have indicated that this mode of operation can yield considerable savings. It appears that continuous runs of four weeks are sufficiently long for the economic advantages to be noticeable [58]. However, experience from industrial microbial cultivation practice show that commercial fermentation technologists actually do not favor this mode of operation as it is considered to be more difficult and risky because of potential contamination problems.

5.3.7
Reliability of the applied AB fermentation

One of the crucial prerequisites for the applied AB fermentation is reliability of operation. Only a completely reliable fermentation process can be expected to be commercially successful. This requires that when the AB fermentation is operated on an industrial scale, the cultivation conditions have to be maintained in such a way that the complete conversion of the substrate can be guaranteed. The sensitivity analyses undertaken by Gapes et al. [33] underline the overriding importance of reliability in production and point out that regardless of the cause, any loss of production can be catastrophic for economic viability.

As currently operated, the applied AB fermentation process requires sterile operating conditions. The need to pre-sterilize the raw materials by exposure to high temperatures and to run the fermentation process under sterile conditions adds considerably to operating costs. In addition, a plant designed for sterile fermentation is significantly and unavoidably more expensive to build than a non-sterile plant, such as those used in ethanol distilleries [33]. The large sterilisable pressure vessels required for the fermenters are expensive and there are additional costs involved in providing dedicated sterilization equipment and the piping, valves, and other fixtures capable of reliably supporting absolute sterility at all times.

The effects of both bacterial and bacteriophage contamination on the reliability in the industrial plants during the early part of the last century were a problem [19]. Contamination by virulent phages usually produced very rapid effects and could halt production within several hours. The effect produced by temperate phages often occurred more gradually and resulted in a reduction in productivity and culture vigor. The importance of phage infection in the applied AB fermentation is reflected in the large amount of time and effort invested in developing phage-resistant strains [19]. Any future developments of the applied AB fermentation will need to ensure that the likelihood of contamination by phages or other microorganisms is kept to a minimum.

Another aspect that was found to effect the reliability and performance of the industrial AB fermentation was variability in the quality of the fermentation substrates. Both annual fluctuations and a general decline in molasses quality over the years caused serious problems for the NCP fermentation process and led to a search for more osmotolerant strains. Attempts to develop small rural-based AB fermentation plants, designed to use a wide range of surplus and low-grade agricultural substrate, would have to content with much greater variability in substrate composition. Such variations are likely to result in sub-optimal growth and solvent production with reduced yields and concentrations of solvents being produced.

Strain degeneration, resulting in a decrease in the ability to produce solvents, was recognized by early workers in the field and is a well documented phenomenon [5, 29]. The potential problem with strain reliability in the industrial batch fermentation process was overcome by the use of starter cultures that were freshly generated from spore stocks. The problem of strain degeneration was encountered in continuous culture, where decreased solvent production was reported to occur over prolonged periods of operation due to culture deterioration [51, 56]. The problems of culture stability and reliability are therefore more likely to be of concern if the applied AB fermentation is operated under continuous conditions.

The commercial batch AB fermentation that operated early last century proved to be a robust and reliable process as illustrated by the analysis of NCP production logs presented in Tables 1 and 2 The importance of reliability is well recognized by industry which has tended to shy away from continuous fermentations in favor of batch forms of production. One of the main reasons that continuous operation in general is not preferred by fermentation technologists, is because of the risks associated with potential contamination and culture stability problems.

5.3.8 Solvent concentration and recovery processes

One of the major limitations in using the fermentation route for the production of solvents is the low final concentration of the end products produced. The low final solvent concentrations achievable are limited by butanol toxicity. This dictates that the AB fermentation will remain a cost-intensive recovery process. The standard method of recovery has been distillation, and although there have been significant improvements in distillation technology, the high energy cost make up a substantial component of the production cost. his constraint has provided an incentive for the development of alternative methods of solvent recovery which are less energy-intensive [30-32].

Many alternative experimental approaches have been investigated. These include the recovery of solvents under vacuum, gas stripping, dialysis, reverse osmosis, liquid-liquid extractions, and the use of two-phase aqueous systems. Absorbents such as activated carbon and synthetic polymeric resins, as well as molecular sieves such as zeolites and chemical extraction using lactones, have also been investigated as potential separation methods [30-32]. All of these approaches suffer limitations and in many cases, although separation of solvents from the fermentation broth can be achieved, no practical methods for the subsequent recovery of solvents and recycling of extractants exist. Although some of these experimental approaches may have potential as adjunct or alternatives to distillation during downstream processing of the spent fermentation broth, most interest has been focused on the *in situ* recovery of solvents from the fermentation broth [30-32].

Extraction methods identified as having potential for *in situ* recovery include gas stripping, vacuum distillation, two-phase aqueous systems, and liquid extractants or absorbents. Developments in semi-permeable membrane technology has provided a variety of novel approaches to the direct recovery of solvents *in situ*. Both cross-flow microfiltration and ultrafiltration have provided feasible technologies for achieving cell recycle for biomass retention while allowing the separation of a solvent stream. Ultrafiltration can be used in conjunction with other extraction processes such as reverse osmosis and adsorption. Pervaporation, utilizing silicone membranes, allows solvents to diffuse from the fermentation broth. These can then be removed by vacuum or a gas stream. Perstraction operates on a similar principle, but utilizes organic solvents or other chemical extracts such as lactones for product removal.

Among the various on-line membrane separation techniques for product removal, pervaporation is considered to provide the most promising approach for the AB fermentation [30, 58, 59].

A number of advantages of on-line product removal can be identified. When the products are separated on-line, the solvent concentration in the fermentation mixture can be maintained at lower levels, thereby reducing the effects of product inhibition. This can contribute to enhanced culture stability and improve substrate utilization. However, before the membrane technologies can gain application in industrial-scale processes, a number of drawbacks will have to be overcome [33]. These include high investment costs, membrane fouling and the blockage of narrow capillaries, as well as contamination risks resulting from membrane perforation. In addition, membrane evaporation techniques remove a large amount of water from the fermentation mixture in addition to the solvents, thereby contributing to the medium concentration and eventually the salt accumulation, both of which can affect the fermentation. Furthermore, pervaporation and membrane evaporation include an energy intensive phase change as the liquids vaporize that results in lost energy which cannot be recovered [33]. Further improvements in membrane fluxes and selectivities could, however, make the application of these technologies more feasible.

Recent analysis indicates that the choice of product separation technology for removal of product from the beer is not likely to be of deciding importance with respect to investment costs for new AB fermentation facilities [33]. Traditional distillation columns incur investment costs that are of a similar magnitude to gas-stripping, extraction, or membrane evaporating equipment. It has been concluded that the use of low flux, highly selective pervaporation membranes could even incur higher investment costs due to the large membrane areas required, and operational problems such as capillary blockage and the compromising of sterility through membrane perforation [33].

Continuous fermentation, coupled with on-line product removal, can provide a way to enhance volumetric productivity of the process. However, on-line removal does involve a compromise because the product separation operates more efficiently when the solvent concentrations are higher. This means that a continuous process of this type is both a conversion and a recovery cost-intensive which detracts from its economic viability. Substantial improvements in both conversion cost-efficiency and the recovery cost-efficiency would be required to attain economy feasibility. In an earlier study, Marlatt and Datta [37] calculated that if improved strains were used which were able to tolerate slightly higher butanol concentrations, and if volumetric productivity could be increased by about 50%, the production costs for the production of butanol by fermentation would be similar to the production costs for synthetic butanol.

5.3.9
Economic perspectives

In addition to the various factors that impinge on the operating costs of the applied AB fermentation covered in the preceding sections, the capital expenditure required to establish a commercial plant is a major consideration in determining commercial feasibility. The substantial investment costs associated with establishing a commercial AB fermentation plant are likely to directly increase production costs through the need to repay interest and financing charges as well as to cover capital depreciation. The requirement for a large capital investment before production can start, coupled with the risk associated with investing in new and innovative technology is likely to constitute a serious hurdle for attracting financial backing.

An analysis of the capital development costs associated with two types of applied AB fermentation plants has been published recently [33]. One plant was designed for batch fermentation followed by distillation as used in the classical industrial AB fermentation process. The other plant was designed for continuous fermentation and on-line product separation using membrane technology. Direct costs were calculated by pricing all major items of equipment. Factors were then applied to estimate the indirect and other costs. It was assumed that the energy requirements would largely be generated internally. The conclusion reached from this analysis was that although batch production plants are usually considered to be more expensive to construct than continuous production facilities, the differential might not be as great as anticipated as there were some additional costs associated with the continuous process, and the capital costs for product separation are comparable for both processes.

This economic analysis indicated that, the costs associated with servicing the capital investment are likely to make up the third largest component of the operating costs, after substrate and energy costs. This could be a major impediment for re-establishing the applied AB fermentation. This cost was identified as a crucial factor in determining the economic viability of the process exploiting the smaller niche market opportunities that have been identified for Europe. Both capital and other costs could also be reduced markedly if an existing plant could be easily modified or extensive use could be made of existing ethanol production facilities, as opposed to the construction of an entirely new plant. The analysis undertaken indicated that this could decrease the cost by up to half [33]. Staggering, or delaying investment costs until after the start of production, might be one way of improving the economics for the AB fermentation process.

The largest operating cost associated with the conventional applied AB fermentation process, is the cost of the fermentation substrate which is estimated to be around 60-65% [34]. At current prices it has been estimated that if purpose-grown crops were to be utilized as substrates, they would make up around three-quarters of production costs [33]. The break-even price for production is therefore very largely dependent on the price that can be obtained for the product, compared to the price paid for the substrate. Analysis of the international product markets show that although the world-wide usage of acetone and butanol is very large, prices can fluctuate considerably from year to year depending, to a large extent, on the price of crude oil [33]. If the applied AB fermentation were to be re-established, it would have to sell into this competitive chemicals market that is currently serviced by a firmly established and highly capitalized petrochemical industry.

The second highest operating cost for the applied AB fermentation is the cost of energy required for substrate preparation and product recovery, which typically makes up between 10% and 20% of production costs [33]. The energy costs of operating small AB fermentation plants of the type considered above were estimated to be low, since energy balances suggested that combustion of the hydrogen and methane from a biogas plant could cover most of the requirements. For small plants of this type, it was suggested that the direct and indirect costs associated with personnel in both production and in administration could be, to some extent reduced by automation and out-sourcing. However, further costs for sales and marketing, as well as research and development would be necessary [33].

5.4 Conclusions and future prospects

The overriding importance of substrate costs for the applied AB fermentation dictates that any future commercial developments will remain dependent on the availability of suitable low-cost raw materials. Justification for the fermentation route for the production of bulk chemicals and fuels has remained contentious, and will remain in competition with the petroleum-based industry, and possibly other processes using other non-renewable reserves such as coal and oil shale. Decisions to utilize the fermentation route have tended to be politically motivated, but emphasis for fuel production has now shifted to reducing pollution and contributing to the Kyoto protocol by limiting global warming [38]. Wheals et al. [38] concluded that the production of fuels by fermentation will only remain a significant, self-sustainable industry in the 21st century if the utilization of lignocellulose becomes a commercial reality. Political, environmental, or socio-economic factors could, however, favor the development of niche markets for low-grade or surplus agricultural products and wastes.

Improvements in productivity and reliability, and increases in yields and concentrations of solvents, will continue to play a key role in any developments to re-establish the applied AB fermentation. Continuing developments in the process technologies to provide greater reliability and operation close to theoretical limits, will be essential. For continuous fermentation, the development of more informative and predictive methods for on-line data-logging and data-processing will be required. Increasing the concentration of the solvents produced, and the development of less energy-intensive methods of solvent extraction will be vital for potential commercial developments. In addition, on-line product separation and removal are predicted to have a major influence on the process economics. The testing of improved process technology in pilot- and demonstration-scale will be essential, to demonstrate that the engineering design innovations and new process technologies can be operated reliably. However, this is likely to remain problematic, requiring governmental support because of the expense and risk involved.

Developments in process technology will need to be supported by laboratory-based studies aimed at strain improvement and increasing strain reliability. Genetic engineering has greatly increased the potential for the development of new or improved biotechnology processes using genetically modified microorganisms. Many of the tools and protocols required for genetic manipulation now exist for at least some of the species of solvent-producing clostridia. Improving polysaccharide degrading ability and the uptake and metabolism of sugars and other nutrients, is one obvious target for strain improvement to expand the range of substrates which can be utilized, and the efficiency with which they can be metabolized. The improvement of solvent yields and concentrations by genetic manipulation are likely to be much more difficult targets to achieve. Increased solvent yields would require the diversion of electron flow away from acid and hydrogen production, to solvent production, without compromising metabolic performance. The solvent concentrations achievable are limited by the toxic effect of butanol but the exact mechanisms involved are not fully understood. Evidence indicates that the major cause of toxicity occurs via damage to the cell membrane, thereby decreasing its efficiency as a barrier, resulting in leakage and disruption of phospholipid and membrane protein organization. Overcoming these intrinsic limitations are likely to present daunting challenges. More amenable targets for the manipulation of solvent-producing strains could be the improvement of culture stability, and the prevention of culture degeneration, as well as the enhancement of resistance to autolysins, bacteriocins, and bacteriophages. Strategies for improving stability and other characteristics of strains for continuous culture and biomass retention could also prove worthwhile. Other possible targets for manipulation could be the improvement of the harvesting ability of the cells, or the improvement of the nutritional characteristics of the cell biomass produced for use as an animal feed or for a source of other products. However, as yet, few practical improvements have been achieved by the genetic manipulation of solvent-producing clostridia and it remains to be seen what impact this approach will have on the applied AB fermentation.

Although the AB fermentation was once commercially successful in many countries, currently these chemicals can be made more cheaply from fossil carbon sources. Since the decline of the industrial AB fermentation a number of economic evaluations have been published [34, 35, 37, 60-62]. All these studies have concluded that, at best, the economics of the applied AB fermentation, using conventional technology and agriculturally-based feedstocks, remain marginal and are currently unlikely to compete commercially with the chemical synthesis of solvents. However, in the longer term it can be predicted that a combination of increasing petroleum prices, environmental constraints, and the availability of suitable low-cost biomass or waste-based raw materials as substrates, coupled with the development of genetically improved strains and continuing advances in process technology, will result in the re-establishment of the applied AB fermentation as a major industry.

Acknowledgements
The NCP company is acknowledged for providing access to archival material and strains. The author would also like to thank Dr Stefanie Keis, Ranad Shaheen, and Hedwig Swoboda for their contributions to the characterization of the NCP strains, and I am grateful to Dr Stefanie Keis for much appreciated editorial assistance.

Note Added in Proof
Since this chapter was written an additional 58 early NCP production strains have been discovered. This new collection of strains came to light during the demolition of the culture laboratory at NCP. These strains have been maintained in sealed glass vials and were kept specifically as backup strains for insurance purposes. Ex-

amples of all 58 of these new strains have now been added to the culture collection maintained in the Department of Microbiology at the University of Otago in New Zealand.

References

[1] Rogers, P., Genetics and biochemistry of *Clostridium* relevant to development of fermentation processes, in: *Advances in Applied Microbiology.* (A. I. Laskin, Ed.), Academic Press, New York, **1984**, pp. 1-89.

[2] Bronnenmeier, K., Staudenbauer, W. L., Molecular biology and genetics of substrate utilisation in clostridia, in: *The Clostridia and Biotechnology* (D. R. Woods, Ed.), Butterworth-Heinemann, Stoneham, MA, **1993**, pp. 261-309.

[3] Canganella, F. and Wiegel, J., The potential of thermophilic clostridia in biotechnology, in: *The Clostridia and Biotechnology* (D. R. Woods, Ed.), Butterworth-Heinemann, Stoneham, MA, **1993**, pp. 393-429

[4] Mitchell, W. J., Physiology of carbohydrate to solvent conversion by clostridia, *Adv. Microbiol. Physiol.* **1998**, 39, 31-130.

[5] Jones, D. T., Woods, D. R., The acetone butanol fermentation revisited, *Microbiol. Rev.* **1986**, 50, 484-524.

[6] Hastings, J. J. H., Development of the fermentation industries in Great Britain, in: *Advances in Applied Microbiology* Vol. 14, (D. Perlman, Ed.), Academic Press, New York, **1971**, pp. 1-45.

[7] Jones, D. T., Keis, S., Origins and relationships of industrial solvent-producing clostridial strains, *FEMS Microbiol. Rev.* **1995**, 17, 223-232.

[8] E. McCoy, E. B. Fred, W. H. Peterson, E. G. Hastings, A cultural study of the acetone butyl alcohol organism, *J. Infect. Dis.* **1926**, 39, 457-484.

[9] Shaheen, R., Shirley, M., Jones, D. T., Comparative fermentation studies of industrial strains belonging to four species of solvent-producing clostridia, *J. Mol. Microbiol. Biotechnol.* **2000**, 2, 115-124.

[10] Hongo, M., Process for producing butanol by fermentation, *US Patent* 2, 945, 786, 1960.

[11] Keis, S., Bennett, C. F., Ward, V. K., Jones, D. T., Taxonomy and phylogeny of industrial solvent-producing clostridia, *Int. J. Syst. Bacteriol.* **1995**, 45, 693-705.

[12] Wilkinson, S. R., Young, M., Goodacre, R., Morris, J. G. , Farrow, J. A. E., Collins, M. D., Phenotypic and genotypic differences between certain strains of *Clostridium acetobutylicum, FEMS Microbiol. Lett.* **1995**, 125, 199-204.

[13] Johnson, J. L., Toth, J., Santiwatanakul, S., Chen, J. S., Cultures of "*Clostridium acetobutylicum*" from various collections comprise *Clostridium acetobutylicum, Clostridium beijerinckii*, and two other distinct types based on DNA–DNA reassociation, *Int. J. Syst. Bacteriol.* **1997**, 47, 420-424.

[14] Keis, S., Taxonomy and phylogeny of industrial solvent-producing clostridia, Department of Microbiology, *Thesis*, University of Otago, Dunedin, New Zealand, **1996**.

[15] Spivey, M. J., The acetone/butanol/ethanol fermentation, *Proc. Biochem.* **1978**, 13, 2-5.

[16] Robson, P. M., Jones, D. T., Production of acetone-butanol by industrial fermentation, in: *Industrielle et Biotechnologie. Production d'Intermédiares Industriels par Culture Anaerobie* (O. Chaude, G. Durand, Eds.), Proceedings Société Française de Microbiologie, Paris, **1982**, pp. 169-213.

[17] Shaheen, R., Phenotypic and fermentation characterisation of industrial solvent-producing clostridia, *Thesis*, University of Otago, Dunedin, New Zealand, **1997**.

[18] Jones, D. T., van der Westhuizen, A., Long, S., Allcock, E. R., Reid, S. J., Woods, D. R., Solvent production and morphological changes in *Clostridium acetobutylicum, Appl. Environ. Microbiol.* **1982**, 43, 1434-1439.

[19] Jones, D. T., Shirley, M., Wu, X., Keis, S., Bacteriophage infections in the industrial acetone butanol (AB) fermentation process, *J. Mol. Microbiol. Biotechnol.* **2000**, 2, 21-26.

[20] Jones, D. T., Woods, D. R., Solvent Production, in: *Clostridia, Handbooks* Vol. 3 (N. P. Minton, D. J. Clarke, Eds.), Plenum Press, New York, **1989**, pp. 105-144.

[21] Young, M., Staudenbauer, W. L., Minton, N. P., Genetics of *Clostridium*, in: *Clostridia, Biotechnology Handbooks* Vol. 3 (N. P. Minton, D. J. Clarke, Eds.), Plenum Press, New York, **1989**, pp. 63-103.

[22] Papoutsakis, E. T., Bennett, G. N., Cloning, structure, and expression of acid and solvent pathway genes of *Clostridium acetobutylicum*, in: *The Clostridia and Biotechnology*, (D. R. Woods, Ed.), Butterworth-Heinemann, Stoneham, MA, **1993**, pp. 157-199.

[23] Blaschek, H. P., White, B. A., Genetic systems development in the clostridia, *FEMS Microbiol. Rev.* **1995**, 17, 349-356.

[24] Dürre, P., Bahl, H., Microbial production of acetone/butanol/isopropanol, in: *Biotechnology* 2nd Edn. Vol. 6 (M. Roehr, Ed.), VCH, Weinheim, **1996**, pp. 229-268.

[25] Dürre, P., New insights and novel developments in clostridial acetone/butanol/isopropanol fermentation, *Appl. Microbiol. Biotechnol.* **1998**, 49, 639-648.

[26] Wilkinson, S. R., Young, M., Physical map of the *Clostridium beijerinckii* (formerly *Clostridium acetobutylicum*)NCIMB 8052 chromosome, *J. Bacteriol.* **1995**, 177, 439-448.

[27] Cornillot, E., Croux, C., Soucaille, P., Physical and genetic map of the *Clostridium acetobutylicum* ATCC 824 chromosome, *J. Bacteriol.* **1997**, 179, 7426-7434.

[28] Formanek, J., Mackie, R., Blaschek, H. P., Enhanced butanol production by *Clostridium beijerinckii* BA 101 grown in semidefined P2 medium containing 6 percent maltodextrin or glucose, *Appl. Environ. Microbiol.* **1997**, 63, 2306-2310.

[29] Kashket, E. R., Cao, Z.-Y., Clostridial strain degeneration, *FEMS Microbiol. Rev.* **1995**, 17, 307-315.

[30] Ennis, B. M., Marshall, C. T., Maddox, I. S., Paterson, A. H. J., Continuous product recovery by in-situ gas stripping/condensation during solvent production from whey permeate using *Clostridium acetobutylicum*, *Biotechnol. Lett.* **1986**, 8, 725-730.

[31] Maddox, I. S., The acetone-butanol-ethanol fermentation: Recent progress in technology, *Biotechnol. Gen. Eng. Rev.* **1989**, 7, 189-220.

[32] Maddox, I. S., Qureshi, N., Gutierrez, N. A., Utilization of whey by clostridia and process technology, in: *The Clostridia and Biotechnology* (D. R. Woods, Ed.), Butterworth-Heinemann, Stoneham, MA, **1993**, pp. 343-369.

[33] Gapes, J. R., The economics of acetone-butanol fermentation: theoretical and market considerations, *J. Mol. Microbiol. Biotechnol.* **2000**, 2, 27-32.

[34] Lenz, T. G., Moreira, A. R., Economic evaluation of the acetone-butanol fermentation, *Ind. Eng. Chem. Prod. Res. Dev.* **1980**, 19, 478-483.

[35] Gapes, J. R., The history of the acetone-butanol project in Austria, *J. Mol. Microbiol. Biotechnol.* **2000**, 2, 5-8.

[36] Lovitt, R. W., Kim, B. H., Shen, G. J., Zeikus, J. G., Solvent production by microorganisms, *CRC Crit. Rev. Biotechnol.* **1988**, 7, 107-186.

[37] Marlatt, J. A., Datta, R., Acetone-butanol fermentation process development and economic evaluation, *Biotechnol. Prog.* **1986**, 2, 23-28.

[38] Wheals, A. E., Basso, L. C., Alves, D. M. G., Amorim, H. V., Fuel ethanol after 25 years, *Trends Biotechnol.* **1999**, 17, 482-487.

[39] Fond, O., Petitdemange, E., Petitdemange, H., Engasser, J. M., Cellulose fermentation by a coculture of a mesophilic cellulolytic *Clostridium* and *Clostridium acetobutylicum*, *Biotechnol. Bioeng. Symp.* **1983**, 13, 217-224.

[40] E. K. C., Yu, Chan, M. K. H., Saddler, J. N., Butanol production from cellulosic substrates by sequential co-culture of *Clostridium thermocellum* and *C. acetobutylicum*, *Biotechnol. Lett.* **1985**, 7, 509-514.

[41] Marchal, R., Rebeller, M., Fayolle, F., Pourquie, J., Vandecasteele, J. P., Acetone butanol fermentation of hydrolysates obtained by enzymatic hydrolysis of agricultural lignocellulosic residues, in: *Energy from Biomass*, (W. Palz, J. Coombs, D. O. Hall, Eds.), Elsevier Applied Science, London, **1985**, pp. 292-696.

[42] Lloyd, A., French push ahead with biomass conversion, *New Scientist* **1984**, 1407, 21.

[43] Nativel, F., Pourquie, J., Ballerini, D., Vandecasteele, J. P., Renault, P., The biotechnology facilities at Soustons for biomass conversion, *Int. J. Solar Energy* **1992**, 11, 219-229.

[44] Desmarquest, J. P., Requillart, V., Modelling simultaneous ABE and ethanol production from various biomass raw materials, in: *Biomass for Energy, Industry and Environment* (G. Grassi, Ed.), Elsevier Applied Science, London, **1992**, pp. 1167-1171.

[45] Marchal, R., Ropars, M., Pourquie, J., Fayolle, F., Vandecasteele, J. P., Large-scale enzymatic hydrolysis of agricultural lignocellulosic biomass, *Bioresource Technol.* **1992**, 42, 205-217.

[46] Datta, R., Zeikus, J. G., Modulation of acetone-butanol-ethanol fermentation by carbon monoxide and organic acids, *Appl. Environ. Microbiol.* **1985**, 49, 522-529.

[47] Parrer, G., Schroll, G., Gapes, J. R., Lubitz, W., Schuster, K. C., Conversion of solvent evaporation residues from the AB- (Acetone-Butanol) bioprocess into bacterial cells accumulating thermoplastic polyesters, *J. Mol. Microbiol. Biotechnol.* **2000**, 2, 81-86.

[48] Girbal, L., Soucaille, P., Regulation of solvent production in *Clostridium acetobutylicum*, *Trends Biotechnol.* **1998**, 16, 11-16.

[49] Clarke, K. G., Hansford, G. S., Jones, D. T., The nature and significance of oscillatory behaviour during solvent production by *Clostridium acetobutylicum* in continuous culture, *Biotechnol. Bioeng.* **1988**, 31.

[50] Qureshi, N., Maddox, I. S., Continuous production of acetone-butanol-ethanol using immobilized cells of *Clostridium acetobutylicum* and integration with product removal by liquid-liquid extraction, *J. Ferment. Bioeng.* **1995**, 80, 185-189.

[51] Wheeler, M. C., Goodale, C. D., Continuous butyl alcohol fermentation process, *US Patent* 1,875,536, **1932**.

[52] McNeil, B., Kristiansen, B., The acetone butanol fermentation, in: *Advances in Applied Microbiology*, Vol. 31, (A. I. Laskin, Ed.), Academic Press, New York, **1986**, pp. 61-92.

[53] Dyr, J. Protiva, J., Praus, R., Formation of neutral solvents in continuous fermentation by means of *Clostridium acetobutylicum*, in: *Continuous Cultivation of Microorganisms. A Symposium* (I. Malek, Ed.), Czechoslovakian Academy of Sciences, Prague, **1958**, pp. 210-226.

[54] Yarovenko, V. L., Principles of the continuous alcohol and butanol-acetone fermentation processes, in: *Continuous Cultivation of Microorganisms. 2nd Symposium* (I. Malek, Ed.), Czechoslovakian Academy of Sciences, Prague, **1964**, pp. 205-217.

[55] Ierusalimsky, N. D., Use of continuous culture method for analysing cell functions, in: *Continuous Cultivation of Microorganisms* (I. Malek, Ed.), Czechoslovakian Academy of Sciences, Prague, **1964**, pp. 83-93.

[56] Hospodka, J., Production of acetone and butanol by continuous fermentation, in: *Theoretical and Methodological Basis of Continuous Culture of Microorganisms* (I. Malek, Z. Fencl, Eds.), Academic Press, New York, **1966**, pp. 611-613.

[57] Gapes, J. R., Nimcevic, D., Friedl, A., Long-term continuous cultivation of *Clostridium beijerinckii* in a two-stage chemostat with on-line solvent removal, *Appl. Environ. Microbiol.* **1996**, 62, 3210-3219.

[58] Nimcevic, D., Gapes, J. R., The acetone-butanol fermentation in pilot plant and pre-industrial scale, *J. Mol. Microbiol. Biotechnol.* **2000**, 2, 15-20.

[59] Qureshi, N., Blaschek, H. P., Production of acetone butanol ethanol (ABE) by a hyper-producing mutant of strain *Clostridium beijerinckii* BA101 and recovery by pervaporation, *Biotechnol. Prog.* **1999**, 15, 594-602.

[60] Volesky, B., Mulchandani, A., Williams, J., Biochemical production of industrial solvents (acetone-butanol-ethanol) from renewable resources, *Ann. N. Y. Acad. Sci.* **1982**, 369, 205-218.

[61] Gibbs, D. F., The rise and fall (....and rise?) of acetone/butanol fermentations, *Trends Biotechnol.* **1983**, 1, 12-15.

[62] Roffler, S., Blanch, H. W., Wilke, C. R., Extractive fermentation of acetone and butanol: process design and economic evaluation, *Biotechnol. Prog.* **1987**, 3, 131-140.

6
Clostridial Toxins Involved in Human Enteric and Histotoxic Infections

Bruce A. McClane and Julian I. Rood

The genus *Clostridium* includes several infamous human and veterinary pathogens that cause such dreaded diseases as tetanus, botulism, gas gangrene (clostridial myonecrosis), pseudomembranous colitis, and food poisoning. The defining hallmark of all pathogenic clostridia is their ability to produce potent protein toxins and degradative enzymes. This chapter focuses on five well-studied clostridial toxins that contribute to enterotoxic or histotoxic infections of humans. The clostridial neurotoxins responsible for botulism and tetanus are discussed separately in chapter 7 of this book. Information regarding clostridial toxins important for veterinary disease can be obtained from a recent review [1].

6.1
Clostridial enterotoxins

Many clostridial toxins are active on the gastrointestinal (GI) tract of humans or domestic animals (see Table 1). Due to page limitations, the first section of this chapter describes only *Clostridium perfringens* enterotoxin (CPE) and *Clostridium difficile* toxins A and B, which are the enterotoxins responsible for the most common human enteric diseases in industrialized countries. Several clostridial toxins listed in Table 1 also contribute to histotoxic infections of humans, so they will be discussed later in section 2 of this chapter. For other GI-active clostridial toxins listed in Table 1, but not discussed in this chapter, consult recent reviews [1, 2].

6.1.1
Clostridium perfringens enterotoxin (CPE)

C. perfringens isolates are commonly classified into one of five types (A-E) [3], based upon their ability to produce alpha, beta, epsilon, and iota toxins (Table 2). The toxin types shown in Table 2 simply reflect the presence (or absence) of plasmids carrying the beta, epsilon, or iota toxin genes [4]. Surveys have shown that about 2-5 % of the global *C. perfringens* population carries the gene (*cpe*) encoding another

Table 1. Clostridial toxins active on the GI tract

Clostridial	*Toxin species*	*Molecular action*	*Associated diseases*
C. botulinum	C2	ADP-Ribosylation of actin	?
C. difficile	Toxin A	Monoglucosyltransferase	Pseudomembranous colitis and antibiotic-associated diarrhea
	Toxin B	"	"
	CDT[a]	ADP-Ribosylation of actin	?
C. perfringens	α toxin	Phospholipase C	Gas gangrene, Animal enterotoxemias
	β toxin	Pore-former?	Necrotizing enteritis
	β-2 toxin	Cytotoxin (mechanism?)	Animal enterotoxemias
	I toxin	ADP-Ribosylates Actin	Animal enterotoxemias
	ε toxin	Cytotoxin (mechanism?)	Animal enterotoxemias
	Enterotoxin (CPE)	Pore-former? Affects Tight Junctions?	*C. perfringens* type A Food poisoning; Nonfoodborne human GI disease; animal enterotoxemias
C. septicum	α toxin	Pore-former	Gas gangrene, Neutropenic enterocolitis, Animal enterotoxemias
C. spiroforme	α toxin	ADP-Ribosylation of actin	Animal enterotoxemias

[a] CDT= *C. difficile* actin-specific, ADP-ribosylstransferase

Table 2. Toxin typing of *C. perfringens* isolates

***C. perfringens* type**	***Toxin(s) produced:* Alpha**	***Beta***	***Epsilon***	***Iota***
A	+	–	–	–
B	+	+	+	–
C	+	+	–	–
D	+	–	+	–
E	+	–	–	+

toxin of major biomedical importance, i. e., *C. perfringens* enterotoxin (CPE) [5]. The vast majority of these *cpe*-positive isolates belong to type A, explaining why virtually all cases of CPE-associated GI illnesses are caused by type A, CPE-producing *C. perfringens* strains [5, 6]. Studies conducted in the 1970s characterized CPE as a heat-labile, single polypeptide of ~35 kDa [3]. Nucleotide sequencing analyses of the *cpe* gene later revealed that the CPE protein consists of 319 amino acids [7]. Those sequencing analyses also indicated that CPE lacks primary sequence homology with other known proteins, except for some limited homology with Antp70/C1, a non-neurotoxic protein made by *C. botulinum* [5]. The significance (if any) of this limited homology is unclear.

The role of CPE in disease

Clostridium perfringens type A food poisoning: For over 30 years, CPE-producing strains of *C. perfringens* type A have been associated with *C. perfringens* type A food poisoning, which currently ranks as the third most commonly identified foodborne illness in the USA [5]. As recently reviewed [5], this foodborne illness is acquired when people consume a food (typically a beef or poultry product) contaminated with large numbers of vegetative cells of a *C. perfringens* type A strain carrying a chromosomal *cpe* gene. Once present in the intestines, the ingested bacteria initially multiply, before sporulating. It is during this *in vivo* sporulation that CPE is produced. The newly synthesized enterotoxin accumulates in the cytoplasm of the sporulating mother cell, until that cell lyses to release its mature endospore. CPE is then released into the intestinal lumen, where it binds to the intestinal epithelium, exerts its molecular action, and produces the histopathologic damage responsible for initiating intestinal fluid and electrolyte losses (discussion later). Most people suffering from *C. perfringens* type A food poisoning are sickened for 12-24 h with diarrhea and abdominal cramps, but then recover without lasting affects. However, this food poisoning can be fatal in elderly or debilitated individuals.

Considerable evidence now implicates CPE as the toxin responsible for most, if not all, symptoms of *C. perfringens* type A food poisoning. Epidemiologic studies established that CPE is present in the feces of virtually all *C. perfringens* type A food poisoning victims, often at levels known to produce GI effects in animal models [5]. Furthermore, human volunteer feeding experiments demonstrated that ingestion of highly purified CPE is sufficient to reproduce the characteristic diarrheal and cramping symptoms of *C. perfringens* type A food poisoning [8].

Studies fulfilling molecular Koch's postulates have recently confirmed the importance of CPE for the GI pathogenesis of *C. perfringens* type A food poisoning isolates [9]. Those studies showed (Figure 1) that lysates prepared from sporulating, but not vegetative, cultures of SM101 (an electroporatable derivative of *C. perfringens* food poisoning isolate NCTC8798 [10]) cause both histopathologic damage and fluid accumulation in rabbit ileal loops, which is consistent with CPE, whose expression is sporulation-associated (discussion later), causing or contributing to the GI pathology of *C. perfringens* type A food poisoning. More importantly, sporulating culture lysates prepared from MRS101, an isogenic SM101 mutant with its *cpe* gene specifically inactivated by allelic exchange, failed to cause either

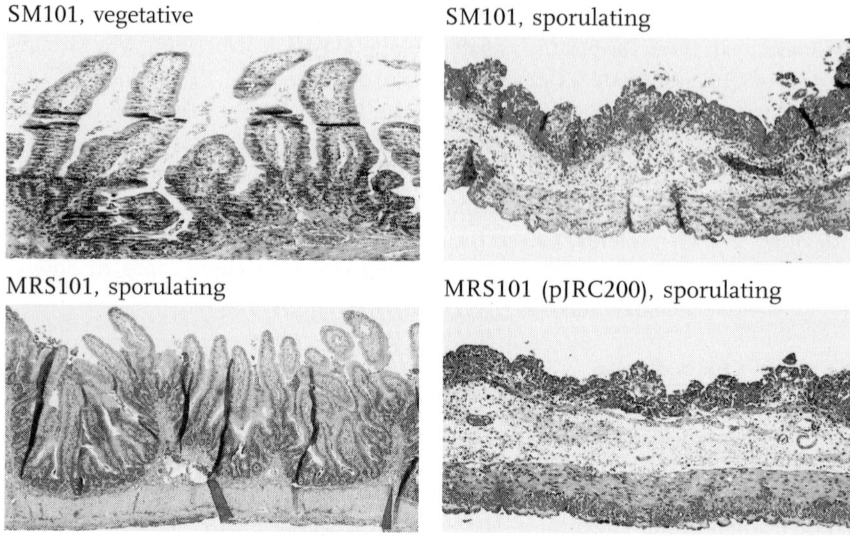

Figure 1. Histopathologic damage induced by sporulating culture lysates of *C. perfringens* isolate SM101 and derivatives. Tissue specimens shown were collected from rabbit ileal loops treated with either concentrated vegetative (FTG) or concentrated sporulating (DS) culture lysates of wild-type *C. perfringens* isolate SM101 (a CPE-positive, electroporatorable derivative of food poisoning isolate NCTC8798 [10]), the isogenic *cpe* knock-out mutant MRS101, or the *cpe*-complementing strain MRS101(pJRC200). Note that extensive epithelial desquamation is present in loop samples treated with DS lysates of SM101 or MRS101(pJRC200) but absent from loop samples treated with FTG lysates of SM101 or DS lysates of MRS101. Tissue specimens treated with concentrated FTG lysates of MRS101 or MRS101(pJRC200) also failed to show histopathologic damage (with permission from [9]).

histopathologic effects or fluid accumulation in rabbit ileal loops. However, full GI virulence was restored when the avirulent MRS101 *cpe* knock-out mutant was complemented with a shuttle plasmid carrying the cloned wild-type *cpe* gene.

CPE-associated nonfoodborne human GI diseases: During the past 15 years, it has become clear that CPE-producing *C. perfringens* type A strains also cause nonfoodborne human GI illnesses [11]. Perhaps 10-15 % of all cases of human antibiotic-associated diarrhea, and 5-20 % of all cases of sporadic diarrhea, involve CPE-producing, type A strains of *C. perfringens* [11]. Transmission of these CPE-associated nonfoodborne GI illnesses is not well understood, but may involve person-to-person transmission of cells or spores [11].

It has recently been established that the *cpe*-positive type A strains of *C. perfringens* causing nonfoodborne GI diseases differ from those causing food poisoning [6, 12-14]. Specifically, nonfoodborne GI disease isolates have been shown to carry their *cpe* gene on a plasmid, while the *cpe* gene has a chromosomal location in food poisoning isolates [12-14]. Recent data suggest at least two explanations for the strong association between each *cpe* genotype (plasmid vs. chromosomal) and a specific CPE-associated human GI disease (foodpoisoning vs. nonfoodborne GI disease). First, studies have now demonstrated that vegetative cells and spores of the chromosomal *cpe* isolates are more heat resistant than those of plasmid *cpe* isolates [15]. Since *C. perfringens* type A food poisoning almost always involves cooked foods, the greater heat resistance of cells/spores of chromosomal *cpe* isolates favors their survival in inadequately cooked or held foods. Second, other recent studies have shown that the *cpe* plasmid can be conjugatively transferred to *cpe*-negative isolates [16]. Conjugative transfer of the *cpe* gene could prove important for transmission of the CPE-associated nonfoodborne GI diseases which probably involve much lower infecting doses of vegetative cells/spores than does *C. perfringens* type A food poisoning. For example, it could be envisioned that *in vivo* conjugative transfer of the *cpe* plasmid from the relatively few infecting *cpe*-positive cells/spores to the many *C. perfringens* normal flora isolates present in the human GI tract converts those *cpe*-negative normal flora isolates to GI virulence, helping to establish the CPE-associated nonfoodborne GI diseases.

Epidemiologic studies first suggested that CPE is an important virulence factor in the pathogenesis of CPE-associated nonfoodborne GI illnesses [11]. For example, CPE has been detected at high levels in the feces of many individuals suffering from antibiotic-associated diarrhea or sporadic diarrhea [11]. Furthermore, recent studies have formally established that CPE expression is necessary for the GI virulence of F4969, a *C. perfringens* type A nonfoodborne human GI disease isolate [9]. Those studies demonstrated that lysates from sporulating, but not vegetative, cultures of F4969 induce fluid accumulation and histopathologic damage in rabbit ileal loops, which is consistent with sporulation-associated CPE expression being important for the GI pathogenesis of F4969. Next, it was shown that sporulating culture lysates of MRS101, an isogenic F4969 mutant with its *cpe* gene specifically inactivated by allelic exchange, failed to cause either histopathologic damage or ileal loop fluid accumulation. However, full GI virulence was restored when MRS4969 was complemented with a shuttle plasmid carrying the cloned wild-type *cpe* gene.

CPE genetics and synthesis

It was mentioned earlier that *C. perfringens* type A nonfoodborne human GI disease isolates carry their *cpe* gene on a plasmid and that the *cpe* plasmid is transferable at high frequency via conjugation, with that genetic exchange possibly contributing to establishment of the CPE-associated nonfoodborne GI diseases. The fact that the plasmid *cpe* gene is located on a highly mobile genetic element also becomes interesting given that the CPE-associated nonfoodborne GI diseases caused by plasmid *cpe* isolates usually involve more severe and longer lasting symptoms compared to *C. perfringens* type A food poisoning [11]. These clinical differences are not explainable by differences in CPE expression levels between plasmid vs. chromosomal *cpe* strains or by the plasmid *cpe* gene encoding a hypertoxic CPE variant [6]. However, it is possible that these symptomology differences could be explainable by high frequency, conjugative transfer of the *cpe* plasmid to naturally *cpe*-negative *C. perfringens* normal flora strains. Since normal flora strains of *C. perfringens* are presumably under selection for the ability to colonize and persist in the GI tract), their acquisition of the *cpe* plasmid might significantly prolong the presence of *cpe*-positive isolates in the GI tract, thereby worsening and increasing the duration of CPE-associated nonfoodborne GI disease symptoms.

Some evidence suggests that the chromosomal *cpe* gene may also be associated with a mobile genetic element (although actual movement of the chromosomal *cpe* gene has not yet been demonstrated). The *cpe* gene of food poisoning isolate NCTC8798 maps to a highly variable region of the *C. perfringens* chromosome, which is consistent with its presence on an integrated phage or transposon [17]. IS*1470* sequences have been detected upstream and downstream of the chromosomal *cpe* gene [18], suggesting that the chromosomal *cpe* gene may be present on a 6.3 kb transposon with terminal IS*1470* repeats (interestingly, different IS elements are associated with the plasmid *cpe* gene [14]). Recent studies suggest this putative 6.3 kb *cpe*-containing transposon, named Tn*5555,* may have a circular intermediate form [19].

The open reading frame (ORF) of the *cpe* gene is highly conserved in type A strains, whether that gene is present on a plasmid or on the chromosome [6]. Interestingly, *C. perfringens* type E isolates, which cause veterinary enterotoxemias, carry a plasmid containing both functional iota toxin genes (see Table 1) and silent *cpe* sequences [20]. It was determined by Billington et al. that the *cpe* sequences of type E isolates are not expressed due to the presence of mutations in their *cpe* promoter(s), ribosome binding site, and ORF sequences [20]. These silent *cpe* sequences of type E isolates are likely to have originated from an interspecies gene transfer event, whereby a mobile genetic element carrying functional iota toxin genes inserted into the promoter region of a plasmid *cpe* gene in a type A isolate. That insertional event probably eliminated *cpe* promoter function, favoring further accumulation of mutations in the, now silent, *cpe* sequences. Interestingly, these silent *cpe* sequences are even highly conserved among type E isolates lacking a clonal relationship [20], suggesting that the plasmid carrying both functional iota toxin genes and silent *cpe* sequences may have recently become mobilizable, with acquisition of that plasmid converting the recipient *C. perfringens* type A isolates to type E.

The linkage between CPE expression and sporulation was first identified by the pioneering studies of Duncan's group conducted in the 1960s-1970s [3, 21]. More recently, highly-sensitive Western blot analyses confirmed the strong association between CPE expression and sporulation by demonstrating that the food poisoning isolate NCTC8239 produces about 1500-fold more CPE during sporulation vs. vegetative growth [7]. CPE synthesis starts soon after sporulation is induced and then continues for at least the next 6-8 hours [22, 23].

The molecular basis for the sporulation-associated expression of CPE has come under investigation. RNA slot blot and Northern blot studies [22, 24] established that CPE expression is regulated at the transcriptional level, i. e., *cpe* mRNA is produced during sporulation, but not during vegetative growth. Northern blot results also strongly suggest that CPE is transcribed as a monocistronic message of ~1.2 kb [24], which is fully consistent with:

- primer extension analyses, RNase T2 protection assays, and deletion mutagenesis studies that identified three transcriptional start sites in the -200 to -50 region of the functional *cpe* gene of type A isolates [10, 22], and
- the presence of a stem-loop and oligo dT tract, with characteristics of a rho-independent transcriptional terminator, located 36 bp downstream of the *cpe* ORF termination codon [7].

Recombinant *Escherichia coli* transformants carrying even 50 copies/cell of pJRC200, a pJIR418 *C. perfringens-E. coli* shuttle plasmid with a cloned wild-type *cpe* gene insert, fail to express detectable levels of CPE [24]. However, sporulating (but not vegetative) cultures of naturally *cpe*-negative *C. perfringens* type A, B, and C strains do express CPE when transformed with pJRC200 [24]. Collectively, these observations suggest that CPE expression may involve positive sporulation-associated regulators expressed by most, or all, *C. perfringens* isolates. These putative positive regulators appear to function globally (i. e., they positively modulate expression of other genes besides *cpe*), since they are present in many naturally *cpe*-negative strains.

A possible insight into the identity of these putative positive regulators was provided by sequence analyses that detected homology between *cpe* transcriptional start sites and SigE-dependent and SigK-dependent promoters (SigE and SigK are sporulation-associated sigma factors active in mother cells of *Bacillus subtilis* during sporulation) [10]. This homology suggests that alternative sigma factor(s) represent at least one type of global regulator contributing to the positive regulation of CPE expression during sporulation.

Another global regulator(s) may repress CPE expression during vegetative growth. Hpr-(hyperprotease-producing) like binding sequences have been identified both upstream and downstream of the *cpe* gene [25]. Furthermore, DNA from several *C. perfringens* strains has been shown to hybridize a *hpr*-specific gene probe. Since Hpr is known to negatively regulate the expression of many proteins during exponential growth of *B. subtilis,* the presence of both Hpr-binding and *hpr*-encoding sequences in *C. perfringens* suggests that Hpr negatively regulates CPE expression during vegetative growth.

A final interesting aspect of CPE expression is the very large quantities of CPE produced by many *C. perfringens* cells during sporulation. CPE often represents 15 %, or more, of the total protein inside a sporulating *C. perfringens* cell [5]. This high-level CPE expression is not dependent on whether an isolate carries a plasmid or chromosomal *cpe* gene [6] and is not a gene dosage effect, since most or all *cpe*-positive isolates apparently carry only a single copy of the *cpe* gene. Some evidence suggests that unusual message stability may contribute to this abundant CPE expression, i. e., the functional half-life of the *cpe* message is reportedly ~58 min [26], which is exceptionally long for a prokaryotic message. Since stem-loop structures contribute to message stability, the putative stability of *cpe* mRNA might result, at least in part, from the presence of a stem-loop structure lying 36 bp downstream of the *cpe* ORF [7].

As mentioned earlier, CPE is not a secreted protein but, instead, accumulates in the cytoplasm of the sporulating mother cell, where it often reaches sufficiently high concentrations to form inclusion bodies [5]. CPE is eventually released into the intestinal lumen when the mother cell lyses at the completion of sporulation.

Intestinal effects of CPE

In animal models, CPE affects all segments of the small intestine, but is most active on the ileum [3]. In rabbit ileum, CPE can alter fluid and electrolyte transport and induce histopathologic damage within 15-30 min [3, 27]. Although CPE binds to rabbit colonic cells, it does not significantly affect fluid/electrolyte transport in the rabbit colon [3]; whether the human colon is affected by CPE remains unclear.

Two observations strongly suggest that CPE-induced histopathologic damage plays a major role in initiating small intestinal fluid/transport alterations. First, CPE-induced histopathologic damage develops concurrently with the onset of small intestinal fluid/electrolyte losses [27]. Second, only those CPE doses capable of inducing histopathologic damage will alter small intestinal fluid/electrolyte transport [28].

Recent studies have shown that, in polarized MDCK cells, a C-terminal CPE fragment slowly induces tight junction rearrangements that ultimately affect paracellular permeability [29]. Other studies using rat liver demonstrated that native CPE also induces tight junction changes [30], which apparently develop more quickly than those caused by C-terminal CPE fragments. If similar CPE-induced tight junction changes also occur in the small intestine, they might lead to paracellular permeability changes that could contribute to the diarrheal symptoms of CPE-associated GI diseases. However, such CPE-induced intestinal paracellular permeability changes (if they occur) are probably only a secondary contributor to the diarrhea of CPE-associated GI disease since both native CPE and C-terminal CPE fragments only induce tight junction rearrangements when applied to the basolateral surface of MDCK cells or rat liver. This limitation is very important given that CPE is initially released into the intestinal lumen, where it interacts with the apical surface of the intestinal epithelium. In this regard, note that CPE is not quickly internalized inside epithelial cells, which inhibits the toxin from gaining access to the basolateral membrane via transcytosis across the epithelium [31]. CPE induces release of proinflammatory cytokines [32, 33], which can promote inflammatory cell infiltration of CPE-treated intestinal tissue [3]. Therefore, inflammation could also be a secondary contributor to CPE-induced gastrointestinal effects.

CPE action: molecular events

The initial step in CPE action (Figure 2) involves the enterotoxin's binding to mammalian cells. CPE binding has the specificity and saturability expected of a receptor-mediated process [3, 34-37]. The specific binding of CPE to mammalian receptors is influenced by temperature and is necessary for the toxin's efficient cytotoxic action [3, 34-37].

It has been clear for twenty years that the CPE receptor(s) is proteinaceous [36]. The identity of CPE receptor protein(s) was first explored by affinity chromatogra-

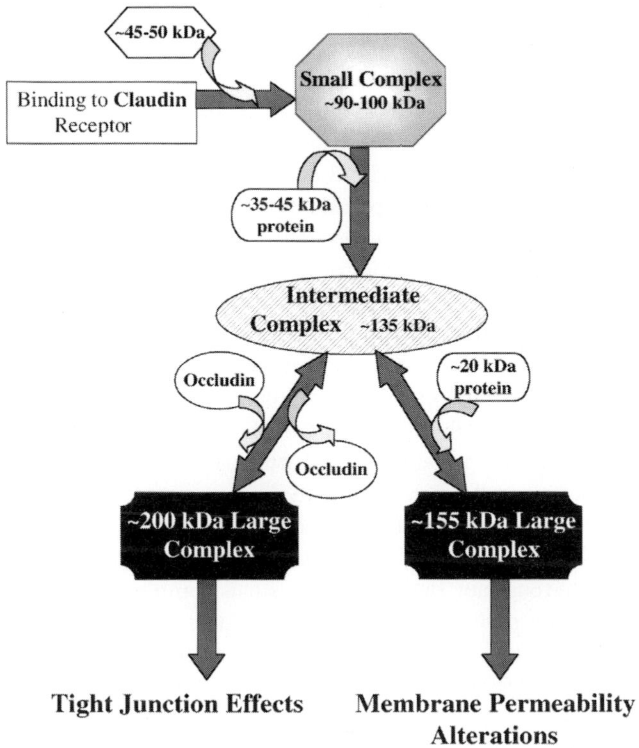

Figure 2. Current model for CPE's molecular action. This model depicts the four known early events in CPE action, including: (i) binding of CPE to a claudin, and/or a ~45 kDa membrane protein receptor, (ii) formation of a small complex that probably includes CPE, a claudin, and the ~45-50 kDa membrane protein; during formation of that small complex, CPE becomes trapped on the membrane surface, (iii) formation of CPE-containing intermediate and large complexes, resulting from interaction between small complex and other eukaryotic proteins, and (iv) consequences of large complex formation, which include cytotoxic membrane permeability alterations (mediated by the ~155 kDa large complex) and tight junction effects (possibly mediated by the ~200 kDa large complex containing occludin). Note: Steps i and ii (but not iii and iv) occur at 4 °C (with permission from [5]).

phy experiments using immobilized CPE, which identified an ~45-50 kDa CPE-binding protein in mammalian plasma membranes [38-40]. Later, native gel electrophoresis studies demonstrated that bound CPE quickly localizes in an ~90 kDa complex [41], now named small complex. Immunoprecipitation analyses showed that small complex contains an ~45-50 kDa protein [41], presumably the same ~45-50 kDa protein identified by affinity chromatography studies. Based upon these biochemical data, it was suggested that the CPE receptor is an ~45-50 kDa protein [41].

However, recent expression cloning studies [42, 43] demonstrated that transfection of human cDNAs encoding either of two related 22 kDa proteins into mouse L cells (which do not naturally bind CPE) converts those transfectants to CPE sensitivity. These two CPE receptors were later named claudins-3 and −4 when they were found to share homology with the multi-member claudin family of tight junction proteins [44]. Very recent studies indicate that the second extracellular loop of claudin-3 is important for CPE binding [45]. Those same studies also reported that some, but not all, other claudins besides claudin-3 and −4 can also bind CPE.

Given the apparent discrepancy between expression cloning vs. biochemical results regarding the identity of CPE receptor(s), several possibilities might explain formation of the ~90 kDa small complex (Figure 2). First, CPE might initially bind only to a claudin receptor, with that claudin-CPE complex then interacting with an ~45-50 kDa protein to form small complex. Second, CPE might initially bind to either a claudin receptor or an ~45-50 kDa protein receptor, with those CPE-receptor complexes then interacting with the other "receptor" protein to form the small complex. A final possibility is that CPE binds simultaneously to both a claudin and the ~45-50 kDa protein as co-receptors, with CPE binding to these co-receptors representing small complex formation.

CPE remains fully accessible to both external antibody and protease probes when present in small complex [46, 47]. indicating that the enterotoxin remains localized on the membrane surface throughout the process of small complex formation. However, despite its membrane surface location, CPE sequestered in small complex dissociates poorly from membranes [47], which suggests that one or more small complex protein(s) undergoes a conformational change during small complex formation, trapping CPE on the membrane surface.

Recent studies suggest that, under physiologic conditions, CPE-containing small complex may interact with additional membrane proteins to form an intermediate complex of ~135 kDa [48]. That ~135 kDa complex may then serve as a precursor for formation of two large (i. e., ~155 kDa and ~200 kDa) CPE-containing complexes (see Figures 2 and 3). At physiologic temperatures, the ~155 kDa large CPE complex is the most abundant CPE complex present in both Caco-2 and Vero cell membranes (see Figure 2).

The eukaryotic protein constituents of the intermediate-large CPE complexes are now being identified. The ~200 kDa complex (but no other CPE complexes) contains occludin [48], a major structural protein of tight junctions. Some evidence [40, 42] suggests that both a claudin and a ~45-50 kDa protein (presumably the same protein found in small complex) may also be present in one, or more, of the intermediate-large CPE complexes.

Formation of these intermediate-large CPE complexes appears to be critical for CPE-induced cytotoxicity. For example, although CPE binds and forms small complex at 4 °C, Vero cells treated with CPE at that low temperature fail to form intermediate-large complexes and they remain viable [48a]. However, if those cells containing enterotoxin bound at 4 °C are then shifted to 37 °C, intermediate-large complex formation closely coincides with onset of the small molecule membrane-permeability alterations that are responsible for CPE-induced cytotoxicity (discussion below). The importance of the intermediate-large CPE complexes for cytotoxicity receives further support from studies using CPE deletion and

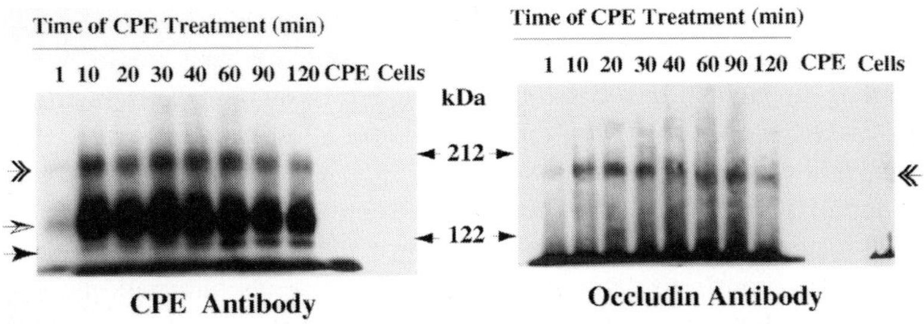

Figure 3. Western Immunoblot Analysis Demonstrates the Presence of Multiple CPE-Containing Intermediate-Large Complexes in CaCo-2 cells Treated with CPE at 37 °C. After CPE treatment for the indicated times (in minutes), CaCo-2 cells were washed to remove unbound toxin and lysed with SDS. The resultant lysates were analyzed by SDS-PAGE (no sample boiling) using 4% acrylamide gels, and those gels were Western blotted with either CPE antibodies or occludin antibodies, as indicated. Migration of myosin (212 kDa) and β galactosidase (122 kDa) markers are indicated. The double, open, and closed arrows indicate the migration of the ~200 kDa large complex, ~155 kDa large complex, and ~135 kDa large complex, respectively (with permission from [48]).

point mutants [49, 50], which established a tight correlation between a CPE mutant's ability to form intermediate-large CPE complexes and its cytotoxic properties.

Recent studies (Singh and McClane, unpublished observations) strongly suggest that formation of the ~155 kDa, large CPE complex is sufficient for cell killing. Those studies showed that treatment of Transwell cultures of CPE-sensitive Caco-2 cells with the enterotoxin at 37 °C causes development of cytotoxic membrane permeability alterations (see below) and formation of the ~155 kDa large complex, but not appreciable formation of the ~200 kDa large complex.

CPE kills mammalian cells by increasing the permeability of their plasma membranes to small molecules [31]. Those CPE-induced permeability alterations collapse the cellular colloid-osmotic equilibrium, causing death of the mammalian cell from either cell lysis or metabolic disturbances [31]. In the small intestine, CPE-induced enterocyte death results in histopathologic damage, which (as mentioned in earlier) initiates the ion/electrolyte transport alterations that manifest clinically as diarrhea. The mechanism by which the ~155 kDa CPE complex induces small molecule plasma membrane permeability changes is not yet completely clear, but a recent study indicates that, when localized in large complexes, CPE becomes closely associated with eukaryotic plasma membranes [47]. That observation suggests the ~155 kDa complex may insert into membranes as a pore-like structure, a possibility receiving support from electrophysiology studies that detected the presence of pores in the apical membranes of CPE-treated CaCo-2 cells [51].

As mentioned earlier, the ~200 kDa CPE complex contains occludin (a major structural protein of tight junctions), and may also contain claudins, which are also major structural proteins of tight junctions [42-44]. Therefore, formation of the ~200 kDa complex might explain the faster ability of CPE (which forms the ~200 kDa complex) vs. C-terminal CPE fragments (which lack the N-terminal sequences necessary for forming the ~200 kDa complex) to induce tight junction rearrangements in rat liver [30]. As indicated earlier, tight junction changes leading to paracellular permeability alterations in the CPE-treated intestine might contribute to diarrhea during CPE-associated GI disease, but those tight junction effects are unlikely to be the initiating factor for CPE-induced intestinal effects since CPE only induces tight junction rearrangements when applied to the basolateral surface of rat liver and MDCK cells [29, 30]. Therefore, the enterotoxin probably must first depend upon histopathologic damage (mediated by CPE's cytotoxic activity on apical membranes) to provide access to the basolateral surface of epithelial cells so it can cause tight junction rearrangements.

CPE structure/function relationships

Since the 3-D structure of the CPE protein has not yet been solved, CPE structure/function relationships have been examined primarily by genetic approaches. Results from those studies (depicted in Figure 4) indicate that receptor binding activity maps to the extreme C-terminus of the CPE protein. For example, both a recombinant CPE fragment and a synthetic peptide corresponding to the thirty C-terminal amino acids of CPE exhibit similar receptor binding properties as native CPE [52]. Furthermore, removing the last five amino acids from the C-terminus of CPE has been shown to abrogate all receptor binding activity [50].

The extreme N-terminal sequences of CPE have no known functional role (this region does not contain signal sequences, since the enterotoxin is not secreted). In fact, cytotoxic activity actually increases when up to the first 45 N-terminal amino acids are removed from native CPE [50]. Trypsin and chymotrypsin can proteolytically remove extreme N-terminal CPE sequences and can produce an activated CPE [53-55]. The presence of those two proteases in the intestinal lumen suggests that CPE may be proteolytically activated during GI disease.

Removal of extreme N-terminal sequences appears to activate CPE's cytotoxic activity by specifically enhancing the ability of the toxin to form intermediate-large CPE complexes [49]. That result suggests amino acids located in the N-terminal half of the CPE protein mediate formation of the intermediate-large complexes, which is fully consistent with the observed inability of receptor binding, but

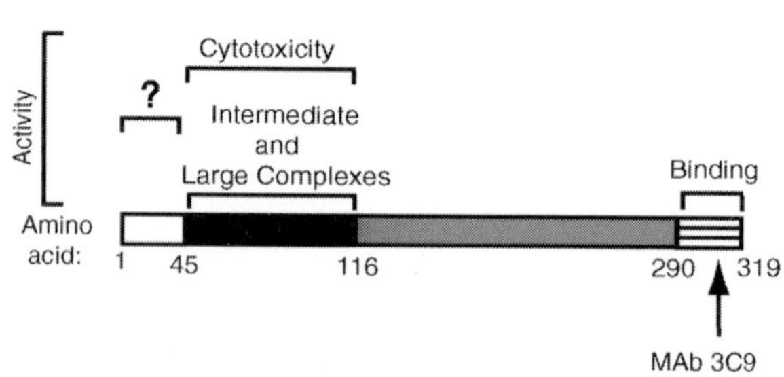

Figure 4. Mapping of CPE functional regions. CPE regions involved in the formation of large complexes, receptor binding, and cytotoxicity are shown on this map. The Figure also shows the location of the neutralizing linear epitope encoding MAb 3C9.

non-cytotoxic, C-terminal CPE fragments to form intermediate-large CPE complexes. [50]. The presence of an intermediate-large complex forming region between amino acids 45 and 116 of native CPE has recently been confirmed by studies using CPE point mutants [49].

CPE vaccines

The linear epitope recognized by MAb 3C9, a neutralizing monoclonal antibody that blocks CPE binding to receptors, maps to the extreme C-terminus of the enterotoxin [55]. The presence of a neutralizing linear epitope in non-cytotoxic C-terminal CPE fragments suggested those fragments could be potential CPE vaccine candidates. That hypothesis receives support from a pilot study that prepared a conjugate consisting of the thirty C-terminal CPE amino acids coupled to a thyroglobulin carrier [56]. When mice were administered that conjugate i. p., they developed a serum IgG antibody response capable of neutralizing the cytotoxic activity of native CPE against Vero cells. The possibility of obtaining a CPE-neutralizing IgA response in the intestines has not yet been explored.

6.1.2 *Clostridium difficile* toxins A and B

C. difficile toxins A and B are extremely large (308,000 M_r and 269,000 M_r, respectively), single polypeptides. They share extensive homology with each other, as well as with several toxins produced by *C. sordelli* and *C. noyvi* (Table 3). Collectively, this homologous toxin family is now referred to as the "large clostridial toxins".

The role of toxins A and B in *C. difficile*-induced gastrointestinal disease

Somewhat surprisingly, *C. difficile* only became recognized as a major human enteric pathogen in the late 1970s [57-59]. However, it is now appreciated that this bacterium is the single most common cause of nosocomial diarrhea in industrialized countries [57-59], causing virtually all cases of pseudomembranous colitis and ~25 % of antibiotic-associated diarrhea cases.

C. difficile pathogenesis begins with the ingestion of spores, which are often present throughout the hospital environment and can be transmitted from patient-to-patient or from health care worker-to-patient [57-59]. Following their ingestion, *C. difficile* spores germinate in the lower GI tract, giving rise to vegetative cells that compete poorly against normal colonic flora. Consequently, *C. difficile* vegetative cells only colonize and proliferate in people whose normal colonic flora has been suppressed by antibiotic therapy or chemotherapy [57-59].

Pseudomembranous colitis is the most serious form of *C. difficile* enteric disease, involving fever, abdominal cramps, diarrhea, and mucosal damage that can lead to colonic perforation or toxic megacolon [57-59]. Without therapy, this illness is potentially life-threatening. Fortunately, pseudomembranous colitis usually responds to vancomycin or metronidazole treatment. However, relapses are common, since antimicrobial agents often fail to completely eradicate *C. difficile* (particularly its spores) from the colon and/or institutionalized patients can reacquire *C. difficile*

Table 3. The large clostridial cytotoxins

Clostridial species proteins	*Toxin*	*M_r (kDa)*	*% Homology with toxin B*	*Co-substrate used*	*Target*
C. difficile:	A	308	48	UDP-Glc[a]	Rho (Thr 37), Rac, Cdc 42 (Thr 35)
	B	270	100	UDP-Glc	Rho (Thr 37) Rac, Cdc 42 (Thr 35)
	B-1470	270	93	UDP-Glc	Rac, Cdc 42 (Thr 35) Ras proteins
C. sordelli:	HT	300	ND[c]	UDP-Glc	Rho (Thr 37) Rac, Cdc 42 (Thr 35)
		LT 270	76	UDP-Glc	Rac, Cdc 42 (Thr 35) Ras Proteins
C. noyvi:	α	250	32	UDP-GlcNAc[b]	Rho (Thr 37) Rac, Cdc 42 (Thr 35)

[a] UDP-Glu= uridine diphosphate-glucose.
[b] UDP-GlcNAc= uridine diphosphate-N-acetyl-glucosamine.
[c] ND, not determined (the HT gene has not yet been cloned and sequenced).
Modified from [74].

spores from their hospital environment. *C. difficile* also causes antibiotic-associated diarrhea, a milder but much more common disease than full-blown pseudomembranous colitis.

Because genetic manipulations of *C. difficile* remain very difficult, molecular Koch's postulate analyses have not yet formally confirmed the importance of toxin A and B for the GI virulence of *C. difficile*. However, a wealth of other evidence directly implicates toxins A and B in the pathogenesis of *C. difficile*-induced enteric disease [57-59]. For example, *C. difficile* strains that produce neither toxin A nor toxin B are avirulent in animal models. Furthermore, immunization of hamsters against toxins A and B provides complete protection against challenge with virulent strains of *C. difficile*.

Several observations from animal model studies [57-60] suggest that toxin A plays a more important role than toxin B in *C. difficile* GI disease. First, the amount of toxin A produced by a *C. difficile* strain closely correlates with that strain's virulence. Second, in the absence of toxin A, toxin B is unable to cause GI effects. Finally, serogroup F strains of *C. difficile*, which produce toxin B but not toxin A, do not cause either diarrhea or lethality. However, toxin A and toxin B probably act synergistically during most human infections [57-59]. Toxin A is likely to initiate diarrhea by damaging the intestinal mucosa (discussion below); that toxin A-induced damage may then provide toxin B access to (otherwise masked) toxin B-sensitive cells.

Besides its cytotoxic properties, toxin A also is a potent chemotactic agent that induces substantial leukocyte infiltration into the colonic mucosa. Those leukocytes release inflammatory mediators produce an intense inflammation, which is now considered a major contributor to the extensive tissue destruction that occurs during fulminant pseudomembranous colitis.

Genetics and Synthesis of *C. difficile* toxins A and B

The genes (*toxA* and *toxB*) encoding toxin A and toxin B are both associated with a 19.6 kb pathogenicity islet (see Figure 5) referred to as "PaLoc" (short for pathogenicity locus) [61, 62]. PaLoc is located on the *C. difficile* chromosome and contains a total of five genes, designated *txeR, toxB, txe2, toxA,* and *txe3* (also known as *tcdD, tcdB, tcdE, tcdA,* and *tcdC*). Four PaLoc genes (including the *toxA* and *toxB* genes) are transcribed in the same direction, while *txe3* (*tcdC*) is transcribed in the opposite direction. The PaLoc sequence is highly conserved among both strong and weak toxigenic strains of *C. difficile* [63]. In non-toxigenic strains of *C. difficile,* the PaLoc chromosomal site is occupied by a small (~120 bp) element [61, 62].

PaLoc may correspond to an inserted transposable element. For example, sequencing studies have shown that the PaLoc borders are highly conserved among different toxigenic *C. difficile* strains [61], which is consistent with PaLoc representing a defined genetic element that has inserted into the *C. difficile* chromosome. However, the genetic organization of the PaLoc region does not resemble a classical pathogenicity island or a transposable element. For example, no insertion sequences, site-specific recombinase genes, or transposase genes appear to be present.

Interestingly, a recent study identified a 1975 bp DNA insertion, named Cd/St1, within the *tcdA* gene of the clinical *C. difficile* isolate C34 [64]. Cd/St1 appears to be a chimeric ribozyme with features of both group 1 introns and insertion elements.

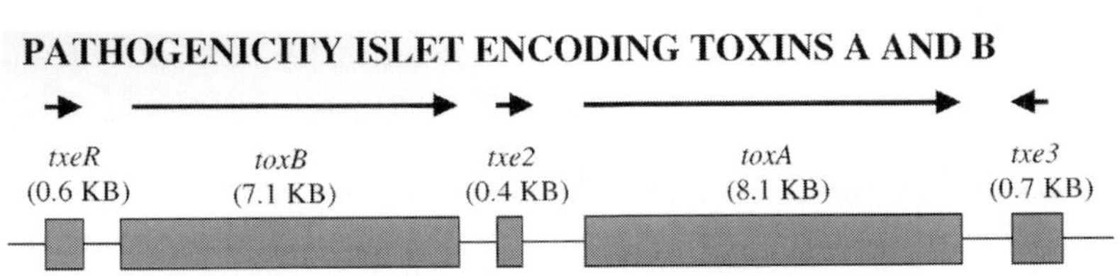

Figure 5. The 19.6kb "PaLoc" pathogenicity islet of toxigenic *C. difficile* strains. The genes encoding toxin A and toxin B are *toxA* and *toxB*, respectively. The *txeR*, *toxB*, *txe2*, *toxA*, and *txe3* genes are also known as *tcdD*, *tcdB*, *tcdE*, *tcdA*, and *tcdC*, respectively (modified from [59], with permission).

Cd/ST1 is spliced from primary *tcdA* transcripts, so its presence does not affect toxin A expression by strain C34. When over 200 *C. difficile* isolates were screened, Cd/ST1-related sequences were universally detected in both toxigenic and non-toxigenic isolates; however, no isolate other than C34 was found to carry a Cd/ST1 insertion in its *tcdA* gene.

About 80 % of *C. difficile* strains are toxigenic [59]; those toxigenic strains usually produce both toxin A and toxin B. However, strains with a toxin A-/toxin B+ phenotype have been described [65, 66]. Recent studies indicate that the toxin A-/toxin B+ phenotype results from production of a truncated (and nontoxic) toxin A fragment [67-70]. These toxin A truncations can result from either the introduction of a nonsense mutation into the 5′ half of *toxA*, or from the deletion of 3′ *toxA* sequences [67-70].

Toxin A+/toxin B+ strains of *C. difficile* vary by as much as 5 orders of magnitude in their expression of toxins A and B [59]. Expression of toxins A and toxin B is glucose-repressible and occurs during stationary, but not exponential, growth [71]. It remains unclear whether this stationary phase expression of toxins A and B is linked to the onset of sporulation.

The expression of toxins A and B during stationary phase apparently involves transcriptional regulation, since toxin mRNA levels closely correlate with toxin A and B production [71]. During stationary phase, toxigenic strains of *C. difficile* typically produce more *toxA* transcript than *toxB* transcript, helping to explain why toxigenic strains usually produce higher levels of toxin A vs. toxin B [59, 63]. Interestingly, *txeR*, *toxR*, *txe2*, and *toxA* transcripts are produced during stationary phase, but the other PaLoc gene, *txe3*, is transcribed only during exponential growth [72]. That (and other) observation suggests that *txe3* is a repressor that negatively regulates expression of toxins A and B during exponential growth of *C. difficile*. Some evidence suggests that *txeR*, which has some homology with a stress response sigma factor, may be a positive regulator of toxin A and B expression during stationary phase [73].

Molecular analyses indicate that the *toxA* and *toxB* genes each have their own promoters, but can also be transcribed from promoters of upstream genes [63, 71, 72]. However, it appears that most *toxA*/*toxB* transcripts are monocistronic and directly produced from the *toxA and toxB* promoters [71]. Some recent evidence suggests that multiple promoters may lie immediately upstream of the *toxB* gene [73a].

After synthesis, toxins A and B accumulate inside a *C. difficile* vegetative cell until that cell undergoes autolysis [59], i. e., toxins A and B, like *C. perfringens* enterotoxin, are not secreted.

Molecular and intestinal action of *C. difficile* toxins A and B

The binding of toxins A and B to cells is the initial, but perhaps least understood, step in the action of these two toxins. Competitive binding studies indicate that toxins A and B do not share the same receptor(s) [74]. Binding of toxin A appears to be specific and stronger at 4 °C vs. 37 °C [75]. A protein of ~160 kDa has been suggested as a/the toxin A receptor [76]. Pre-treating mammalian cells with α- or β-galactosidases was shown to decrease toxin A binding, suggesting that galactose moieties are present on the toxin A receptor [77]. Since Lewis X, Y, and I antigens contain galactose and are present on intestinal epithelial cells, it was once proposed that those proteins might be the toxin A receptor(s); however, data arguing against that hypothesis have now been reported [78].

After binding, toxins A and B are internalized by receptor-mediated endocytosis [75, 79]. This entry process is thought to involve endosomal acidification which probably induces conformational changes that facilitate entrance of part or all of the *C. difficile* toxins into the cytoplasm [79]. Some evidence suggests toxins A and B may also enter (although less efficiently) host cells by generalized pinocytosis [59, 74].

Elegant biochemical studies have recently established that, once present in the host cell cytoplasm, toxins A and B exert the same enzymatic action [80-82]. Both *C. difficile* toxins are monoglucosyl-transferases that transfer the glucose moiety from UDP-glucose onto several members of the Rho family, including Rho, Rac, and Cdc 42 (see Table 2). This transferred glucose is added onto residue Thr 37 of Rho, or onto residue Thr 35 of Rac or Cdc 42.

Rho proteins are small GTP-binding proteins that function as molecular switches for important cellular processes, such as cytotoskeletal assembly and signal transduction pathways [83]. The glucosylation of Rho family proteins by toxins A and B inactivates these mammalian targets by decreasing their ability to interact with downstream effectors. That disruption causes a cascade of consequences, including a cytoskeletal collapse that results in morphologic alterations (cell rounding), as well as alterations in signal transduction pathways that trigger such cytotoxic processes as apoptosis [75, 80-83]. It has also been shown that toxin A- and B-altered signal transduction pathways affect tight junctions [84, 85] which could increase paracellular permeability in the colonic mucosa and contribute to diarrhea [80].

Interestingly, the hemorrhagic toxin (HT) and lethal toxin (LT) of *C. sordelli* (a wound pathogen) are also monoglucosyltransferases that use UDP-glucose as a cosubstrate [[75, 80-82] and Table 3). However, while HT shares the same Rho family targets as *C. difficile* toxins A and B, LT has a somewhat different target protein profile (Table 3), i.e., LT monoglucosylates Rac, Cdc 42, and Ras (instead of Rho protein itself). A variant toxin B produced by *C. difficile* strain B-1470 has recently been shown [86] to share the same Rho family protein targets as LT. A final member of the large clostridial cytotoxin family, the α toxin of *C. noyvi* (another wound pathogen), targets the same Rho family proteins as *C. difficile* toxins A and B, but uses UDP-N-acetyl-glucosamine (instead of UDP-glucose) as a cosubstrate [87].

The recently discovered similarities in the action and sequence homology of the large clostridial cytotoxins raise fascinating, but still largely unexplored, questions: Have large clostridial toxin genes been transferred between different clostridial species? If so, using what genetic mechanism? What is the evolutionary relationship between these toxins? Furthermore, additional studies are also needed to better connect the enzymatic actions of toxin A and B to the intestinal pathophysiology observed during *C. difficile* GI infections.

Structure/function relationships of C. *difficile* toxins A and B

Consistent with the shared primary sequence homology of toxins A and B, the structure/function relationships of these two toxins are also similar, as depicted in Figure 6. The C-terminal region of both toxins contains a series of repeating units, called CROPS (for clostridial repetitive oligopeptides). CROPS share homology with regions of glucosyltransferases produced by *Streptococcus mutants* and *Streptococcus sobrinus* [74, 75, 80, 83]. Experimental evidence suggests that CROPS mediate binding of toxin A to galactose moieties on cell receptors. For example, monoclonal antibodies directed against epitopes located in the CROPS region of toxin A can interfere with receptor binding of that toxin, thereby blocking enterotoxigenic effects [75, 88]. Furthermore, isolated C-terminal fragments of toxin A can block binding of native toxin A to mammalian cells [78]. Presumably, the CROPS of toxin B also mediate binding of that toxin to its receptor, but this process has not been well studied.

Toxins A and B also contain a central hydrophobic region thought to be involved in toxin uptake and/or intracellular processing [59, 89]. In addition, both *C. difficile* toxins A and B contain four conserved cysteine residues, which also apparently contribute to toxicity by facilitating endocytosis or translocation [89, 90].

The monoglucosyltransferase activity of toxins A and B has recently been localized [90-92] to their N-terminus (i. e. to amino acids 1-546 for toxin B, or 1-659 for toxin A). For example, the presence of a toxin $A_{1\text{-}546}$ fragment in the cytoplasm of a mammalian cell is sufficient to produce both monoglucosyltransferase activity and cytotoxic effects. Additionally, the toxin B region containing amino acids ~360-520 has now been shown to recognize specific Rho protein substrates [93]. Meanwhile, a DXD motif located between amino acids 285 and 290 of toxins A and B binds Mn^{2+} and appears to be necessary for those toxins to bind UDP-glucose [94].

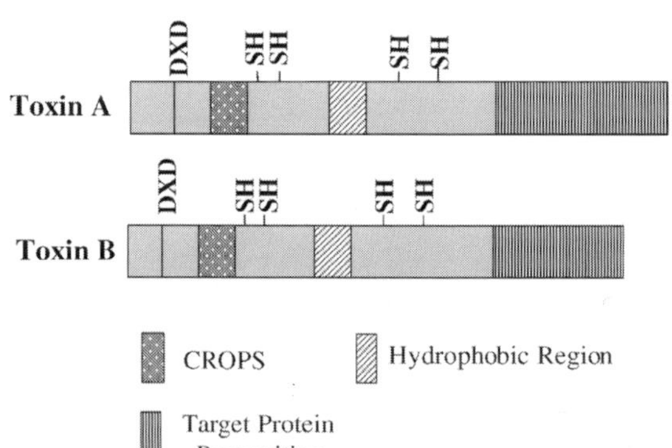

Figure 6. Map of structure/function relationships for *C. difficile* toxins A and B. Shown are: the C-terminal CROPS region that has putative receptor binding activity and contains repeating sequences; the central hydrophobic region and four conserved cysteine residues (marked SH) thought to be important for translocation/internalization; the N-terminal region responsible for recognizing specific Rho protein targets; and the DXD site which binds the Mn^{+2} ion required for UDP-glucose binding.

Toxin A and B vaccines

The development of vaccines to protect against *C. difficile* toxins A and/or B is under active investigation. Initial progress towards such vaccines involved mouse experiments demonstrating that intranasal administration of a nontoxic C-terminal fragment of toxin A elicits local and systemic antibodies capable of neutralizing native toxin A [95]. This approach was taken further by constructing attenuated *Salmonella* carriers expressing a C-terminal toxin A fragment; when those recombinant vaccine strains were administered intranasally to mice, significant levels of mucosal and serum antibodies capable of neutralizing native toxin A were detected [95a].

6.2 Clostridial toxins involved in histotoxic infections

Gas gangrene or clostridial myonecrosis is an invasive, histotoxic infection. This disease primarily results from the traumatic contamination of healthy tissues by intestinal or soil bacteria and is facilitated by extensive wounds that damage the blood supply to the tissues, thereby providing the anaerobic conditions required for clostridial growth. Gas gangrene is characterized by extensive tissue invasion and necrosis; unless surgical intervention occurs, this disease rapidly progresses to systemic shock and death. The microorganism responsible for most human cases of gas gangrene is *C. perfringens* type A, although other clostridia such as *C. septicum, C. novyi* and *C. histolyticum* cause up to 20 % of all cases [96]. There is also a nontraumatic form of this disease, which most commonly occurs as a consequence of GI lesions. *C. septicum* is the most frequent cause of that nontraumatic gas gangrene [96].

C. perfringens produces many extracellular toxins and enzymes (Table 1), several of which have been proposed to contribute to gas gangrene. However, only the α toxin and, to a lesser extent, perfringolysin O (also known as θ toxin) have been conclusively shown to play a role in the disease process [97]. The major toxin implicated in the *C. septicum*-mediated disease is also called α toxin [98]. Note that the α toxins from *C. perfringens* and *C. septicum* are distinct toxins that have no sequence similarity and have very different modes of action. In this chapter only the α toxin and perfringolysin O from *C. perfringens* and the α toxin from *C. septicum* will be discussed in detail. Other clostridial toxins and extracellular toxins of possible relevance to histotoxic infections are listed in Table 4.

6.2.1 The α toxin from *C. perfringens*

The role of *C. perfringens* α toxin in disease

There is clear immunological and genetic evidence that the ability to produce α toxin is an essential component of the pathogenesis of *C. perfringens*-mediated gas gangrene. The *C. perfringens* α toxin has both phospholipase C and sphingomyelinase activity, is hemolytic and lethal [99]. The N-terminal domain (aa 1-249) has the phospholipase C active site, but is not immunoprotective. However, immu-

Table 4. Toxins and extracellular enzymes produced by histotoxic clostridia

Species and toxin	*Mode of action*	*References*
C. perfringens		
α toxin	Phospholipase C and sphingomyelinase	[99]
β toxin	Lethal pore-forming toxin	[166,167]
β2 toxin	Cytotoxin	[168, 169]
ε toxin	Lethal cytotoxin	[170]
perfringolysin O (θ toxin)	Pore-forming cytolysin	[138]
sialidase	Both the large (NanH) and small (NanI) enzymes are sialic acid-specific glycohydrolases	[171]
hyaluronidase (μ toxin)	N-acetyl glucosaminidase (NagH)	[172]
collagenase (κ toxin)	Zinc-metalloprotease (ColA)	[173, 174]
λ toxin	Zinc-metalloprotease	[175]
ι toxin	Binary toxin, actin-specific ADP-ribosyl-transferase	[176, 177]
C. septicum		
α toxin	Pore-forming cytotoxin	[158, 164]
sialidase	Sialic acid-specific glycohydrolase	[171]
C. histolyticum		
collagenase	Zinc-metalloproteases (ColG and ColH)	[178]
C. novyi		
α toxin	N-acetyl-glucosaminyltransferase	[87]
phospholipase C	Phospholipase C	[179]
C. sordellii		
lethal toxin (LT)	Monoglucosyltransferase	[180, 181]
hemorrhagic toxin (HT)	Monoglucosyltransferase	[180]
sialidase	Sialic acid-specific glycohydrolase	[171]

nization with the nontoxic C-terminal domain (aa 249-370) of the mature α toxin protein will protect mice injected with a normally lethal *C. perfringens* dose [100]. In addition, insertional inactivation of the α toxin structural gene, *plc*, has been carried out by homologous recombination. Virulence testing of the resultant mutants in a mouse myonecrosis model showed that, unlike the isogenic wild-type control, the *plc* mutants were unable to cause disease to any significant extent [101]. Mice infected with the wild-type strain exhibited considerable muscle swelling, extensive muscle necrosis, and a 30% survival rate after 18 h. In contrast, mice infected with the *plc* mutants appeared healthy, had minimal muscle destruction and necrosis, and exhibited a 100% survival rate. These effects on virulence were reversed when the wild-type *plc* gene was introduced into the *plc* mutant on a multicopy plasmid, fulfilling molecular Koch's postulates and providing genetic proof for the essential role of α toxin in the disease.

One of the classical features of *C. perfringens*-mediated gas gangrene is the absence of polymorphonuclear leukocytes (PMNLs) at the site of infection, which is exactly the opposite of what is observed in most bacterial infections]102]. Comparative analyses of the wild-type and *plc* mutants in the mouse model revealed that animals infected with the *plc* mutants show an inflammatory response with a significant PMNL infiltration (Figure 7) compared to that of mice infected with the wild-type, which exhibited the typical histopathology of myonecrosis and an attenuated inflammatory response [102]. The effect of the *plc* mutation on the resultant inflammatory response was reversed when the mutation was complemented by the introduction of a wild-type *plc* gene. Using these complemented strains, little or no PMNL influx to the lesion was once more observed. These results provide good genetic evidence to support previous observations made using purified α toxin [103].

It was suggested that α toxin, and perfringolysin O, directly inhibit PMNL influx into the active lesion by increasing the adhesion of PMNLs to the vascular epithelium [102]. These conclusions were supported by a subsequent series of experiments in which leukocyte accumulation within the blood vessels of the infected lesions was determined quantitatively [104]. The percentage of vessels with severe leukocyte accumulation dropped from 8.6% in mice infected with the wild-type strain to 3.0% with the α toxin mutant and 0.6% with the perfringolysin O mutant. These two values were significantly different ($P < 0.0001$). Significantly reduced thrombosis (0.4% c.f. 14.2% thrombotic vessels) was also observed on histopathological examination of mice infected with the *plc* mutant compared to the

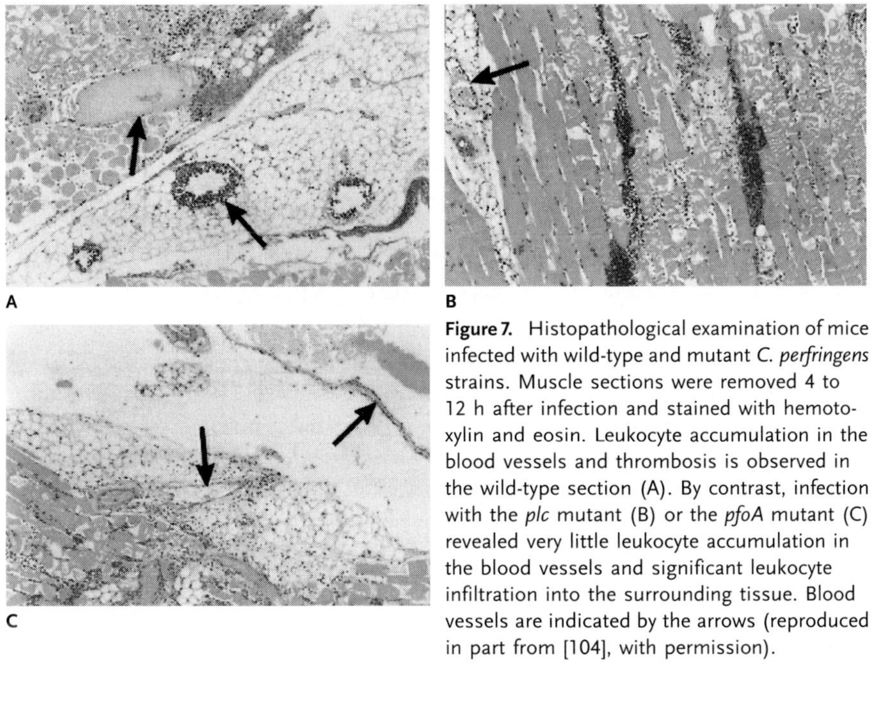

Figure 7. Histopathological examination of mice infected with wild-type and mutant *C. perfringens* strains. Muscle sections were removed 4 to 12 h after infection and stained with hemotoxylin and eosin. Leukocyte accumulation in the blood vessels and thrombosis is observed in the wild-type section (A). By contrast, infection with the *plc* mutant (B) or the *pfoA* mutant (C) revealed very little leukocyte accumulation in the blood vessels and significant leukocyte infiltration into the surrounding tissue. Blood vessels are indicated by the arrows (reproduced in part from [104], with permission).

wild-type strain. Therefore, α toxin not only appears to play an important role in reducing the ability of the host's inflammatory response to control the infection, but is also responsible for reducing the flow of blood to the infected tissues, thereby helping to enhance the anaerobic environment that is essential for the growth of *C. perfringens* cells in the lesion [104].

Genetics and synthesis of *C. perfringens* α toxin

The *plc* gene is located on the chromosome in a region within close proximity to the origin of replication [17, 105]. It is expressed from a single promoter that is regulated at the transcriptional level by a temperature-dependent mechanism involving three phased poly(dA)$_{5\text{-}6}$ tracts located immediately upstream of the σ^{70}-like *plc* promoter. At low temperatures, DNA fragments containing this region exhibit anomalous electrophoretic behavior, suggesting that the A-tracts impart DNA bending on the *plc* promoter [106]. Since progressive deletion of these tracts, or insertion of 5 to 21 nucleotides between the A-tracts and the −35 box, leads to decreased transcription from the *plc* promoter, there is good experimental evidence that this region of DNA curvature is involved in the regulation of α toxin production [107]. The stimulatory effect of the A-tracts on *plc* transcription and α toxin production was more pronounced at 25 °C than at 45 °C, in agreement with the temperature-dependent nature of DNA bending.

Subsequent studies have examined the mechanism by which this transcriptional activation occurs [108]. *In vitro* transcription studies carried out with RNA polymerase from *C. perfringens* showed that, at 25 °C, the level of transcription was dependent on the number of A-tracts present in the template DNA. However, a greatly reduced dependence on these A-tracts was observed at 45 °C. Quantitative gel shift and DNA footprinting analysis showed that the increased level of transcription was a direct reflection of the ability of RNA polymerase to bind in a stable manner to the promoter region. It is intriguing that the production of α toxin, the major toxin involved in gas gangrene, is specifically downregulated at normal body temperatures. It suggests that α toxin production is more critical to the survival of the bacterium when it is growing outside a mammalian host, such as in the soil, where the temperature is lower and nutrient availability is reduced [108]. Therefore, the key role of α toxin may not be tissue destruction in the living host, but the degradation of membrane-related material and the resultant release of nutrients from decomposing animal matter.

Expression of the *plc* gene is also controlled by a regulatory cascade involving the VirS/VirR two-component signal transduction system [109, 110]. Interaction of the VirS sensor histidine kinase with an unknown growth phase or environmental stimulus leads to its autophosphorylation and subsequent phosphotransfer to the cognate response regulator, VirR. Phosphorylation of the VirR protein leads to transcriptional activation of several extracellular toxin genes, including *plc* [111]. Recent studies have shown that VirR does not directly bind to the *plc* gene ([112] but activates a secondary regulatory locus, the *hyp7* gene [113]. The *hyp7* gene product then appears to activate transcription of the *plc* gene. Note that since there is always a constitutive level of *plc* gene expression, the production of α toxin is not totally dependent upon activation of the VirS/VirR/*hyp7* regulatory cascade. It is not known how this cascade interacts with the temperature-dependent regulatory process or what physiological changes induce the VirS/VirR phosphorelay [97].

Action of *C. perfringens* α toxin

C. perfringens α toxin was the first bacterial toxin shown to have enzymatic activity [114]. The enzyme is a zinc-metallophospholipase C that binds three zinc ions and is activated by calcium ions [99, 115]. It has both phospholipase C, or lecithinase, activity and sphingomyelinase activity [116], hydrolyzing glycerophospholipids such as lecithin to produce diacylglycerol and phosphorylcholine and hydrolyzing sphingomyelin to produce ceramides. Since phosphatidylcholine and sphingomyelin are the major components of the outer leaflets of human erythrocytes, it is not surprising that α toxin is also hemolytic and toxic [117]. Toxic activity is related to its sphingomyelinase rather than its phospholipase C activity, although it is clear that the enzyme has only one active site that can catalyze the hydrolysis of both phosphatidylcholine and sphingomyelin [99]. Loss of phospholipase C activity results in loss of all biological activity [118, 119].

It is clear that the action of α toxin on the host cell membrane induces a regulatory cascade that affects host cell metabolism, without necessarily lysing the cell. The production of diacylglycerol leads to, (1) activation of protein kinase C (PKC) and the production of endogenous phospholipases [120, 121], (2) activation of the arachidonic acid cascade [122, 123], leading to pro-inflammatory changes [124] and (3) potential activation of the platelet aggregation pathway [99].

Exposure of human endothelial cell monolayers to recombinant α toxin has been shown to activate the PKC pathway, with a resultant production of platelet-activating factor (PAF) [124]. PAF signaling appears to be involved in the P-selectin-mediated adhesion of PMNLs to endothelial cells, which may explain why, in infected tissues, PMNLs are sequestered within the blood vessels at the periphery of the site of infection. Other studies have shown that α toxin also induces synthesis of the immune modulator interleukin-8, the cell surface receptors intercellular adherence molecule 1 and E-selectin in cultured epithelial cells [125]. However, the precise role of these cytokines and receptors in the α toxin-induced inflammatory response *in vivo* remains to be determined.

Structure function/relationships of *C. perfringens* α toxin

Determination of the structure of *C. perfringens* α toxin by X-ray crystallography has shown that the toxin has two distinct domains [126] an N-terminal domain (residues 1-246) that has phospholipase C activity but is not hemolytic, as well as a C-terminal domain (residues 257-370) that has no enzyme activity, but confers sphingomyelinase activity, hemolytic activity, and toxicity on the N-terminal domain [127]. These domains are connected by a flexible linker region (residues 247-255). The N-terminal domain is comprised almost entirely of α-helices (Figure 8), has three Zn^{2+} binding sites, and has sequence and structural similarity to other phospholipase C enzymes such as PC-PLC from *Bacillus cereus*. The N-terminal tryptophan residue, several histidine residues, and Glu-152 are all

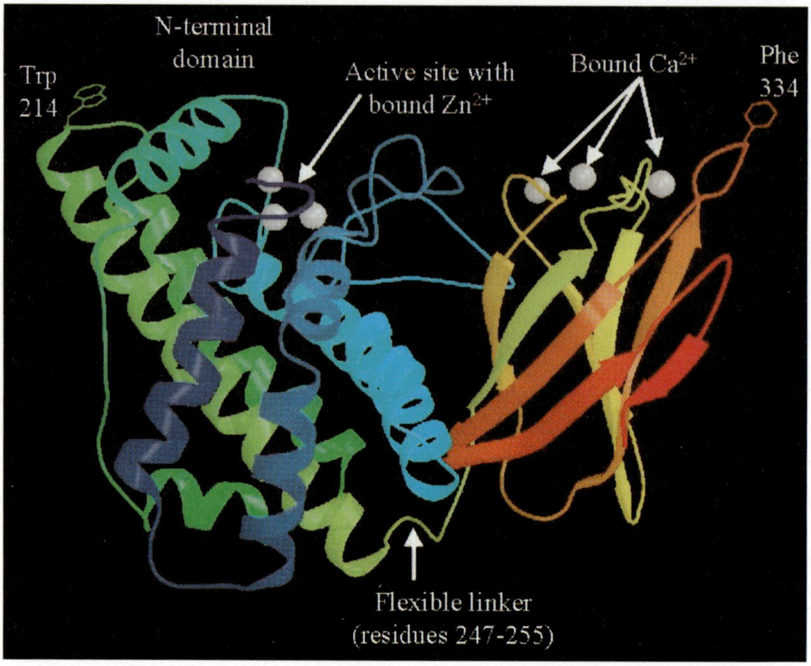

Figure 8. Crystal structure of *C. perfringens* α toxin. The crystal structure is represented as a ribbon diagram, with a color transition from the N-terminus (blue) to the C-terminus (red). As shown, the active site is in the N-terminal domain and the calcium-binding site is in the C-terminal domain (reproduced from Figure 2 of [99], with permission).

involved in the α-helical Zn^{2+}-binding domain [126]. Site-directed mutagenesis confirms that these residues, which are conserved in other phospholipase C enzymes, are essential for biological activity [118, 119, 128].

The common features of *C. perfringens* α toxin, the γ toxin from *Clostridium novyi* and phospholipase C from *Clostridium bifermentans* are that they are all hemolytic and all have an additional C-terminal domain when compared to non-hemolytic PLCs such as PC-PLC [99]. The C-terminal α toxin domain consists of an eight-stranded antiparallel β-sandwich that has structural similarity to eukaryotic calcium-dependent phospholipid-binding C2 domains such as the C-terminal domains of pancreatic lipase and synaptotagmin I and the N-terminal domain of soybean lipoxygenase-1 [126, 129]. C2 domains bind calcium ions and/or phospholipids and are commonly found in eukaryotic signaling proteins. The unexpected finding of a C2 domain in the α toxin suggests that α toxin binds mammalian cell membranes in a similar manner to the process by which host cell proteins bind to cell membranes [126]. The structural data suggest that α toxin binds in a C2 domain mediated manner, with calcium ions moderating interactions between the phospholipid headgroups. Surface exposed hydrophobic residues such as Trp-214, Tyr-331, and Phe-334 appear to be involved in membrane penetration.

The location of the calcium-binding sites could not be determined with certainty from the initial X-ray crystallography data, but it appeared that the C-terminal domain had the ability to bind three calcium ions close to the surface of the enzyme, at sites postulated to be in close proximity to the cell membrane. Based on this structural analysis and other mutagenesis studies [129], it has been proposed that Asp-269 and Asp-336, both of which have been shown to be required for Ca^{2+}-mediated activation of α toxin activity [129], and Gly-271 and Ala-337, comprise a calcium-binding site [126]. The other sites were proposed to be located in loops that are topologically equivalent to Ca^{2+} binding sites of other C2 domain proteins [126]. In subsequent studies, both crystallographic and spectroscopic methods were used to further define these binding sites [130]. Determination of the structure of calcium-bound α toxin identified three calcium ions at the putative membrane-binding surface. The first site, Ca1, was identical to that previously identified [126, 129]. The other sites, Ca2 and Ca3, had not been identified previously but involved residues that were also present in the C-terminal domains of the *C. bifermentans* and *C. novyi* enzymes [130]. It was proposed that, when α toxin binds to the cell membrane, coordination of the three calcium ions is completed by phospholipid head groups, resulting in the enhancement of membrane binding. Furthermore, modeling suggests that once the toxin is bound to the membrane via these calcium ions, the zinc ions at the active site are suitably positioned for the toxin to bind phosphatidylcholine, without that phospholipid having to be withdrawn from the membrane [130]. Note that the binding of calcium ions by the toxin does not appear to induce major conformational changes.

Finally, based upon structural comparisons between α toxin and other zinc metallophospholipase C enzymes, it is possible to understand why only the α toxin is hemolytic and toxic [130]. Clearly, phospholipase C enzymes such as PC-PLC are not non-hemolytic because they lack a C-terminal membrane binding domain and do not have any of the exposed hydrophobic domains that are involved in membrane binding. Although the nontoxic but weakly hemolytic phospholipase C from *C. bifermentans* has a C-terminal C2-domain, it does not have the Tyr-331 and Phe-334 residues that appear to be required for effective membrane insertion. Recently, a hybrid protein, which consisted of the N-terminal domain of phospholipase C from *C. bifermentans* and the C-terminal domain of *C. perfringens* α toxin, was constructed [131]. The resultant protein was toxic and ten times more hemolytic that the *C. bifermentans* enzyme, confirming the key role of the C-terminal α toxin domain in potentiating its full biological activity.

C. perfringens α toxin vaccines

By use of α toxin variants that were overexpressed from recombinant plasmids in *E. coli*, it was shown that the C-terminal α toxin domain is immunoprotective in mice [100]. Immunization with an α toxin-derived peptide that consists of amino acids 247-370 was sufficient to completely protect mice (6/6 survived) against a challenge of 10^9 *C. perfringens* cells, a challenge that killed all six mice in the control group within approximately seven hours. No protection was observed when mice were immunized with the N-terminal α toxin domain, even though equivalent antibody responses were observed [100]. The latter antibodies were effective in inhibiting phospholipase C activity but, unlike antibodies raised against the C-terminal domain, did not block the hemolytic activity of α toxin. These findings are consistent with the conclusions drawn in the previous section regarding the essential role of the C-terminal domain in lethal and hemolytic α toxin activity. These studies shown that an effective nontoxic vaccine with potential for use in high risk groups, such as the elderly, diabetics and surgical patients, could be prepared from the C-terminal α toxin domain. Subsequent studies have shown that recombinant vaccinia viruses that express this domain can also be used to protect mice against a lethal dose of α toxin [132].

6.2.2 Perfringolysin O from *C. perfringens*

The role of perfringolysin O in disease

The role of perfringolysin O, or θ toxin, in *C. perfringens*-mediated gas gangrene is less certain than that of α toxin [133]. The major pathophysiologic effects of this toxin do not seem solely attributable to its cytolytic activity. Perfringolysin O appears to be involved in the interaction of leukocytes and endothelial cells [125], vascular dysfunction, and shock [134]. Homologous recombination has been used to insertionally inactivate the perfringolysin O structural gene, *pfoA*, and the resultant mutants tested in the mouse myonecrosis model [101]. The results showed that these *pfoA* mutants, which did not produce any detectable perfringolysin O, were still able to cause fulminant myonecrosis. Animals infected with the *pfoA* mutants exhibited significant muscle destruction, unlike the equivalent *plc* mutants. These data provide solid evidence that perfringolysin O is not an essential toxin in the pathogenesis of gas gangrene. However, it still appears to play some role in the disease process. Histopathological examination of the tissues of mice infected with the *pfoA* mutants revealed a significant increase in PMNL influx into the lesion, an effect that was reversed when the mutant was complemented with a multicopy plasmid carrying the wild-type *pfoA* gene [102]. These results suggest that perfringolysin O is important for preventing PMNL influx into the developing myonecrotic lesion. Subsequent studies showed that the number of blood vessels in which severe leukocyte accumulation occurred dropped from 8.6 % in the wild-type strain to 0.6 % in the *pfoA* mutant, providing evidence that perfringolysin O is important for the vascular accumulation of leukocytes [104].

Genetics and synthesis of perfringolysin O

The *pfoA* gene is located on the *C. perfringens* chromosome, within close proximity (12 kb) of the collagenase structural gene, *colA* [135, 136]. Upstream of the *pfoA* gene is a putative regulatory gene, *pfoR* [137]. On multicopy plasmids in *E. coli*, deletion of an internal region of *pfoR* leads to reduced production of perfringolysin O, suggesting that the PfoR protein is a positive regulator of *pfoA* expression. Confirmation of this hypothesis awaits the construction of *pfoR* mutants in *C. perfringens*.

There is clear evidence that perfringolysin O production is regulated at the level of transcription by the VirS/VirR system in *C. perfringens* [109-111]. Mutation of either the *virS* [109] or *virR* [110] genes completely eliminates perfringolysin O production. Hemolytic activity is restored when these mutants are complemented by multicopy plasmids carrying the respective wild-type genes. It was postulated that the VirR-mediated activation of *pfoA* may have occurred indirectly, by activation of *pfoR* expression [109, 111]. However, recent evidence has shown that purified VirR binds directly to two imperfect directly repeated sequences, or VirR boxes, which are located immediately upstream of the *pfoA* promoter [112]. Mutation of these sites eliminates VirR binding. These studies suggest that VirR activates *pfoA* transcription by binding directly to the promoter region and stimulating RNA polymerase binding or activity. This mechanism is in contrast to that observed for the VirR-mediated activation of the *plc* and *colA* genes, where transcriptional activation appears to involve a regulatory cascade induced by the product of the VirR-regulated *hyp7* transcript [113].

Action of Perfringolysin O

Perfringolysin O is a member of a large family of pore-forming, cholesterol-binding cytolysins that are active against mammalian cells that contain cholesterol in their cell membranes [138]. These proteins have at least 40 % amino acid sequence identity and contain a highly conserved sequence, ECTGLAWEWWR, near their C-terminus. Although they were once known as thiol-activated cytolysins, it has now been shown that the thiol-active reagents that inhibit activity do so by binding to the cysteine residue in the conserved undecapeptide, which leads to steric hindrance of the membrane binding site [138, 139]. Site-directed mutagenesis of the same cysteine residue has shown that it is not essential for cytolytic activity [138].

The mode of action of perfringolysin O involves the formation of large oligomers on the membrane of the target cell [140]. The toxin binds to cholesterol residues in the plasma membrane with high affinity ($K_d \sim 2$ nM) [141]. The end result is the formation of a large protein pore or channel in the membrane. The toxin is inhibited by cholesterol; while the exact mechanism of this inhibition is not known, it does involve the binding of cholesterol to the tryptophan-rich C-terminal domain of perfringolysin O [138].

As already discussed, perfringolysin O also partially inhibits the influx of PMNLs into myonecrotic lesions [96, 104]. Therefore, at sublytic concentrations the toxin clearly must modulate intracellular signaling pathways, especially in endothelial cells, which leads to the accumulation of PMNLs in the blood vessels. The mechanism by which these effects are mediated is also not known. Treatment of cultured human umbilical vein endothelial cells with purified perfringolysin O leads to a small but significant increase in ICAM-1 expression but, unlike α toxin, has no effect on ELAM-1 or IL-8 production [125]. In other experiments, purified perfringolysin O has been shown to upregulate leukocyte CD11b/CD18 [142] and the synthesis of PAF by endothelial cells [143, 143a]. Studies carried out *in vivo* with *pfoA* mutants suggest that upregulation of ICAM-1 expression is not the mechanism by which perfringolysin O mediates its effects on PMNL influx [104].

Structure/function relationships of perfringolysin O

The C-terminal domain of perfringolysin O contains the cholesterol binding site and the conserved tryptophan-rich undecapeptide and is therefore essential for activity [139]. Spectroscopic analysis revealed that binding of the toxin to the membrane involved conformation changes around the tryptophan residues [144, 145]. Furthermore, mutation of all seven tryptophan residues in perfringolysin O to phenylalanine showed that only the three tryptophan residues located in the conserved region are required for binding to erythrocyte membranes and for hemolytic activity. However, these residues are not necessary for cholesterol binding or oligomerization on the membranes [146]. Subsequent studies have shown that the two C-terminal amino acids of perfringolysin O are required for correct protein folding, thereby ensuring cholesterol binding and hemolytic activity. Deletion of the C-terminal tyrosine and asparagine residues had no effect on biological activity, whereas deletion of an additional threonine residue significantly reduces hemolytic activity and the ability to bind to sheep erythrocyctes [147].

Perfringolysin O was the first cholesterol-binding cytolysin for which the crystal structure was determined [148]. The molecule has an elongated structure consisting of four discrete domains that are all rich in β-sheet (Figure 9A). Domain 1 has a seven-stranded antiparallel β-sheet and is connected to domain 4 by the elongated domain 2. Domain 3 has a three layered α/β/α structure that has a five-stranded antiparallel sheet that continues the structure found in domain 1. The tryptophan-rich undecapeptide is located in a loop at the most elongated point of the molecule in domain 4, which is a compact β sandwich domain. The analysis of this region explains why modification of Cys-459 with bulky sulphhydryl reagents inhibits activity. This modification results in a conformational change that disturbs the tight packing of the tryptophan-rich loop, which is essential for cholesterol binding.

Based upon electron microscopy data and evidence that domain 4 is involved in cell binding and spans the phospholipid bilayer, a model for the membrane-bound oligomeric state of perfringolysin O was proposed [148]. In this model the monomers fit together to form an L-shaped repeating unit, with domains 1,2, and 4 forming the stem and domain 3 the flange of the structure. Domain 4 is positioned at the base of the structure, in the membrane-spanning region. The model predicts that domain 1 of each monomer forms tight contacts with its neighboring domain 1 region. It was proposed that membrane insertion involves the formation of oligomers on the membrane surface. These oligomers are then activated by the binding of cholesterol to the Trp-464 pocket of domain 4, displacing the Trp-rich loop. The movement of this hairpin loop is proposed to be the trigger for insertion of the toxin, but it is too short to span the lipid bilayer. Instead its movement is proposed to lead to the formation of a hydrophobic region that enables domain 4 to penetrate into the membrane and ultimately form a pore [148].

Subsequent studies have shown that it is a conformational change in domain 3 that leads to the insertion of perfringolysin O oligomers into the membrane [149]. Substitution of the amino acids between 189 and 218 with cysteine, and subsequent fluorescence spectroscopy studies carried out in the absence and presence

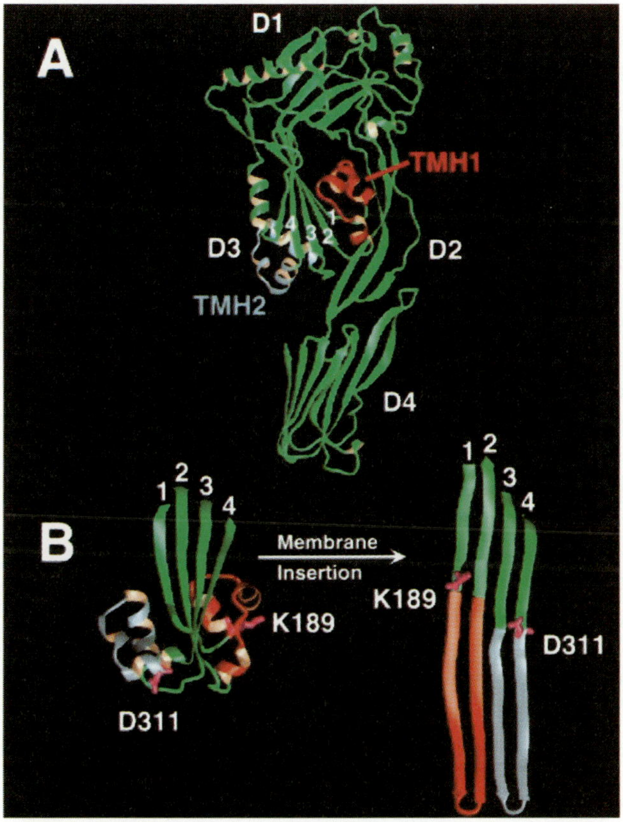

Figure 9. Crystal structure of perfringolysin O. Panel A is a representation of the crystal structure of perfringolysin O. Domains 1-4 are labelled as D1-D4. Transmembrane hairpins (TMH) 1 (aa 190-217) and 2 (aa 288-311) are shown in red and blue, respectively. The core β strands of domain 3 are labelled 1-4 for reference. Panel B shows the putative transition from the helical domains in the soluble monomer (left) to the dual β hairpins (right) that are proposed to span the membrane (reproduced from Fig 4 of [150], with permission).

of liposomes that contain cholesterol, have shown that these amino acids alternate between an aqueous and hydrophobic environment. These studies indicate that, although these residues form three short α helices in the soluble perfringolysin O monomer, they form a two-stranded amphipathic β sheet in the membrane bound form of the toxin. Therefore, it was postulated that the formation of oligomers is a cholesterol-dependent process that involves a major structural transition in which α helices in domain 3 unfold to form a membrane-spanning β sheet. The implication was that the association of perfringolysin O monomers with the mammalian cell membrane induces major conformation changes in the structure of the protein [149].

This model has been further refined by additional cysteine-scanning mutagenesis and fluorescence microscopy studies which have shown that domain 3 contains a second series of α helices (aa 288 to 311) that undergo a conformational change to an amphipathic β sheet [150]. Both this region and the domain 3 region previously demonstrated to undergo an α helix to β strand transition [149] were shown to completely span the membrane [150]. In conclusion, insertion of perfringolysin O into the cell membrane appears to occur by a unique mechanism that involves the conformational change of six α helices in domain 3 to form two amphipathic β hairpins that completely span the cell membrane (Figure 9B). Since each perfringolysin O monomer would contribute four β strands to an oligomer that may contain up to 50 monomers, the oligomeric pore may consist of a 200 stranded β barrel transmembrane domain [151].

Is the oligomeric complex formed at the membrane surface before penetration of the membrane or does insertion occur first, forming a small pore that grows as the toxin oligomerizes? Recent studies have shown that, at 4 °C, oligomerization occurs on the surface of liposomes, without significant penetration of the membrane by the β-hairpins [151a]. Conductivity data were also consistent with the hypothesis that perfringolysin O forms an oligomeric complex on the membrane prior to the insertion of the transmembrane domains into the lipid bilayer.

Perfringolysin O vaccines

Vaccine studies on *C. perfringens*-mediated gas gangrene have centered upon use of α toxin variants. However, passive immunization with monoclonal antibodies specific for perfringolysin O has been shown to enhance survival rates in experimentally infected mice [142].

6.2.3
The α toxin from *C. septicum*

The role of *C. septicum* α toxin in disease

C. septicum is a major cause of spontaneous nontraumatic gas gangrene, which often occurs in patients with predisposing diseases such as colonic cancer, leukemia or diabetes [152]. The pathogenesis of *C. septicum* disease is related to the production of α toxin, which is a pore-forming hemolytic toxin related to aerolysin from *Aeromonas* spp. [153]. Although little is known about the precise role of α toxin in virulence, it does appear to be the major toxin involved in the disease process [98, 154].

Genetics and synthesis of *C. septicum* α toxin

In the absence of defined systems for genetic analysis in *C. septicum,* little is known about the genetics of α toxin production or its expression and regulation. The α toxin structural gene has been cloned and sequenced [155, 156]; those studies indicated that this toxin gene encodes a single 443 amino acid preproprotein containing a 32 amino acid signal sequence that is cleaved after secretion, forming an inactive proprotein [156].

Action of *C. septicum* α toxin

The α toxin is secreted as an inactive protoxin monomer. It is then activated by trypsin- or furin-mediated cleavage of a 5.1 kDa peptide from the C-terminus [157, 158]. The resultant active toxin molecule forms stable oligomers that insert into the cell membrane to form a pore or channel [159]. The cleaved propeptide has been proposed [160] to function as an intramolecular chaperone that remains associated with the active toxin after tryptic cleavage and stabilizes it until the toxin oligomerizes on the cell surface to form a pre-pore complex [159, 160].

Furin is a serine protease found in the Golgi apparatus and on the plasma membrane surface. It plays a role in the processing of proreceptors and other proproteins. Cells that express furin on their surface are lysed more rapidly by α toxin than cells without this cell-surface protease [158]. α Toxin contains a potential furin recognition sequence, RXXR, at its C-terminal cleavage site. Mutation of this site to SGSR yielded a protein that could still be cleaved and activated by trypsin, but not by furin. The mutant protein was no longer able to kill CHO cells, unless it was pretreated with trypsin. These data provide good evidence that in the absence of other serine proteases furin-mediated cleavage of α toxin is required for activation at the cell surface [158].

Structure/function relationships of *C. septicum* α toxin

C. septicum α toxin has 27% identity and 72% similarity to aerolysin, a toxin expressed by *Aeromonas hydrophila* [156]. The crystal structure of aerolysin has been determined, showing that toxin consists of two major lobes [153, 161]. The larger C-terminal lobe of aerolysin, but not the N-terminal lobe, is responsible for oligomerization and activation. That C-terminal lobe has extensive sequence similarity to *C. septicum* α toxin suggesting that the toxins have a similar structure and mechanism of action. The smaller N-terminal lobe, or domain 1, of aerolysin is not present in α [156, 162]. Other studies involving the construction of aerolysin-α toxin hybrid proteins have shown that aerolysin contains two receptor binding sites, including a glycosylphosphatidylinositol (GPI) receptor binding site located in the larger lobe and a second carbohydrate binding site located in the smaller lobe [163]. In agreement with these observations, recent studies have shown that, like aerolysin, α toxin binds to C-terminal GPI-anchored protein receptors. Chinese Hamster Ovary (CHO) cells unable to produce GPI were more resistant to both α toxin and aerolysin as were CHO cells that had been pretreated with PI-PLC, which cleaves terminal GPI residues, or other cell lines unable to synthesize GPI anchors [164].

It has been proposed that, once cleaved and oligomerized on the cell surface, aerolysin forms an amphipathic β barrel capable of penetrating the cell membrane and forming a pore. Despite the absence of the smaller N-terminal lobe, α toxin can be modeled on the aerolysin structure and may well penetrate the cell membrane by forming a similar oligomeric β barrel structure [138].

C. *septicum* α toxin vaccines

Toxoids prepared from culture supernatants are used to vaccinate animals against *C. septicum* infections, often as components of multivalent clostridial vaccines [165, 165a, 166]. In the only vaccine study to be carried using purified α toxin, the survival time in a mouse myonecrosis model was increased when the mice were immunized with α toxin [98].

Acknowledgements

The preparation of this chapter was supported, in part, by U. S. Public Health Service Grant AI19844-18 from the National Institute of Allergy and Infectious Diseases and grant 9802822 from the Ensuring Food Safety Program of the United States Department of Agriculture (both to BAMcC) and by grants from the Australian National Health and Medical Research Council (to JIR).

We thank Mahfuzur Sarker for the preparation of Figures 1,4, and 5; Usha Singh for the preparation of Figures 3 and 6; Darren Ellemor for the preparation of Figure 7; Richard Titball for the preparation of Figure 8; and Rod Tweten for the the preparation of Figure 9.

References

[1] Songer, J. G., Clostridial enteric diseases of domestic animals, *Clin. Microbiol. Rev.* **1996**, 9, 216-234.

[2] Hatheway, C., Toxigenic clostridia, *Clin. Microb. Rev.* **1990**, 3, 66-76.

[3] McDonel, J. L., Toxins of *Clostridium perfringens* types A, B, C, D, and E, in: *Pharmacology of Bacterial Toxins* (F. Dorner and H. Drews, Eds.), Oxford: Pergamon Press, **1986**, pp. 477-517.

[4] Cole, S. T. and Canard, B. T., Structure, organization, and evolution of the genome of *Clostridium perfringens*, in: *The Clostridia: Molecular Biology and Pathogenesis* (J. I. Rood, B. A. McClane, J. G. Songer and R. W. Titball, Eds.), Academic Press, London, **1997**, pp. 49-64.

[5] McClane, B. A. *Clostridium perfringens*, in: *Food Microbiology: Fundamentals and Frontiers*. (M. P. Doyle, L. R. Beuchat and T. J. Montville, Eds.), ASM Press, Washington, DC, **2001**, pp. 351-372.

[6] Collie, R. E., Kokai-Kun, J. F. and McClane, B. A., Phenotypic characterization of enterotoxigenic *Clostridium perfringens* isolates from nonfoodborne human gastrointestinal diseases, *Anaerobe* **1998**, 4, 69-79.

[7] Czeczulin, J. R., Hanna, P. C. and McClane, B. A., Cloning, nucleotide sequencing, and expression of the *Clostridium perfringens* enterotoxin gene in *Escherichia coli*, *Infect. Immun.* **1993**, 61, 3429-3439.

[8] Skjelkvale, R. and Uemura, T., Experimental diarrhea in human volunteers following oral administration of *Clostridium perfringens* enterotoxin, *J. Appl. Bacteriol.* **1977**, 46, 281-286.

[9] Sarker, M. R., Carman, R. J. and McClane, B. A., Inactivation of the gene (*cpe*) encoding *Clostridium perfringens* enterotoxin eliminates the ability of two *cpe*-positive *C. perfringens* type A human gastrointestinal disease isolates to affect rabbit ileal loops, *Mol. Microbiol.* **1999**, 33, 946-958.

[10] Zhao, Y. and Melville, S. B., Identification and characterization of sporulation-dependent promoters upstream of the enterotoxin gene (*cpe*) of *Clostridium perfringens*, *J. Bacteriol.* **1998**, 180, 136-142.

[11] Carman, R. J., *Clostridium perfringens* in spontaneous and antibiotic-associated diarrhoea of man and other animals, *Rev. Med. Microbiol.* **1997**, 8, supplement 1, S43-S45.

[12] Collie, R. E. and McClane, B. A., Evidence that the enterotoxin gene can be episomal in

Clostridium perfringens isolates associated with nonfoodborne human gastrointestinal diseases, *J. Clin. Microbiol.* **1998**, 36, 30-36.

[13] Cornillot, E., Saint-Joanis, B., Daube, G., Katayama, S., Granum, P. E., Carnard, B. and Cole, S. T., The enterotoxin gene (*cpe*) of *Clostridium perfringens* can be chromosomal or plasmid-borne, *Mol. Microbiol.* **1995**, 15, 639-647.

[14] Katayama, S. I., Dupuy, B., Daube, G., China, B. and Cole, S. T., Genome mapping of *Clostridium perfringens* strains with *I-Ceu* I shows many virulence genes to be plasmid-borne, *Mol. Gen. Genet.* **1996**, 251, 720-726.

[15] Sarker, M. R., Shivers, R. P., Sparks, S. G., Juneja, V. K. and McClane, B. A., Comparative experiments to examine the effects of heating on vegetative cells and spores of *Clostridium perfringens* isolates carrying plasmid versus chromosomal enterotoxin genes, *Appl. Environ. Microbiol.* **2000**, 66, 3234-3240.

[16] Brynestad, S., Sarker, M. R., McClane, B. A., Granum, P. E. and Rood, **J. I.** Conjugative transfer of the plasmid carrying the *Clostridium perfringens* enterotoxin gene, *Infect. Immun.* **2001**, 69, 3483-3487.

[17] Canard, B., Saint-Joanis, B. and Cole, S. T., Genomic diversity and organization of virulence genes in the pathogenic anaerobe *Clostridium perfringens, Mol. Microbiol.* **1992**, 6, 1421-1429.

[18] Brynestad, S., Synstad, B. and Granum, P. E., The *Clostridium perfringens* enterotoxin gene is on a transposable element in type A human food poisoning strains, *Microbiology* **1997**, 143, 2109-2115.

[19] Brynestad, S. and Granum, P. E., Evidence that Tn*5565*, which includes the enterotoxin gene in *Clostridium perfringens*, can have a circular form which may be a transposition intermediate, *FEMS Microbiol. Lett.* **1999**, 170, 281-286.

[20] Billington, S. J., Wieckowski, E. U., Sarker, M. R., Bueschel, D., Songer, J. G. and McClane, B. A., *Clostridium perfringens* type E animal enteritis isolates with highly conserved, silent enterotoxin sequences, *Infect. Immun.* **1998**, 66, 4531-4536.

[21] Duncan, C. L., Strong, D. H. and Sebald, M., Sporulation and enterotoxin production by mutants of *Clostridium perfringens, J. Bacteriol.* **1972**, 110, 378-391.

[22] Melville, S. B., Labbe, R. and Sonenshein, A. L., Expression from the *Clostridium perfringens cpe* promoter in *C. perfringens* and *Bacillus subtilus, Infect. Immun.* **1994**, **62**, 5550-5558.

[23] Smith, W. P. and McDonel, J. L., *Clostridium perfringens* type A: *in vitro* systems for sporulation and enterotoxin synthesis, *J. Bacteriol.* **1980**, 144, 306-311.

[24] Czeczulin, J. R., Collie, R. E. and McClane, B. A., Regulated expression of *Clostridium perfringens* enterotoxin in naturally *cpe*-negative type A, B, and C isolates of *C. perfringens, Infect. Immun.* **1996**, 64, 3301-3309.

[25] Brynestad, S., Iwanejko, L. A., Stewart, G. S. A. B. and Granum, P. E., A complex array of Hpr consensus DNA recognition sequences proximal to the enterotoxin gene in *Clostridium perfringens* type A, *Microbiology* **1994**, 140, 97-104.

[26] Labbe, R. G. and Duncan, C. L., Evidence for stable messenger ribonucleic acid during sporulation and enterotoxin synthesis by *Clostridium perfringens* type A, *J. Bacteriol.* **1977**, 129, 843-849.

[27] Sherman, S., Klein, E. and McClane, B. A., *Clostridium perfringens* type A enterotoxin induces concurrent development of tissue damage and fluid accumulation in the rabbit ileum, *J. Diarrheal Dis. Res.* **1994**, 12, 200-207.

[28] McDonel, J. L. and Duncan, C. L., Histopathological effect of *Clostridium perfringens* enterotoxin in the rabbit ileum, *Infect. Immun.* **1975**, 12, 1214-1218.

[29] Sonoda, N., Furuse, M., Sasaki, H., Yonemura, S., Katahira, J., Horiguchi, Y. and Tsukita, S., *Clostridium perfringens* enterotoxin fragments removes specific claudins from tight junction strands: evidence for direct involvement of claudins in tight junction barrier, *J. Cell Biol.* **1999**, 147, 195-204.

[30] Rahner, C., Mitic, L. L., McClane, B. A. and Anderson, J. M., *Clostridium perfringens* enterotoxin impairs bile flow in the isolated perfused rat liver and induces fragmentation of tight junction fibrils, *Hepatology* **1999**, 30, 326A.

[31] McClane, B. A., *Clostridium perfringens* enterotoxin acts by producing small molecule permeability alterations in plasma membranes, *Toxicology* **1994**, 87, 43-67.

[32] Krakauer, T., Fleischer, B., Stevens, D. L., McClane, B. A. and Stiles, B. G., *Clostridium perfringens* enterotoxin lacks superantigenic activity but induces an interleukin-6 response

from human peripheral blood mononuclear cells, *Infect. Immun.* **1997**, 65, 3485-3488.

[33] Wallace, F. M., Mach, A. S., Keller, A. M. and Lindsay, J. A., Evidence for *Clostridium perfringens* enterotoxin inducing a mitogenic and cytokine response *in vitro* and a cytokine response *in vivo*, *Curr. Microbiol.* **1999**, 38, 96-100.

[34] Horiguchi, Y., Uemura, T., Kozaki, S. and Sakaguchi, G., The relationship between cytotoxic effects and binding to mammalian cultures cells of *Clostridium perfringens* enterotoxin, *FEMS Microbiol. Lett.* **1985**, 28, 131-135.

[35] McClane, B. A., Hanna, P. C. and Wnek, A. P., *Clostridium perfringens* type A enterotoxin, *Microb. Pathogen.* **1988**, 4, 317-323.

[36] McDonel, J. L., Binding of *Clostridium perfringens* ^{125}I- enterotoxin to rabbit intestinal cells, *Biochemistry* **1980**, 21, 4801-4807.

[37] McDonel, J. L. and McClane, B. A., Binding vs. biological activity of *Clostridium perfringens* enterotoxin in Vero cells, *Biochem. Biophys. Res. Commun.* **1979**, 87, 497-504.

[38] Sugii, S. and Horiguchi, Y., Identification and isolation of the binding substance for *Clostridium perfringens* enterotoxin on Vero cells, *FEMS Microbiol. Lett.* **1988**, 52, 85-90.

[39] Wnek, A. P. and McClane, B. A., Identification of a 50,000 M_r protein from rabbit brush boarder membranes that binds *Clostridium perfringens* enterotoxin, Biochem. *Biophys. Res. Commun.* **1983**, 112, 1099-1105.

[40] Wnek, A. P. and McClane, B. A., Preliminary evidence that *Clostridium perfringens* type A enterotoxin is present in a 160,000-M_r complex in mammalian membranes, *Infect. Immun.* **1989**, 57, 574-581.

[41] Wieckowski, E. U., Wnek, A. P. and McClane, B. A., Evidence that an ~50kDa mammalian plasma membrane protein with receptor-like properties mediates the amphiphilicity of specifically-bound *Clostridium perfringens* enterotoxin, *J. Biol. Chem.* **1994**, 269, 10838-10848.

[42] Katahira, J., Inoue, N., Horiguchi, Y., Matsuda, M. and Sugimoto, N., Molecular cloning and functional characterization of the receptor for *Clostridium perfringens* enterotoxin, *J. Cell Biol.* **1997**, 136, 1239-1247.

[43] Katahira, J., Sugiyama, H., Inoue, N., Horiguchi, Y., Matsuda, M. and Sugimoto, N., *Clostridium perfringens* enterotoxin utilizes two structurally related membrane proteins as functional receptors *in vivo*, *J. Biol. Chem.* **1997**, 272, 26652-26658.

[44] Furuse, M., Fujita, K., Hiiragi, T., Fujumoto, K. and Tsukita, S., Claudin-1 and -2: Novel integral membrane proteins localizing at tight junctions with no sequence similarity to occludin, *J. Cell Biol.* **1998**, 141, 1539-1550.

[45] Fujita, K., Katahira, J., Horiguchi, Y., Sonoda, N., Furuse, M. and Tskuita, S., *Clostridium perfringens* enterotoxin binds to the second extracellular loop of claudin-3, a tight junction membrane protein, *FEBS Lett.* **2000**, 476, 258-261.

[46] Kokai-Kun, J. F. and McClane, B. A., Evidence that region(s) of the *Clostridium perfringens* enterotoxin molecule remain exposed on the external surface of the mammalian plasma membrane when the toxin is sequestered in small or large complex, *Infect. Immun.* **1996**, 64, 1020-1025.

[47] Wieckowski, E., Kokai-Kun, J. F. and McClane, B. A., Characterization of membrane-associated *Clostridium perfringens* enterotoxin following Pronase treatment, *Infect. Immun.* **1998**, 66, 5897-5905.

[48] Singh, U., Van Itallie, C. M., Mitic, L. L., Anderson, J. M. and McClane, B. A, CaCo-2 cells treated with *Clostridium perfringens* enterotoxin form multiple large complex species, one of which contains the tight junction protein occludin, *J. Biol. Chem.* **2000**, 275, 18407-18417.

[48a] McClane, B. A. and Wnek, A. P., Studies of *Clostridium perfringens* enterotoxin action at different temperatures demonstrate a correlation between complex formation and cytotoxicity, *Infect. Immun.* **1990**, 58, 3109-3115.

[49] Kokai-Kun, J. F., Benton, K., Wieckowski, E. U. and McClane, B. A., Identification of a *Clostridium perfringens* enterotoxin region required for large complex formation and cytotoxicity by random mutagenesis, *Infect. Immun.* **1999**, 67, 6534-6541.

[50] Kokai-Kun, J. F. and McClane, B. A., Deletion analysis of the *Clostridium perfringens* enterotoxin, *Infect. Immun.* **1997**, 65, 1014-1022.

[51] Hardy, S. P., Denmead, M., Parekh, N. and Granum, P. E., Cationic currects induced by *Clostridium perfringens* type A enterotoxin in human intestinal Caco-2 cells, *J. Med. Microbiol.* **1999**, 48, 235-243.

[52] Hanna, P. C., Mietzner, T. A., Schoolnik, G. K. and McClane, B. A., Localization of the

receptor-binding region of *Clostridium perfringens* enterotoxin utilizing cloned toxin fragments and synthetic peptides. The 30 C-terminal amino acids define a functional binding region, *J. Biol. Chem.* **1991**, 266, 11037-11043.

[53] Granum, P. E. and Richardson, M., Chymotrypsin treatment increases the activity of *Clostridium perfringens* enterotoxin, *Toxicon* **1991**, 29, 445-453.

[54] Granum, P. E., Whitaker, J. R. and Skjelkvale, R., Trypsin activation of enterotoxin from *Clostridium perfringens* type A, *Biochim. Biophys. Acta* **1981**, 668, 325-332.

[55] Hanna, P. C., Wieckowski, E. U., Mietzner, T. A. and McClane, B. A., Mapping functional regions of *Clostridium perfringens* type A enterotoxin, *Infect. Immun.* **1992**, 60, 2110-2114.

[56] Mietzner, T. A., Kokai-Kun, J. F., Hanna, P. C. and McClane, B. A., A conjugated synthetic peptide corresponding to the C-terminal region of *Clostridium perfringens* type A enterotoxin elicits an enterotoxin-neutralizing antibody response in mice, *Infect. Immun.* **1992**, 60, 3947-3951.

[57] Johnson, S. and Gerding, D. N., Enterotoxemic Infections in *The Clostridia: Molecular Biology and Pathogenesis* (J. I. Rood, B. A. McClane, J. G. Songer and R. W. Titball, Eds.), Academic Press, London, **1997**, pp. 117-140.

[58] Kelly, C. P., Immune responses to *Clostridium difficile* infection, *Eur. J. Gastroenterol. Hepatol.* **1996**, 8, 1048-1053.

[59] McClane, B. A., Lyerly, D. M., Moncrief, J. S. and Wilkins, T. D., Enterotoxic clostridia: *Clostridium perfringens* type A and *Clostridium difficile,* in: *Gram-Positive Pathogens* (V. A. Fischetti, R. P. Novick, J. J. Ferretti, D. A. Portnoy, and J. Rood, Eds.), ASM Press, Washington, DC, **2000**, pp. 551-562.

[60] Moncrief, J. S., Barroso, L. A. and Wilkins, T. D., Positive regulation of *Clostridium difficile* toxins, *Infect. Immun.* **1997**, 65, 1105-1108.

[61] Braun, V., Hundsberger, T., Leukel, P., Sauerborn, M. and von Eichel-Streiber, C., Definition of the single integration site of the pathogenicity locus in *Clostridium difficile, Gene* **1996**, 18, 29-38.

[62] Hammond, G. A. and Johnson, J. L., The toxigenic element of *Clostridium difficile* strain VPI 10463, *Microb. Pathogen.* **1995**, 19, 203-213.

[63] Hammond, G. A., Lyerly, D. A. and Johnson, J. L., Transcriptional analysis of the toxigenic element of *Clostridium difficile, Microb. Pathogen.* **1997**, 22, 143-154.

[64] Braun, V., Mehlig, M., Moos, M., Rupnik, M., Kalt, B., Mahony, D. E. and von Eichel-Streiber, C., A chimeric ribozyme in *Clostridium difficile* combines features of group I introns and insertion elements, *Mol. Microbiol.* **2000**, 36, 1447-1459.

[65] Depitre, C., Delmee, M., Avesani, V., L' Haridon, R., Roels, A., Popoff, M. and Corthier, G., Serogroup F strains of *Clostridium difficile* produce toxin B but not toxin A. *J. Med. Microbiol* **1993**, 38, 434-441.

[66] Torres, J. F., Purification and characterization of toxin B from a strain of *Clostridium difficile* that does not produce toxin A, *J. Med. Microbiol.* **1991**, 35, 40-44.

[67] Kato, H., Kato, N., Katow, S., Maegawa, T., Nakamura, S. and Lyerly, D. M., Deletions in the repeating sequences of the toxin A gene of toxin A-negative, toxin B-positive *Clostridium difficile* strains, *FEMS Microbiol. Lett.* **1999**,175, 197-203.

[68] Soehn, F., Wagenknecht-Wiesner, A., Leukel, P., Kohl, M., Weidmann, M., von Eichel-Streiber, C. and Braun, V., Genetic rearrangements in the pathogenicity locus of *Clostridium difficile* strain 8864-implications for transcription, expression and enzymatic activity of toxins A and B, *Mol. Gen. Genet.* **1998**, 258, 222-232.

[69] Song, K. P., Bai, X. L. and Chang, S. Y., Nucleotide and peptide sequences of the open reading frame encoding a truncated toxin A gene of *Clostridium difficile* strain CCUG 20309, *DNA Sequence* **1999**, 10, 93-96.

[70] von Eichel-Streiber, C., Zec-Pirnat, I., Grabnar, M. and Rupnik, M., A nonsense mutation abrogates production of a functional enterotoxin A in *Clostridium difficile* toxinotype VIII stains of serogroups F and X, *FEMS Microbiol. Lett.* **1999**, 178, 163-168.

[71] Dupuy, B. and Sonenshein, A. L., Regulated transcription of *Clostridium difficile* toxin genes, *Mol. Microbiol.* **1998**, 27, 107-120.

[72] Hundsberger, T., Braun, V., Weidmann, M., Leukel, P., Sauerborn, M. and von Eichel-Streiber, C., Transcription analysis of the genes *tcdA-E* of the pathogenicity locus of

Clostridium difficile, Eur. J. Biochem. **1997**, 244, 735-742.
[73] Moncrief, J. S., Lyerly, D. M. and Wilkins, T. D., Molecular biology of the *Clostridium difficile* toxins, in: *The Clostridia: Molecular Biology and Pathogenesis* (J. I. Rood, B. A. McClane, J. G. Songer, R. W. Titball, Eds.), Academic Press, London, **1997**, pp. 369-392.
[73a] Song, K. P. and Faust, C., Molecular analysis of the promoter region of the *Clostridium difficile* toxin B gene that is functional in *Escherichia coli, J. Med. Microbiol.* **1998**, 47, 309-316.
[74] von Eichel-Streiber, C., Boquet, P., Sauerborn, M. and Thelestam, M., Large clostridial cytotoxins- a family of glycosyltransferases modifying small GTP-binding proteins, *Trends Microbiol.* **1996**, 4, 375-382.
[75] Smith, J. A., Cooke, D. L., Hyde, S., Borriello, S. P. and Long, R. G., *Clostridium difficile* toxin A binding to human intestinal epithelial cells, *J. Med. Microbiol.* **1997**, 46, 953-958.
[76] Rolfe, R. D. and Song, W., Purification of a functional receptor for *Clostridium difficile* toxin A from intestinal brush border membranes of infant hamsters, *Clin. Inf. Dis.* **1993**, 16 (suppl.4), S219-227.
[77] Lyerly, D. M. and Wilkins, T. D., *Clostridium difficile*, in: *Infections of the Gastrointestinal Tract* (M. J. Blaser, P. D. Smith, J. I. Ravdin, H. B. Greenberg, R. L. Guerrant, Eds.), Raven Press, New York, **1995**, pp. 867-891.
[78] Sauerborn, M., Leukel, P. and von Eichel-Streiber, C., The C-terminal ligand-binding domain of *Clostridium difficile* toxin A (TcdA) abrogates TcdA-specific binding to cells and prevents mouse lethality, *FEMS Lett.* **1997**, 155, 45-54.
[79] Qa'Dan, M., Spyres, L. M. and Ballard, J. D., pH-Induced conformational changes in *Clostridium difficile* toxin B, *Infect. Immun.* **2000**, 68, 2470-2474.
[80] K. Aktories, Bacterial toxins that target Rho proteins, *J. Clin. Invest.* **1997**, 99, 827-829.
[81] Just, I., Selzer, J., Wilm, M., von Eichel-Streiber, C., Mann, M. and Aktories, K., Glucosylation of Rho proteins by *Clostridium difficile* toxin B, *Nature* **1995**, 375, 500-503.
[82] Just, I., Wilm, M., Selzer, J., von Eichel-Streiber, C., Mann,. M. and Aktories, K., The enterotoxin from *Clostridium difficle* (ToxA) monoglucosylates the Rho proteins, *J. Biol. Chem.* **1995**, 270, 13932-13936.
[83] Aktories, K. and Just, I., Monoglucosylation of low-molecular-mass GTP-binding Rho proteins by clostridial cytotoxins, *Trends Cell Biol.* **1995**, 5, 441-443.
[84] Hecht, G., Pothoulakis, C., LaMont, J. T., and Madera, J. L., *Clostridium difficile* toxin A perturbs cytoskeletal structure and tight junction permeability of cultured intestinal epithelial monolayers, *J. Clin. Invest.* **1988**, 82, 1516-1524.
[85] Hecht, G., Koutsouris, A., Pothoulakis, C., LaMont, J. T. and Madara, J. L., *Clostridium difficile* toxin B disrupts the barrier function of T84 monolayers, *Gastroenterology* **1992**, 102, 416-423.
[86] Chaves-Olarte, E., Low, P., Freer, E., Norlin, T., Weidmann, M., von Eichel-Streiber, C. and Thelestam, M., A novel cytotoxin from *Clostridium difficile* serogroup F is a functional hybrid between two other large clostridial cytotoxins, *J. Biol. Chem.* **1999**, 274, 11046-11052.
[87] Selzer, J., Hofmann, F., Rex, G., Wilm, M., Mann, M., Just, I. and Aktories, K., *Clostridium novyi* α-toxin-catalyzed incoproration of GlcNAc into Rho subfamily proteins, *J. Biol. Chem.* **1996**, 271, 25173-25177.
[88] Frey, S. M. and Wilkins, T. D., Localization of two epitopes recognized by monoclonal antibody PCG-4 on *Clostridium difficile, Infect. Immun.* **1992**, 60, 2488-2492.
[89] Barroso, L. A., Moncrief, J. S., Lyerly, D. M. and Wilkins, T. D., Mutagenesis of the *Clostridium difficile* toxin B gene and effect on cytotoxic activity, *Microb. Pathogen.* **1994**, 16, 297-303.
[90] Hofmann, F., Busch, C., Prepens, U., Just, I. and Aktories, K., Localization of the glucosyltransferase activity of *Clostridium difficile* toxin B to the N-terminal part of the holotoxin, *J. Biol. Chem.* **1997**, 272, 11074-11078.
[91] Faust, C., Ye, B. and Song, K. P., The enzymatic domain of *Clostridium difficile* toxin A is located within its N-terminal region, *Biochem. Biophys. Res. Commun.* **1998**, 25, 100-105.
[92] von Eichel-Streiber, C., Heringdorf, D. M., Habermann, E. and Sartingen, S., Closing in on the toxic domain through analysis of a variant *Clostridium difficile* toxin B, *Mol. Microbiol.* **1995**, 17, 313-321.
[93] Hofmann, F., Busch, C. and Aktories, K., Chimeric clostridial cytotoxins: identification of the N-terminal region involved in protein

substrate recognition, *Infect. Immun.* **1998**, 66, 1076-1081.

[94] Busch, C., Hofmann, F., Selzer, J., Munro, S., Jeckel, D. and Aktories, K., A common motif of eukaryotic glycosyltransferases is essential for the enzyme activity of large clostridial cytotoxins, *J. Biol. Chem.* **1998**, 273, 19566-19572.

[95] Ward, S. J., Douce, G., Dougan, G. and Wren, B. W., Local and systemic neutralizing antibody responses induced by intranasal immunization with the nontoxic binding domain of toxin A from *Clostridium difficile*, *Infect. Immun.* **1999**, 67, 5124-5132.

[95a] Ward, S. J., Douce, G., Figueiredo, D., Dougan, G. and Wren, B., Immunogenicity of a *Salmonella typhimurium aroAaroD* vaccine expressing a nontoxic domain of *Clostridium difficile* toxin A, *Infect. Immun.* **1999**, 67, 2145-2152.

[96] Stevens, D. L., Necrotizing clostridial soft tissue infections, in: *The Clostridia: Molecular Biology and Pathogenesis* (J. I. Rood, B. A. McClane, J. G. Songer, R. W. Titball, Eds.), Academic Press, London, **1997**, pp. 141-151.

[97] Rood, J. I., Virulence genes of *Clostridium perfringens*, *Ann. Rev. Microbiol.* **1998**, 52, 333-360.

[98] Ballard, J., Bryant, A., Stevens, D. and Tweten, R., Purification and characterization of the lethal toxin (alpha toxin) of *Clostridium septicum*, *Infect. Immun.* **1992**, 60, 784-790.

[99] Titball, R. and Rood, J., Bacterial phospholipases, in: *Bacterial Protein Toxins* (K. Aktories, I. Just, Eds.), Springer-Verlag, Heidelberg, **2000**, pp. 529-556.

[100] Williamson, E. D. and Titball, R. W., A genetically engineered vaccine against the alpha-toxin of *Clostridium perfringens* protects against experimental gas gangrene, *Vaccine* **1993**, 11, 1253-1258.

[101] Awad, M. M., Bryant, A. E., Stevens, D. L. and Rood, J. I., Virulence studies on chromosomal α-toxin and θ-toxin mutants constructed by allelic exchange provide genetic evidence for the essential role of α-toxin in *Clostridium perfringens*-mediated gas gangrene, *Mol. Microbiol.* **1995**, 15, 191-202.

[102] Stevens, D. L., Tweten, R., Awad, M. M., Rood, J. I. and Bryant, A. E., Clostridial gas gangrene: evidence that α and θ toxins differentially modulate the immune response and induce acute tissue necrosis, *J. Infect. Dis.* **1997**, 176, 189-195.

[103] Stevens, D. L., Mitten, J. and Henry, C., Effects of α and θ toxins from *Clostridium perfringens* on human polymorphonuclear leukocytes, *J. Infect. Dis.* **1987**, 156, 324-333.

[104] Ellemor, D., Baird, R., Awad, M., Boyd, R., Rood, J. and Emmins, J., Use of genetically manipulated strains of *Clostridium perfringens* reveals that both alpha-toxin and theta-toxin are required for vascular leukostasis to occur in experimental gas gangrene, *Infect. Immun.* **1999**, 67, 4902-4907.

[105] Canard, B. and Cole, S. T., Genome organization of the anaerobic pathogen *Clostridium perfringens*, *Proc. Natl. Acad. Sci. USA* **1989**, 86, 6676-6680.

[106] Toyonaga, T., Matsushita, O., Katayama, S.-I., Minami, J. and Okabe, A., Role of the upstream region containing an intrinsic DNA curvature in the negative regulation of the phospholipase C gene of *Clostridium perfringens*, *Microbiol. Immunol.* **1992**, 36, 603-613.

[107] Matsushita, C., Matsushita, O., Katayama, S., Minami, J., Takai, K. and Okabe, A., An upstream activating sequence containing curved DNA involved in activation of the *Clostridium perfringens plc* promoter, *Microbiology* **1996**, 142, 2561-2566.

[108] Katayama, S., Matsushita, O., Jung, C.-M., Minami, J. and Okabe, O., Promoter upstream bent DNA activates the transcription of the *Clostridium perfringens* phospholipase C gene in a low teperature-dependent manner, *EMBO J.* **1999**, 18, 3442-3450.

[109] Lyristis, M., Bryant, A. E., Sloan, J., Awad, M. M., Nisbet, I. T., Stevens, D. L. and Rood, J. I., Identification and molecular analysis of a locus that regulates extracellular toxin production in *Clostridium perfringens*, *Mol. Microbiol.* **1994**, 12, 761-777.

[110] Shimizu, T., Ba-Thein, W., Tamaki, M. and Hayashi, H., The *virR* gene, a member of a class of two-component response regulators, regulates the production of perfringolysin O, collagenase, and hemagglutinin in *Clostridium perfringens*, *J. Bacteriol.* **1994**, 176, 1616-1623.

[111] Ba-Thein, W., Lyristis, M., Ohtani, K., Nisbet, I. T., Hayashi, H., Rood, J. I. and Shimizu, T., The *virR/virS* locus regulates the transcription of genes encoding extracellular

toxin production in *Clostridium perfringens, J. Bacteriol.* **1996**, 178, 2514-2520.

[112] Cheung, J. K. and Rood, J. I., The VirR response regulator from *Clostridium perfringens* binds independently to two imperfect direct repeats located upstream of the *pfoA* promoter, *J. Bacteriol.* **2000**, 182, 57-66.

[113] Banu, S., Ohtani, K., Yaguchi, H., Swe, T., Cole, S., Hayashi, H. and Shimizu, T., Identification of novel VirR/VirS-regulated genes in *Clostridium perfringens, Mol. Microbiol.* **2000**, 35, 854-864.

[114] MacFarlane, M. G. and Knight, B. C. J. G., The biochemistry of bacterial toxins I. The lecithinase activity of *Cl. welchii* toxins, *Biochem. J.* **1941**, 35, 884-902.

[115] Moreau, H., Pieroni, G., Joilivet-Reynaud, C., Alouf, J. E. and Verger, R., A new kinetic approach for studying phospholipase C (*Clostridium perfringens* a-toxin) activity on phospholipid monolayers, *Biochemistry* **1988**, 27, 2319-2323.

[116] Saint-Joanis, B., Garnier, T. and Cole, S. T., Gene cloning shows the alpha toxin of *Clostridium perfringens* to contain both sphingomyelinase and lecithinase activities, *Mol. Gen. Genet.* **1989**, 219, 453-460.

[117] Titball, R. W., Bacterial phospholipases C, *Microbiol. Rev.* **1993**, 57, 347-366.

[118] Guillouard, I., Garnier, T. and Cole, S. T., Use of site-directed mutagenesis to probe structure-function relationships of alpha-toxin from *Clostridium perfringens, Infect. Immun.* **1996**, 64, 2440-2444.

[119] Nagahama, M., Ochi, S., Kobayashi, K. and Sakurai, J., The relationship between histidine residues and various biological activities of *Clostridium perfringens* alpha toxin, *Adv. Exp. Med. Biol.* **1996**, 391, 251-5.

[120] Sakurai, J., Ochi, S. and Tanaka, H., Evidence for coupling of *Clostridium perfringens* alpha-toxin-induced hemolysis to stimulated phosphatidic acid formation in rabbit erythrocytes, *Infect. Immun.* **1993**, 61, 3711-3718.

[121] Sakurai, J., Ochi, S. and Tanaka, H., Regulation of *Clostridium perfringens* alpha-toxin-activated phospholipase C in rabbit erythrocyte membranes, *Infect. Immun.* **1994**, 62, 717-721.

[122] Fujii, T. and Sakurai, J., Contraction of the rat isolated aorta caused by *Clostridium perfringens* alpha-toxin (phospholipase C): evidence for the involvement of arachidonic acid metabolism, *Br. J. Pharmacol.* **1989**, 97, 119-124.

[123] Gustafson, C. and Tagesson, C., Phospholipase C from *Clostridium perfringens*-stimulates phospholipase A2-mediated arachidonic acid release in cultured intestinal epithelial cells (INT 407), *Scand. J. Gastroenterol.* **1990**, 25, 363-371.

[124] Bunting, M., Lorant, D. E., Bryant, A. E., Zimmerman, G. A., McIntyre, T. M., Stevens, D. L. and Prescott, S. M., Alpha toxin from *Clostridium perfringens* induces proinflammatory changes in endothelial cells, *J. Clin. Invest.* **1997**, 100, 565-574.

[125] Bryant, A. E. and Stevens, D. L., Phospholipase C and perfringolysin O from *Clostridium perfringens* upregulate endothelial cell-leukocyte adherance molecule 1 and intercellular leukocyte adherance molecule 1 expression and induce interleukin-8 synthesis in cultured human umbilical vein endothelial cells, *Infect. Immun.* **1996**, 64, 358-362.

[126] Naylor, C., Eaton, J., Howells, A., Justin, N., Moss, D., Titball, R. and Basak, A., Structure of the key toxin in gas gangrene has a prokaryotic calcium-binding C2 domain, *Nature Struct. Biol.* **1998**, 5, 738-746.

[127] Titball, R. W., Leslie, D. L., Harvey, S. and Kelly, D., Hemolytic and sphingomyelinase activities of *Clostridium perfringens* alpha-toxin are dependent on a domain homologous to that of an enzyme from the human arachadonic acid pathway, *Infect. Immun.* **1991**, 59, 1872-1874.

[128] Nagahama, M., Okagawa, Y., Nakayama, T., Nishioka, E. and Sakurai, J., Site-directed mutagenesis of histidine residues in *Clostridium perfringens* alpha-toxin, *J. Bacteriol.* **1995**, 177, 1179-1185.

[129] Guillouard, I., Alzari, P. M., Saliou, B. and Cole, S. T., The carboxy-terminal C2-like domain of the α-toxin from *Clostridium perfringens* mediates calcium-dependent membrane recognition, *Mol. Microbiol.* **1997**, 26, 867-876.

[130] Naylor, C., Jepson, M., Crane, D., Titball, R., Miller, J., Basak, A. and Bolgiano, B., Characterisation of the calcium-binding C-terminal domain of *Clostridium perfringens* alpha-toxin, *J. Mol. Biol.* **1999**, 294, 757-770.

[131] Jepson, M., Howells, A., Bullifent, H., Bolgiano, B., Crane, D., Miller, J., Holley, J.,

Jayasekera, P. and Titball, R., Differences in the carboxy-terminal (putative phospholipid binding) domains of *Clostridium perfringens* and *Clostridium bifermentans* phospholipases C influence the hemolytic and lethal properties of thes enzymes, *Infect. Immun.* **1999**, 67, 3297-3301.

[132] Bennett, A., Lescott, T., Phillpotts, R., Mackett, M. and Titball, R., Recombinant vaccinia viruses protect against *Clostridium perfringens* alpha-toxin, *Viral. Immunol.* **1999**, 12, 97-105.

[133] Bryant, A. E. and Stevens, D. L., The pathogenesis of gas gangrene. In *The Clostridia: Molecular Biology and Pathogenesis* (J. I. Rood, B. A. McClane, J. G. Songer, R. W. Titball, Eds.), Academic Press, London, **1997**, pp. 185-196.

[134] Asmuth, D. M., Olson, R. D., Hackett, S. P., Bryant, A. E., Tweten, R. K., Tso, J. Y., Zollman, T. and Stevens, D. L., Effects of *Clostridium perfringens* recombinant and crude phospholipase C and theta-toxin on rabbit hemodynamic parameters, *J. Infect. Dis.* **1995**, 172, 1317-23.

[135] Ohtani, K., Bando, M., Swe, T., Banu, S., Oe, M., Hayashi, H. and Shimizu, T., Collagenase gene (*colA*) is located in the 3'-flanking region of the perfringolysin O (*pfoA*) locus in *Clostridium perfringens, FEMS Microbiol. Lett.* **1997**, 146, 155-159.

[136] Ohtani, K., Takamura, H., Yaguchi, H., Hayashi, H. and Shimizu, T., Genetic analysis of the *ycgJ-metB-cysK-ygaG* operon negatively regulated by the VirR/VirS system in *Clostridium perfringens, Microbiol. Immunol.* **2000**, 44, 525-528.

[137] Shimizu, T., Okabe, A., Minami, J. and Hayashi, H., An upstream regulatory sequence stimulates expression of the perfringolysin O gene of *Clostridium perfringens, Infect. Immun.* **1991**, 59, 137-142.

[138] Rossjohn, J., Tweten, R., Rood, J. and Parker, M., Perfringolysin O, in: *Bacterial Toxins: A Comprehensive Sourcebook* (J. Alouf, J. Freer, Eds.), Academic Press, London, **1999**, pp. 496-510.

[139] Iwamoto, M., Ohno-Iwashita, Y. and Ando, S., Effect of isolated C-terminal fragment of θ-toxin (perfringolysin O) on toxin assembly and membrane lysis, *Eur. J. Biochem.* **1990**, 194, 25-31.

[140] Tweten, R. K., Pore-forming toxins in gram-positive bacteria, in: *Virulence Mechanisms of Bacterial Pathogens* (J. A. Roth, C. A. Bolin, K. A. Brogden, C. Minion, M. J. Wannemuehler, Eds.), American Society for Microbiology, Washington, DC, **1995**, pp. 207-229.

[141] Ohno-Iwashita, Y., Iwamoto, M., Mitsui, K., Ando, S. and Nagai, K., Protease-nicked θ-toxin of *Clostridium perfringens*, a new membrane probe with no catlaytic effect reveals two classes of cholesterol as toxin binding sites on sheep erythrocytes, *Eur. J. Biochem.* **1988**, 176, 95-101.

[142] Bryant, A. E., Bergstrom, R., Zimmerman, G. A., Salyer, J. L., Hill, H. R., Tweten, R. K., Sato, H. and Stevens, D. L., *Clostridium perfringens* invasiveness is enhanced by effects of theta toxin upon PMNL structure and function: the roles of leukocytotoxicity and expression of CD11/CD18 adherence glycoprotein, *FEMS Immunol. Med. Microbiol.* **1993**, 7, 321-336.

[143] Webster, A. C. and Frank, C. L., Comparison of immune response stimulated in sheep, rabbits and guinea pigs by the administration of multi-component clostridial vaccines, *Aust. Vet. J.* **1985**, 62, 112-4.

[143a] Whately, R., Zimmerman, G., Stevens, D., Parker, C., McIntyre, T. and Prescott, S., The regulation of platelet activating factor synthesis in endothelial cells – the role of calcium and protein kinase, *J. Biol. Chem.* **1989**, 11, 6325-6333.

[144] Nakamura, M., Sekino, N., Iwamoto, M. and Ohno-Iwashita, Y., Interaction of theta-toxin (perfringolysin O), a cholesterol-binding cytolysin, with liposomal membranes: change in the aromatic side chains upon binding and insertion, *Biochemistry* **1995**, 34, 6513-20.

[145] Nakamura, M., Sekino-Suzuki, N., Mitsui, K. and Ohno-Iwashita, Y., Contribution of tryptophan residues to the structural changes in perfringolysin O during interaction with liposomal membranes, *J. Biochem.* (Tokyo) **1998**, 123, 1145-55.

[146] Sekino-Suzuki, N., Nakamura, M., Mitsui, K. and Ohno-Iwashita, Y., Contribution of individual tryptophan residues to the structure and activity of theta-toxin (perfringolysin O), a cholesterol-binding cytolysin, *Eur. J. Biochem.* **1996**, 241, 941-947.

[147] Shimada, Y., Nakamura, M., Naito, Y., Nomura, K. and Ohno-Iwashita, Y., C-ter-

minal amino acid residues are required for the folding and cholesterol binding property of perfringolysin O, a pore-forming cytolysin, *J. Biol. Chem.* **1999**, 274, 18536-18542.

[148] Rossjohn, J., Buckley, J. T., Hazes, B., Murzin, A. G., Read, R. J. and Parker, M. W., Aerolysin and pertussis toxin share a common receptor-binding domain, *EMBO J.* **1997**, 16, 3426-34.

[149] Shepard, L. A., Heuck, A. P., Hamman, B. D., Rossjohn, J., Parker, M. W., Ryan, K. R., Johnson, A. E. and Tweten, R. K., Identification of a membrane-spanning domain of the thiol-activated pore-forming toxin *Clostridium perfringens* perfringolysin O: an alpha-helical to beta-sheet transition identified by fluorescence spectroscopy, *Biochemistry* **1998**, 37, 14563-74.

[150] Shatursky, O., Heuck, A., Shepard, L., Rossjohn, J., Parker, M., Johnson, A. and Tweten, R., The mechanism of membrane insertion of a cholesterol-dependent cytolysin: a novel paradigm for pore-foring toxins, *Cell* **1999**, 99, 293-299.

[151] van der Goot, F. G., Plasticity of the transmembrane beta-barrel, *Trends Microbiol.* **2000**, 8, 89-90.

[151a] Shepard, L., Shatursky, O., Johnson, A. and Tweten, R., The mechanism of pore assembly for a cholesterol-dependent cytolysin: formation of a large prepore complex precedes the insertio of the transmembrane b-hairpins, *Biochemistry* (in press).

[152] Stevens, D., Musher, D., Watson, D., Eddy, H., Hamill, R., Gyorkey, F., Rosen, H. and Mader, J., Spontaneous, nontraumatic gangrene due to *Clostridium septicum, Rev. Infect. Dis.* **1990**, 12, 286-296.

[153] Rossjohn, J., Feil, S. C., McKinstry, W. J., Tsernoglou, D., van der Goot, G., Buckley, J. T. and Parker, M. W., Aerolysin – a paradigm for membrane insertion of beta-sheet protein toxins? *J. Struct. Biol.* **1998**, 121, 92-100.

[154] Ballard, J. and Tweten, R., *Clostridium septicum* a-toxin, in: *The Clostridia: Molecular Biology and Pathogenesis* (J. Rood, B. McClane, J. Songer, R. Titball, Eds.), Academic Press, London, **1997**, pp. 251-257.

[155] Imagawa, T., Dohi, Y. and Higashi, Y., Cloning, nucleotide sequence and expression of a hemolysin gene of *Clostridium septicum, FEMS Microbiol. Lett.* **1994**, 117, 287-92.

[156] Ballard, J., Crabtree, J., Roe, B. and Tweten, R., The primary structure of *Clostridium septicum* alpha-toxin exhibits similarity with that of *Aeromonas hydrophila* aerolysin, *Infect. Immun.* **1995**, 63, 340-344.

[157] Ballard, J., Sokolov, Y., Yuan, W.-L., Kagan, B. and Tweten, R., Activation and mechanism of *Clostridium septicum* alpha toxin, *Mol. Microbiol.* **1993**, 10, 627-634.

[158] Gordon, V., Benz, R., Fujii, K., Leppla, S. and Tweten, R., *Clostridium septicum* alpha-toxin is proteolytically activated by furin, *Infect. Immun.* **1997**, 65, 4130-4134.

[159] Sellman, B. and Tweten, R., The propeptide of *Clostridium septicum* alpha toxin functions as an intramolecular chaperone and is a potent inhibitor of alpha toxin-dependent cytolysis, *Mol. Microbiol.* **1997**, 25, 429-440.

[160] Sellman, B. R., Kagan, B. L. and Tweten, R. K., Generation of a membrane-bound, oligomerized pre-pore complex is necessary for pore formation by *Clostridium septicum* alpha toxin, *Mol. Microbiol.* **1997**, 23, 551-558.

[161] Parker, M. W., Buckley, J. T., Postma, J. P., Tucker, A. D., Leonard, K., Pattus, F. and Tsernoglou, D., Structure of the *Aeromonas* toxin proaerolysin in its water-soluble and membrane-channel states, *Nature* **1994**, 367, 292-5.

[162] Rossjohn, J., Feil, S. C., McKinstry, W. J., Tweten, R. K. and Parker, M. W., Structure of a cholesterol-binding, thiol-activated cytolysin and a model of its membrane form, *Cell* **1997**, 88, 685-692.

[163] Diep, D. B., Nelson, K. L., Lawrence, T. S., Sellman, B. R., Tweten, R. K. and Buckley, J. T., Expression and properties of an aerolysin–*Clostridium septicum* alpha toxin hybrid protein, *Mol. Microbiol.* **1999**, 31, 785-94.

[164] Gordon, V., Nelson, K., Buckley, J., Stevens, V., Tweten, R., Elwood, P. and Leppla, S., *Clostridium septicum* alpha toxin uses glycosylphosphatidylinositol-anchored protein receptors, *J. Biol. Chem.* **1999**, 274, 27274-27280.

[165] Roth, F., Jansen, K. and Petzke, S., Detection of neutralizing antibodies against alpha-toxin of different *Clostridium septicum* strains in cell culture, *FEMS Immunol. Med. Microbiol.* **1999**, 24, 353-9.

[165a] Webster, A. C., Frank, C. L., Comparison of immune response stimulated in sheep, rabbits and guinea pigs by the administration

of multi-component clostridial vaccines, *Aust. Vet. J.*, **1985**, 62, 112-114.

[166] Leary, S. E. C. and Titball, R. W., The *Clostridium perfringens* β-toxin, in: *The Clostridia: Molecular Biology and Pathogenesis* (J. I. Rood, B. A. McClane, J. G. Songer, R. W. Titball, Eds.), Academic Press, London, **1997**, pp. 243-250.

[167] Steinthorsdottir, V., Halldorsson, H. and Andrésson, O. S., *Clostridium perfringens* beta-toxin forms multimeric transmembrane pores in human endothelial cells, *Microb. Pathogen.* **2000**, 28, 45-50.

[168] Gibert, M., Jolivet-Renaud, C. and Popoff, M. R., Beta2 toxin, a novel toxin produced by *Clostridium perfringens, Gene* **1997**, 203, 65-73.

[169] Herholz, C., Miserez, R., Nicolet, J., Frey, J., Popoff, M., Gibert, M., Gerber, H. and Straub, R., Prevalence of b-toxigenic *Clostridium perfringens* in horses with intestinal disorders, *J. Clin. Microbiol.* **1999**, 37, 358-361.

[170] Payne, D. and Oyston, P., The *Clostridium perfringens* e-toxin, in: *The Clostridia: Molecular Biology and Pathogenesis* (J. I. Rood, B. A. McClane, J. G. Songer, R. W. Titball, Eds.), Academic Press, London, **1997**, pp. 439-447.

[171] Roggentin, P. and Schauer, R., Clostridial sialidases, in: *The Clostridia: Molecular Biology and Pathogenesis* (J. I. Rood, B. A. McClane, J. G. Songer, R. W. Titball, Eds.), Academic Press, London, **1997**, pp. 423-437.

[172] Canard, B., Garnier, T., Saint-Joanis, B. and Cole, S. T., Molecular genetic analysis of the *nagH* gene encoding a hyaluronidase of *Clostridium perfringens, Mol. Gen. Genet.* **1994**, 243, 215-224.

[173] Okabe, A. and Cole, S., Extracellular enzymes from *Clostridium perfringens* and *Clostridium histolyticum* that damage connective tissue, in: *The Clostridia: Molecular Biology and Pathogenesis* (J. Rood, B. McClane, J. Songer, R. Titball, Eds.), Academic Press, London, **1997**, pp. 411-422.

[174] Awad, M., Ellemor, D., Bryant, A., Matsushita, O., Boyd, R., Stevenes, D., Emmins, J. and Rood, J., Construction and virulence testing of a collagenase mutant of *Clostridium perfringens, Microb. Pathogen.* **2000**, 28, 107-117.

[175] Jin, F., Matsushita, O., Katayama, S., Jin, S., Matsushita, C., Minami, J. and Okabe, A., Purification, characterization, and primary structure of *Clostridium perfringens* lambda-toxin, a thermolysin-like metalloprotease, *Infect. Immun.* **1996**, 64, 230-237.

[176] Carman, R. J., Perelle, S. and Popoff, M. R., Binary toxins from *Clostridium spiroforme* and *Clostridium perfringens*, in: *The Clostridia: Molecular Biology and Pathogenesis* (J. I. Rood, B. A. McClane, J. G. Songer, R. W. Titball, Eds.), Academic Press, London, **1997**, pp. 359-367.

[177] Nagahama, M., Sakaguichi, Y., Kobayashi, K., Ochi, S. and Sakurai, J., Characterization of the enzymatic component of *Clostridium perfringens iota-toxin, J. Bacteriol.* **2000**, 182, 2096-2103.

[178] Matsushita, O., Jung, C.-M., Katayama, S., Minami, J., Takahashi, Y. and Okabe, A., Gene duplication and multiplicity of collagenases in *Clostridium histolyticum, J. Bacteriol.* **1999**, 181, 923-933.

[179] Tsutsui, K., Minami, J., Matsushita, O., Katayama, S.-I., Taniguchi, Y., Nakamura, S., Nishioka, M. and Okabe, A., Phylogenetic analysis of phospholipase C genes from *Clostridium perfringens* types A to E and *Clostridium novyi, J. Bacteriol.* **1995**, 177, 7164-7170.

[180] Boquet, P., Munro, P., Fiorentini, C. and Just, I., Toxins from anaerobic bacteria: specificity and molecular mechanisms of action, *Curr. Opin. Microbiol.* **1998**, 1, 66-74.

[181] Popoff, M. R., Chaves-Olarte, E., Lemichez, E., von Eichel-Streiber, C., Thelestam, M., Chardin, P., Cussac, D., Antonny, B., Chavrier, P., Flatau, G., Giry, M., de Gunzburg, J. and Boquet, P., Ras, Rap, and Rac small GTP-binding proteins are targets for *Clostridium sordellii* lethal toxin glucosylation, *J. Biol. Chem.* **1996**, 271, 10217-10224.

7
Clostridial Neurotoxins

Conrad P. Quinn and Nigel P. Minton

7.1
Introduction

Clostridial neurotoxins (CNTs) are the most potent natural poisons known [1, 2] The CNT family comprises eight structurally related proteins; tetanus toxin (TeNT) and the seven botulinum neurotoxin types (BoNT/A-BoNT/G). The CNTs intoxicate the peripheral (BoNTs) or central (TeNT) nervous systems where, respectively, they inhibit calcium-dependent secretion of acetylcholine at the neuromuscular junction and glycine secretion from Renshaw cells of the spinal cord. The consequences of these neural blockades are the classic flaccid paralyses of botulism and spastic paralyses of tetanus, which may manifest in a variety of clinical conditions.

A tremendous amount of literature has been and continues to be published on the molecular genetics of the organisms and the structure, function and exploitation of the clostridial neurotoxins they produce. This includes the recent excellent and detailed reviews by Popoff and Marvaud [3], Herreros et al. [2] and Schiavo et al. [4]. The focus of the present chapter will be to summarize our increasing understanding of the organism responsible for the production of these remarkable proteins, to review neurotoxin mode of action and to divine their current and future prospects as therapeutic molecules. In this context there will be an emphasis on the genetics of production and on the enzymatic activity and neuronal targeting capabilities of the proteins. In this later respect, particular emphasis is placed on the botulinum neurotoxin types A and B and tetanus toxin.

7.2
General considerations

7.2.1
The disease

Botulism can occur when toxin-producing bacteria infect wounds (wound botulism) or the intestinal tract (intestinal toxaemia), or following the ingestion of contaminated food in which toxin has been produced (foodborne botulism) [5]. Until recently [6] wound botulism has rarely been encountered and the clinical importance of the "organism" relates to intestinal toxaemia (infant/intestinal botulism) and foodborne botulism. In the USA, intestinal toxaemia in the very young (infant botulism) represents the most common manifestation of the disease, and its prevalence has led to suggestions of a link to sudden infant death syndrome [7]. In contrast, in the UK foodborne botulism represents the major concern [5]. Indeed, the organism has been the principal target in food processing for almost a century.

Tetanus ("lockjaw"), occurs when spores of toxinogenic strains of *C. tetani* gain access to a site in the body of reduced oxygen tension, where they germinate to produce vegetative rods. The TeNT thereafter produced may cause localized, or more usually, generalized tetanus. The normal site of entry are wounds or abrasions, but in the developing world neonatal tetanus commonly occurs following infection of the umbilical stump. While tetanus is entirely preventable through immunization, it is endemic in 90 developing countries, where half a million infants die every year due to neonatal tetanus. In contrast, cases of human botulism occur relatively infrequently, with for instance around 200 cases of foodborne botulism within the EU each year, and 930 cases on a world-wide basis. It does, however, remain one of the most emotive of diseases due to its serious nature, including an extremely high fatality rate. The cost (per incident) of foodborne botulism is several orders of magnitude higher than for other forms of foodborne illness (e. g., listeriosis, salmonellosis). Cases of botulism therefore continue to command a high public profile in the western media.

7.2.2
Characteristics and epidemiology of the organisms

Botulinum neurotoxins have been subdivided into seven distinct serotypes (types A to G), although variations within an individual serotype are evident, particularly type F toxins [8]. The ability to produce toxin is confined to the genus *Clostridium*, the members of which are anaerobic, spore-forming, Gram-positive rods. Traditionally, botulinogenic clostridial strains have been classified as *Clostridium botulinum*. However, it has been known for many years that strains of *C. botulinum* are composed of four distinct physiological groupings [9]: Group I (proteolytic *C. botulinum*) strains producing toxin of types A, B or F; Group II (non-proteolytic *C. botulinum*) strains producing toxin of types B, E, or F; Group III strains producing toxin of types C or D, and; Group IV strains producing toxin of type G. In contrast, TeNT appears to show little variation in structure, regardless of source. Similarly, the ability to produce this neurotoxin is confined to a single clostridial species, *C. tetani*. Each of the four *C. botulinum* Groups encompass isolates which resemble their toxinogenic counterparts in all respects bar their ability to produce botulinum toxin. In addition to these four groupings, clostridial strains otherwise indistinguishable from *C. butyricum* and *C. barati* have been isolated from cases of infant botulism which produce BoNT/E and BoNT/F, respectively [10-13]. These six Groups, together with *C. tetani*, comprise all the currently known neurotoxinogenic clostridia (Table 1). The Group in which an isolate falls influences the nucleotide sequence of the cloned gene, the nature of the ancillary genes involved in progenitor toxin formation, and the type of genetic element on which the determinants of botulism are found.

Table 1. Characteristics of the neutotoxinogenic clostridia

Species	*Group*	*Toxin type*	*Complex form*	*Proteolysis*[a]	*Glucose*	*Lipase*	*Related species*
C. botulinum	I	A	M,L,LL	+	+	+	*C. sporogenes*
		B	M,L	+	+	+	*C. sporogenes*
		F	M	+	+	+	*C. sporogenes*
C. botulinum	II	B	M,L	–	+	+	"E-like" organims
		E	M	–	+	+	"E-like" organims
		F	M	–	+	+	"E-like" organims
C. botulinum	III	C	M,L	+/–	–	–	*C. novyi*
		D	M,L	+/–	–	–	*C. novyi*
C. argentinense	IV	G	M,L	+	–	–	*C. argentinense*
C. butyricum	V	E	M	[b]	–	–	*C. butyricum*
C. barati	VI	F	M	–	+	–	*C. barati*
C. tetani		TeTx		[c]	–	–	*C. tetanomorphum*

[a] digestion of milk proteins.
[b] "E-like" organisms refers to clostridial strains that share characteristics with *C. botulinum* group II.
[c] *C. butyricum* partially digests milk protein.
[d] *C. tetani* is obligately proteolytic but cannot degrade complex proteins such as those in milk.

Only Group I (proteolytic *C. botulinum*) and Group II (non-proteolytic *C. botulinum*) are of importance with regard to human disease. The neurotoxins implicated being those of types A, B, E and to a lesser extent type F. Group III strains appear to be associated with botulism in birds and animals, while Group IV has no known association with disease [5]. In more recent years; the application of 16sRNA sequencing technology has unequivocally demonstrated that the Groups I to IV are composed of distinct species ([14, 15] see Figure 1). In the case of those strains important to human botulism, Group I strains, regardless of toxin type, are highly related to one another (99.7 to 100% 16sRNA sequence similarity) and together with *Clostridium sporogenes* form a single phylogenic unit. Group II strains that produce toxin of types B, E, or F (and their non-toxinogenic counterparts) form a distinct line quite separate from other saccharolytic clostridia and phylogenetically far removed from the Group I strains.

In order to further delineate strains, Lin and Johnson, subjected representatives of Groups I and II to analysis by pulsed-field gel electrophoresis (PFGE) [16]. In the case of the former grouping, only a limited number of type A strains were examined, *viz.*, 4 Hall A strain culture deposits (WHO A, ATCC 3502, and those Hall A deposits from the laboratories of E. A. Johnson and H. Sugiyama). Their analyses revealed that although there was some conservation in overall genome size (approximately 4.0 Mb) and restriction profiles with the three enzymes tested (*Mlu*I, *Ksp*I and *Sma*I), significant differences were apparent. These data indicate that substantial genomic variation occurs, even between supposedly identical strains. In contrast, PFGE analysis of Group II strains was undertaken on a much wider selection of strains [17]. In the initial study some 4 type B, 14 type E and 3 type F strains were analyzed. Except for strains of obvious clonal origin, the different strains were easily distinguishable based on dissimilar restriction profiles. This diversity made it difficult to assign strains to common lineages, and indicated a large diversity within Group II strains. Variation was also evident from the estimates of genome size, which ranged from between 3.6 to 4.1 Mb. A wide biodiversity was particularly evident for type E strains.

In a later study 42 further strains were examined, isolated from 21 Finnish trout farms [18]. A total of 24 different restriction profiles were generated using *Xma*I, which could be grouped into 15 clusters, with a similarity index of 76%, which also encompassed the 12 previously characterized North American strains. These clusters could be further subdivided into 22 different pulsotypes through the use of additional restriction enzymes, particularly *Xho*I. The data showed that strains with the same pulsotype were not restricted to samples from the same location, or even the same country. The authors concluded "that (i) *C. botulinum* type E is an organism that is well adapted to northern aquatic environments and occasionally contaminates animal or human food chains at random, and (ii) there is or has been extensive spread and exchange of *C. botulinum* type E strains in the northern temperate regions, the vehicle of which has not been determined". These studies suggest that PFGE has great utility as a typing method for establishing a causal link between patients and incriminated foodstuffs.

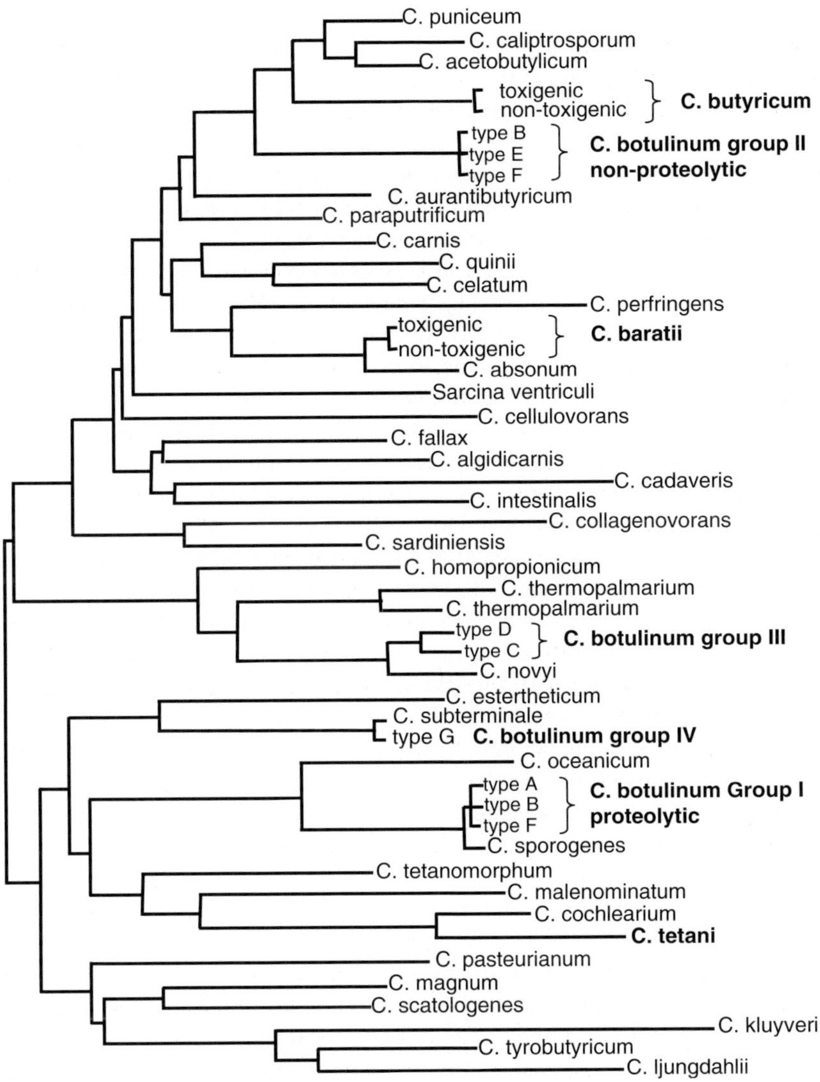

Figure 1. Dendrogram showing the phylogenic position of clostridial species able to produce CNTs.

7.3
Genetic organization of toxin genes

Botulinum toxin found in bacterial cultures and contaminated foodstuffs is not composed of the BoNT alone but is always found complexed with other nontoxic proteins. Three forms of complex, or progenitor toxins, are recognized: M toxin (300 kDa), L toxin (500 kDa), and LL toxin (900 kDa). All three forms contain a protein of equivalent size to the BoNT called the nontoxic, non-hemagglutinating protein (NTNH, 150 kDa). The larger L and LL progenitor toxins additionally contain an undefined number of proteins with hemagglutinin (HA) activity. The form of progenitor toxin found varies between the different toxinogenic types (Table 1). TeNT, on the other hand, is never found in association with any of these proteins. Botulism, unlike tetanus, is principally caused by the ingestion of food in which contaminating clostridial cells have produced BoNT. Thus, as progenitor toxins exhibit greater stability to high temperature and pH extremes than purified BoNT, it has been suggested that the nontoxic proteins serve to aid the neurotoxin in withstanding the acidic and proteolytic environment of the stomach [19, 20].

Analysis of the genes concerned with botulinum toxin production initiated with the cloning and characterization of the genes encoding the neurotoxins. The first CNT gene sequence to be reported was that encoding TeNT [21, 22]. This was closely followed by the determination of the sequence of the CNT gene of a Group I type A strain of *C. botulinum* [23], and thereafter the elucidation of the sequence of representative sequences of all CNT types, including those carried by non-*C. botulinum* species [24]. Thereafter, attention has switched to the analysis of those ancillary genes coding for the non-toxin components of the botulinum toxin complex. As a consequence, considerable information is now available as to the location and structure of genes concerned with toxin production.

7.3.1
The organization of the toxin genes

In all cases examined to date, the structural gene for a given BoNT is found to be immediately preceded by the structural gene encoding the nontoxic non-hemagglutinating protein (NTNH). A NTNH protein is not found in association with tetanus toxin, and no such gene is located within the vicinity of the TeNT gene. Aside from the juxtaposition of these two genes, there is no other feature that is common to all neurotoxinogenic clostridia. Accordingly, four different gene loci (I to IV) have been described [25]. These are reproduced in Figure 2, together with two further recognizable structural arrangements, types V and VI.

In those clostridia known to produce a botulinum toxin complex with hemagglutinating (HA) activity, the genes encoding the protein components responsible reside 5′ to the NTNH gene (gene loci types I, II, and V), and in all cases to date are encoded by the opposite DNA strand (Figure 2). The arrangements in gene loci I and II differ only in the relative position of an open reading frame (ORF) encoding a transcriptional factor (CntR) believed to be responsible for the transcription of the toxin genes. In the case of all type B strains, and certain type A strains, *cntR* resides

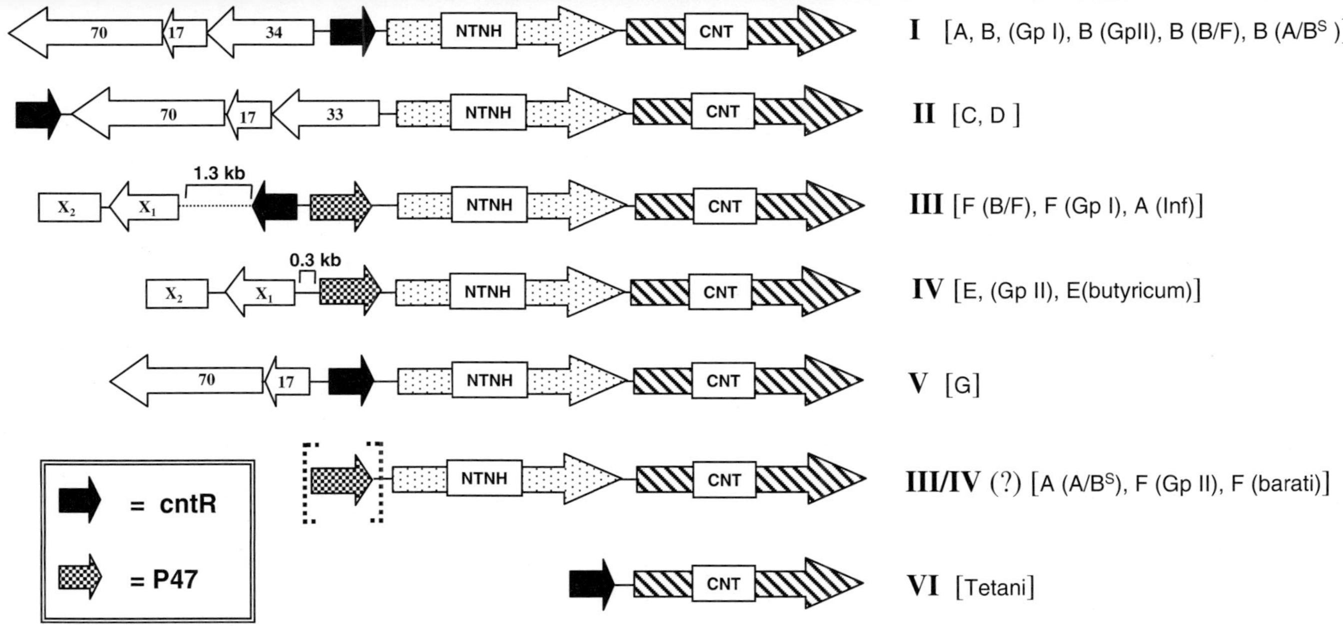

Figure 2. Organization of botulinum toxin genes. Strains have been assumed to carrying a gene locus corresponding to structure I if the presence of the genes for CntR, NTNH and CNT had been established and at least the first gene (i. e., HA33/34) of the upstream hemagglutinin operon. Only in the case of strains ATCC 43757 (type B/F) and NCTC 2916 (type A/ B^S) has the entire operon been shown to be present. In the case, of strains NCTC 7272 (type A), NCTC 7273 (type B), 17B (type A), Lamanna (type B), A-NIH (type A) and 667 (type A/B^S), only part of the hemagglutinin operon has been characterized. Strains possessing gene locus of structure II are the type C strains/phage C-Stockholm, C-6814 and C-468, and the type D strains CB-16, D-1873 and D-4947. Gene locus of structure III are found in strain Langeland and Kyoto-F (infant type A strain). In the case of ATCC 43757, the presence of OrfX1 and OrfX2 has not been established upstream of the type F locus. Structure IV has been established in the type E strains Iwanai and BL6340 (*C. butyricum*). Structure V is unique to the type G strain, ATCC 27322. The assignment of the remaining strains to any structure is not possible, as the sequences either upstream of the NTNH gene (type A strain 62A, type B strain ACTC25765), or upstream of the P47 gene (type A strains NCTC 2916 and 667 (in relation to the BoNT/A gene), the type E strain Mashike, and the type F strains 202F and ATCC 43256 (*C. barati*), have not been determined. Figure compiled using data from GenBank sequences with the following Accession numbers, X79102 (NCTC 7273), X79104 (NCTC 7272), X92973 (62A), D67030 (A-NIH), D84289 (A-Infant 7I03-H), Y14238, Y14239 and L42537 (NCTC 2916), X8748, X87850 and X87849 (667Ab), Y13630 and Y13631 (ATCC 43757), X79103 (ATCC 25765), U63808 (B-Lamanna), X78229 (ATCC 25765), X96494, X99064, L35496 and AB004779 (F-Langeland), S73676, Y12091 and X71086 (202F), X96493, X87974 and AB0044778 (F-Kyoto), Y12091 (*C. barati*), X87972 (ATCC 27322), X72793 (C-468), AB037166 (C6814), AB012111 (D-CB16), AB012112 (D-1873), AB37920 (D-4947), D88419 and D12739 (*C. butyricum* BL6340), D88418 (E-Iwanai) and D12697 (E-Mashike).

between the genes encoding the NTNH/CNT and HA genes. In the case of type C and D strains (genome type III) the *cntR* gene is located 3′ to the HA genes. The *cntR* gene of strains producing BoNT/G (genome type V) is also located between the HA genes and NTNH. However, in this instance the gene encoding HA33 is absent (Figure 2).

Clostridia of the remaining botulinum genome types are unified by the presence upstream of the NTNH gene of an ORF encoding a protein with a molecular weight of 47 kDa, and hence designated as P47. Its function remains unknown, but there have been some suggestions that it may play a regulatory role [25]. There is, however, no evidence to support this view. Furthermore, in gene loci III (Group I type F strains, type F strains which also produce type B toxin, and type A strains associated with infant botulism) the *cntR* gene, which is believed to play a role in toxin gene transcription, is also present immediately upstream of P47. Thus, if P47 is involved in regulation of toxin gene expression, then it must be over and above the regulatory control brought to bear by the CntR transcription factor.

In organisms which produce BoNT/E, it has been clearly established that *cntR* does not reside upstream of P47. Rather, two genes of unknown function are present almost immediately (0.3 kb) upstream of P47 which have been designated orfX1 and orfX2 [25]. These same genes are present some 1.3 kb upstream of *cntR* in genome type III strains. Primers based on conserved DNA regions from the *cntR* genes failed to amplify appropriate fragments from DNA from the genome of the type A strain 62A, the type B strain Lamanna, and the type C strain Stockholm. However, this may not be taken as conclusive evidence for the absence of *cntR* homologues from these, or similar, strains.

In certain instances, the regions upstream of the NTNH gene have not been rigorously analyzed. Thus, while it is possible to position a P47 gene immediately upstream, it is not possible to say whether either *cntR* or orfX1-orfX2 are present in these strains. It is therefore not possible to assign them to any particular gene locus structure (Figure 2).

7.3.2
Strains carrying more than one toxin gene

Certain strains of *C. botulinum* are able to produce more than one toxin serotype. These are confined to strains able to produce either both type B and F CNT (e. g., ATCC 43757), or strains able to produce both type A and B toxins. Other Group I type A strains have been discovered which, while they only produce a type A toxin, also possess defective genetic loci for type B toxin. The presence of these so-called "silent" genes was uncovered following the development of serotype-specific primers for the detection of toxin genes. When such a primer pair specific to the BoNT/B gene were employed against 79 type A strains, positive results were unexpectedly obtained with 43 strains, of which only one was shown to be producing BoNT/B [26]. Subsequent sequence analysis of one of these strains (667) showed that a genetic locus corresponding to structure I (Figure 2) was present in which the CNT gene encoded a protein typical of a type B CNT [27]. Among other changes, this BoNT/B gene contained a proximal stop codon (at codon position 128), two amino acid deletions, and two single nucleotide positions producing two shifts in reading frame. The two deleted amino acids (positions 328-329) were subsequently shown, by appropriate PCR, to be similarly absent in 35 of 37 strains thought to possess the silent BoNT/B gene isolated from diverse US locations [28]. The two strains which did not carry the deletion were shown to produce BoNT/B.

Analysis of the NTNH encoded by the gene preceding the silent BoNT/B gene of strain 667 indicated that it is composed of a chimeric molecule, comprising the N-terminal domain of NTNH/A and the COOH-terminal domain of NTNH/B. Thus, the N-terminus (amino acids 1 to 627) shares 69.4 % identity with the NTNH/A of NTCT 2916 and 97.5 % identity with the NTNH/B of the strain Lamanna, while the COOH-terminal domain (residues 542 to 1194) shares 99.5 % identity with NTNH/A and 64.6 % identity to NTNH/B. Residues 542 to 627 are, with the exception of a single amino acid conserved between all three proteins. These data strongly suggest that the NTNH gene of this strain was generated by recombination between a NTNH/B and NTNH/A gene. Recombination was apparently made possible by the presence of a highly conserved stretch of DNA encoding 76 out of 77 identical amino acids (corresponding to position 519 to 695 in NTNH/A, and residues 551 to 639 in NTNH/B).

The presence of this conserved region of DNA between the two classes of NTNH gene led directly to the erroneous assignment of the NTNH/B gene in strain NCTC 2916 to a position immediately upstream of the BoNT/A gene by Henderson et al. [29]. Thus, the partial nucleotide sequence of NTNH/A previously obtained during the sequencing of the BoNT/A gene [23] was extended through the use of inverse PCR. The use of primers based on the region conserved between NTNH/A and NTNH/B led to the cloning of a fragment which was derived not from the intended NTNH/A-encoding region of the NCTC 2916 genome, but to the isolation of a DNA fragment coding for NTNH/B, and thereafter to adjacent regions of the operon encoding the type B hemagglutinating genes. This incorrect assignment was subsequently pointed out by Hutson et al [27].

This discovery of chimeric genes in neurotoxinogenic clostridia was not the first instance of its kind, as previous workers had demonstrated the presence of mosaic CNT genes encoding chimeric toxins composed of BoNT/D (N-terminus) and BoNT/C (COOH-terminus) domains [30, 31]. The existence of such mosaic molecules provides insight into the mechanism by which the different proteins/genes have evolved, through recombination.

7.3.3 Localization of the genes

In the majority of the neurotoxinogenic clostridia the genes responsible for botulinum toxin production appear to be localized to the chromosome. This certainly seems to be the case for those strains of most importance in human disease, the Group I and Group II strains. However, the presence of BoNT genes within genotypically distinct clostridial groupings indicates that lateral transfer of the genes has occurred. The toxin genes are, or were at some point, localized to a mobile genetic element.

Plasmid-encoded genes

Many Group I and II *C. botulinum* strains have been shown to harbor cryptic plasmids, but curing experiments have failed to prove any association with toxin genes [32-34]. In contrast, the botulinum gene complex of Group IV *C. botulinum* type G strains has been definitively shown to reside on a 81 MDa plasmid [35]. Similarly, the TeNT gene has been shown to be encoded by a large 75 kb plasmid [36]. The large size of both plasmids is consistent with a ability to undergo conjugative transfer, but no evidence for transfer between clostridial strains has been reported.

The demonstration that strains other than *C. botulinum* can produce either BoNT/E (*C. butyricum*) or BoNT/F (*C. barati*) may be indicative of the localization of the genes on a large conjugative plasmid. Data has been presented in support of this view for *C. butyricum*, ATCC 43181, and ATCC 43755 [37]. Thus, two BoNT/E gene-specific oligonucleotide primers were shown to PCR-amplify “larger” quantities of the expected 1.2 kb fragment when the template employed was a “plasmid” preparation than when “chromosomal” DNA was used. These data are clearly subjective, and more substantive data will be required before the CNT genes of these strains can be confidently assigned to an extrachromosomal plasmid. More recently, the restriction fragments derived form the genomes of 11 Chinese isolates of *C. butyricum* producing BoNT/E were separated by PFGE and probed with toxin gene-specific probes. The data obtained indicated that, in all cases, the gene was chromosomally located [38].

Genes encoded by pseudolysogenic phage

While many strains of *C. botulinum* possess bacteriophages [39-41], a direct relationship between bacteriophage and toxinogenicity has only been established in the case of BoNT/C1 and BoNT/D. The localization of the botulinum toxin genes on phage genomes was initially established using classical genetic techniques, involving the curing and re-infection of the resultant toxin minus strains with phage [42]. Thus, a *C. botulinum* type C strain could be cured of its TOX+ phage, and the resultant non-toxigenic derivative converted to a toxinogenic *C. botulinum* type D strain by infection with a type D TOX+ phage. Thereafter, the phage location of toxin genes was established via recombinant methods, when both the BoNT genes were cloned from their respective bacteriophage genomes and their nucleotide sequences determined [43, 44].

As the TOX+ phage is lost relatively easily by Group III clostridia, it would appear that the prophage is not stably integrated into the host chromosome, but exists in an extrachromosomal state termed “pseudolysogeny”. This form of existence falls somewhere between virulence and lysogeny. This location was confirmed by the demonstration that BoNT gene-specific probes hybridized only with phage DNA, and not host chromosome [43, 44]. The phages of Group I and II *C. botulinum* in contrast exist in a lysogenic state. As a consequence, phage-sensitive bacteria are seldom isolated, and when they are, retain the ability to produce botulinum toxin [42]. Despite these findings, the involvement of bacteriophages in the toxigenesis of *C. botulinum* Groups I and II cannot be discounted.

Genes encoded by lysogenic prophage

There is no evidence to definitively show that the botulinum toxin genes are carried by a lysogenic prophage. A number of experiments do, however, suggest that this may be the case in certain clostridial species.

In a study conducted by Zhou et al. the use of BoNT/E gene-specific probes in Southern blots demonstrated that the CNT gene was located in the chromosome of *C. butyricum* strains ATCC 5839 and 5521, but was absent in the non-toxinogenic strain ATCC 18398 [45]. Both the toxinogenic and non-toxinogenic were found to harbor a prophage, which though of a similar genomic size (34 kb), were entirely unrelated [45]. However, because probes derived either from the BoNT/E gene or phage DNA hybridized to similar sized *Nde*I and *Hin*dIII fragments of the ATCC 5839 chromosome, it was suggested [45] that *botE* and the phage genome are in some way linked on the host chromosome. In keeping with this hypothesis, it was shown that the ability to produce BoNT/E could be transferred from *C. butyricum* ATCC 5839 to a non-toxinogenic E-like *C. botulinum* strain by a phage-mediated process. Transfer (at frequencies of 1 in 10^4 recipient cells) was only possible when the recipient was incubated in broth containing phage induced by mitomycin C from the donor strain which had been supplemented with the filter-sterilized culture filtrate of a helper strain (a non-toxinogenic *C. butyricum* strain ATCC 19398). It was concluded that the phage particles produced by ATCC 5839 are defective, and only become infective through the participation of the helper strain.

On the face of it these data indicate that the toxin gene is encoded by a lysogenic phage. However, while purified phage DNA was shown to act as a template for the PCR mediated amplification of a *botE*-specific DNA fragment, it did not hybridize to a radiolabelled BoNT/E gene probe [45]. This implies that only a very minor proportion of the phage genomes contained the BoNT/E gene. The most likely explanation for this observation is that more than one type of phage was present, with the BoNT/E-encoding phage being in the minority. If the attachment site of the identified prophage resides next to the BoNT/E gene in the *C. butyricum* chromosome, then the phage preparation isolated could have consisted of wild-type phage genomes and derivatives carrying host chromosome from adjacent to the prophage attachment site. Such derivatives would arise following imperfect excision, in much the same way as in other specialized transducing phage such as phage lambda. The abundance of phage particles carrying the BoNT/E gene would than be dictated by the frequency of imperfect excision, e.g., 1 in 10^{-6} in the case of phage lambda. The low frequency of these types of event would explain why the gene could not be detected in preparations of phage DNA by standard DNA/DNA hybridization techniques.

Indirect evidence for the involvement of phages in neurotoxin gene transmission has also come from the analysis of DNA regions in the vicinity of the botulinum toxin locus. In certain strains an open reading frame (ORF) has been shown to be present immediately downstream of the CNT gene which encodes a polypeptide with significant homology to the *Charalopsis* group of lysozymes. These include the autolysins of *Clostridium acetobutylicum* and *Streptomyces globisporus* [46, 47], and the bacteriophage encoded lysozymes of *Lactobacillus bulgaricus* and *Strepto-*

coccus pneumoniae [48, 49]. Confirmation of the lytic properties of the encoded Lyc protein was obtained through its recombinant expression in *E. coli*. Thus, production of Lyc resulted in lysis of the *E. coli* host employed, while lysates generated were shown to degrade clostridial cell wall extracts [50].

The gene was initially discovered in the type A strain NCTC 2916, and was located some 1.0 kb 3′ to the *botA* stop codon. Thereafter, oligonucleotide primers, based on two conserved amino acid motifs, were used in PCR to show that *lycA* was also present in five other Group I *C. botulinum* strains which produce BoNT/A (NCTC 2012), BoNT/F (Langeland, ATCC 25764, ATCC 23387) or BoNT/B and BoNT/F (ATCC 43757 & 43758), and in the Group II strain ATCC 23387 which produces BoNT/F. Sequencing of the amplified DNA fragments derived from these six strains showed that encoded proteins exhibited a high degree of identity with the LycA protein of NCTC 2916, with between only 1 and 9 amino acid differences out of 150 amino acids. Furthermore, in three of the six cases it was possible to show, using appropriate PCR primers, that the *lyc* gene resided immediately downstream of the respective CNT gene.

A DNA fragment of the expected size was not generated using the *lycA* gene specific primers when the template DNA was derived from any strain producing either BoNT/B alone, BoNT/C, BoNT/D BoNT/E or BoNT/G. Furthermore, a DNA fragment encompassing the majority of the gene failed to hybridize to chromosomal preparations of the *C. botulinum* type B strains Danish (Group I), and ATCC 43756 (Group II), *C. botulinum* NCTC 11219 (type E), *C. botulinum* 89G (type G), and purified DNA from phage C-St (type C) and DVD/-3 (type D). Thus, to date the gene is confined to strains which produce BoNT/F (either alone or in conjunction with BoNT/B) or BoNT/A. Given that the release of toxin from the cell is intimately linked with autolysis [51], the proximity of these genes seems more than just coincidental. Furthermore, the occurrence of similar genes on bacteriophage genomes raises the possibility that *lycA*, and the adjacent toxin genes, forms part of a prophage genome present in the chromosome of certain *C. botulinum* strains.

7.3.4
Involvement of transposons

The diverse genetic localization of toxin genes (i. e., plasmid, phage, or chromosome) suggests that they themselves are mobile through the agency of transposable elements. As yet there is little direct evidence to support this view. What is clear, however, is the fact that the highly conserved regions of the chromosome which encode the neurotoxin genes themselves can abruptly diverge in homology shortly after the translational stop codons. Thus, considerable homology between the sequence of botulinum locus of the type A strains NCTC 2916 and 62A ends some 100 bp downstream of their respective stop codons. This raises the possibility that these two genes may have transposed to two different genome environments. Conversely, in the case of the genetic loci of strains NCTC 2916 and Langeland, minimal sequence identity is evident between the two sequences until some 600 bp past their CNT coding regions when their sequences become essentially

99 % homologous. This level of homology continues for at least a further 1.2 kb, and encompasses the *lycA* gene described above. This may be the result of the transposition of the BoNT/A and BoNT/F genetic loci to the same region of the genome of a common Group I *C. botulinum* strain.

Indirect evidence for the transposition of the genes encoding exoenzyme C3 has also been presented [52]. These particular genes are found on the same phage genomes that carry the type C and type D botulinum genetic loci. Analysis of a mutant type C phage (CN) which had lost the C3 gene demonstrated that a 21.5 kb fragment had been deleted. The sequenced ends of this 21.5 kb fragment (along with the encompassed C3 genes) from a type C (C-Stockholm) and a type D (D-1873) phage were found to be essentially identical. However, the nucleotide sequences of the C-St and D-1873 genomes flanking the 21.5 kb region were unrelated. The presence of a common 6 bp core sequence (5′-AAGGAG-3′) at either end of this fragment has parallels in the Tn554 family of site-specific transposons. However, although sequence divergence between the two phage begins immediately past the core motif at one end of the 21.5 kb fragment, homology continues for a further 61 nucleotides past the core sequence at the opposite end, before diverging.

Finally, TOX^- derivatives of the type A strain 62A have been isolated following the introduction of the conjugative transposon Tn*916* [53]. This loss in activity was due to the deletion of a 12 kb region of the chromosome which appeared, from DNA/DNA hybridization data, to encompass the entire botulinum toxin locus. However, neither the precise nature of the deletion, nor the mechanism by which Tn*916* induced the loss of this fragment has been elucidated. Nevertheless, its occurrence was suggested to indicate the presence of the botulinum locus as part of a prophage.

7.4
Regulatory control of toxin gene expression

Microorganisms have developed tightly managed regulatory systems to precisely control the circumstances under which virulence/toxin genes are expressed. It is clear that complex schemes involving regulatory cascades, acting predominantly at the level of transcription, control virulence gene expression. The so-called two-component systems play a crucial role in such regulation, allowing the organism to sense changes in the external environment, such as nutrient availability, and modulate expression of gene sets accordingly. Classical physiological studies have established that both BoNT and TeNT production are significantly affected by the availability of exogenous nitrogen [54-56]. However, the mechanism by which such signals are transmitted to toxin gene expression are unknown. Elucidation of these mechanisms by which environmental factors affect the ability of the organism to grow and/or elaborate toxin, either when present in food or when present in the GI tract, are essential if we are to prevent foodborne botulism or infant/intestinal botulism. Accordingly, a number of research groups have initiated studies aimed at understanding how expression of the genes in the botulinum toxin locus are regulated.

7.4.1
CntR is a transcription factor

There is now convincing evidence that the *cntR* gene (Figure 2) encodes a transcription factor which positively regulates expression of the toxin complex genes. Prior to any experimental evidence to support this fact it had been noted that the encoded protein shared significant homology with other known proteins involved in transcription, such as the plasmid-encoded UviA protein of the *C. perfringens* plasmid pIP404 [57], the *txeR* gene of *C. difficile* [58] and the MsmR protein of *Streptococcus mutans* [59]. The former has an unspecified role in regulating the production of bacteriocin, while the latter is presumed to contribute to the transcriptional regulation of a streptococcal operon involved in sugar transport. The conclusion that MsmR is a DNA binding protein is based on its amino acid sequence similarity with known transcriptional activators, such as MelR and AraC of *E. coli*, and VirF of *Yersinia*. All these proteins contain helix-turn-helix motifs presumed to bind to DNA operator sites. The CntR proteins also contain amino acid sequence motifs with a high probability of being a helix-turn-helix (Figure 3).

To test this assumption, the *cntR-A* gene of the type A strain 62A was cloned on a multicopy plasmid and re-introduced into this *Clostridium* by electroporation. Subsequent dot blot analysis of mRNA levels indicated that in addition to a rise in *cntR-A* mRNA levels, there was also a significant increase in the level of both BoNT/A and NTNH/A, as measured by toxicity and western blotting respectively [60]. The increase in toxin activity was estimated to be some 16 to 32-fold in magnitude. An 8 to 16-fold increase in the levels of HA proteins was also detected. Furthermore, inhibition of *cntR-A* activity through the expression of *cntR*-specific antisense resulted in a 4 to16-fold decrease in levels of BoNT/A and NTNH/A mRNA, as measured by dot blot analysis.

A comparable effect was observed when the *cntR-T* gene of *C. tetani* was similarly introduced into *C. tetani* CN655, where a 16-fold increase in toxin levels was estimated on the basis of immunoblot assay [61]. Furthermore, despite sharing only 60 % identity with CntR-T, CntR-A was also found to stimulate TeNT production some 8-fold when the encoding gene was introduced into *C. tetani*. The CntR-C protein also stimulated TeNT production, albeit at a reduced level. These experiments demonstrated the close functional similarity between clostridial CntR proteins, and indicate that the genes involved in toxin production share common DNA regulatory elements.

Having demonstrated the product of the *cntR* gene control gene transcription it was of interest to obtain evidence for any interaction of the encoded transcription factor with the promoter elements identified above. Accordingly, an approximately 1.2 kb fragment was amplified from the NCTC 2916 genome using primers based on sequences residing within the 5′ end of the HA34 gene and the 5′ end of the NTNH/B gene entire. It therefore encompassed the entire *cntR* coding sequence and the non-coding regions between *cntR* and the adjacent HA34 and NTNH/B genes. This fragment was shown to be retarded in electrophoretic mobility when incubated with cell extracts of strain 62A which was overproducing CntR-A. A sim-

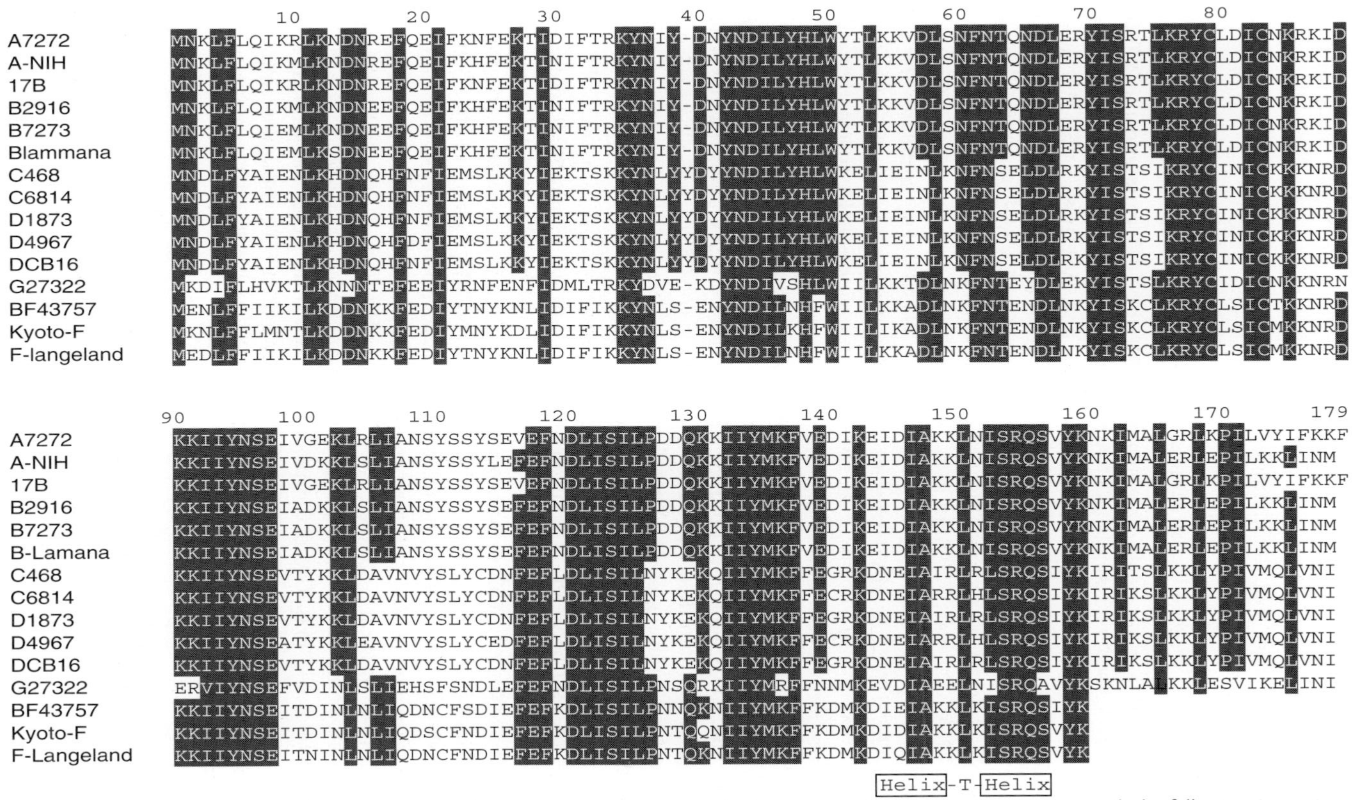

Figure 3. Multiple alignment of all known clostridial CntR proteins. Figure compiled using data from GenBank sequences with the following Accession numbers, X79104 (NCTC 7272), D67030 (A-NIH), L42537 (NCTC 2916), X79103 (ATCC 25765), U63808 (B-Lamanna), AB004779 (F-Langeland), AB0044778 (F-Kyoto), X72793 (C-468), AB037166 (C6814), AB012111 (D-CB16), AB012112 (D-1873), AB37920 (D-4947), Y13631 (ATCC 43757) and X87972 (ATCC 27322).

ilar shift in mobility was not observed when a control DNA fragment, derived from the BoNT/A coding region was employed. Furthermore, a greater degree of retardation was observed when polyclonal antibody raised against CntR-A was included in the assay. These data provide direct evidence that the CntR factor binds to DNA fragments carrying the HA34 and NTNH promoter regions, but of course does not conclusively prove that the identified motifs are involved in this interaction.

7.4.2
Gene transcription in gene loci structures I and II

Experimental data amassed to date suggests that the CNT gene and the NTNH gene are transcribed as a single mRNA species, although the number of strains analyzed is small. This conclusion was first reached by Hauser et al. [62] using RT-PCR analysis of mRNA isolated from a *C. botulinum* type C strain. A similar conclusion was reached by Henderson et al. [29] for the type A locus of strain NCTC 2916, where it was shown that a NTNH-derived gene specific probe hybridized to mRNA of at least 7.5 kb in size. In both studies, primer extension studies identified a transcript start point upstream of the NTNH gene. Similarly, the HA genes (HA33, HA17 & HA7) present in the *C. botulinum* type C strain and NCTC 2916 were also shown to be specified by a single RNA species [29, 62]. Thus, the botulinum gene cluster is transcribed as two divergent polycistronic messages. Intriguingly, as discussed above, the genetic locus analyzed by Henderson et al. [29] corresponds to the locus later identified as incorporating the silent BoNT/B gene. Thus, failure to produce type B toxin by NCTC 2916 may be solely attributable to the identified translational defects.

Examination of the regions of DNA within the vicinity of the determined transcriptional start points has revealed the presence of characteristic sequence motifs which are highly conserved, both between the regions upstream of BoNT and NTNH genes, and between different neurotoxic clostridia (Figure 4). These comprise a highly conserved 17 bp motif (Box 1) separated by 12 nt from a conserved 13 bp sequence (Box 2). This same sequence motif immediately precedes the TeNT gene in *C. tetani* [21]. In the case, of the HA34 and NTNH/B genes of NCTC 2916, and the TeNT gene of *C. tetani*, the transcriptional start points occur almost immediately after the end of Box 2. This is consistent with sequences within Box 1 and Box 2 acting as classical -35 and -10 promoter elements. In contrast, the transcription start point of the NTNH/C gene of strain C468 was mapped to the initial nucleotide of Box 2 [62]. Given the position of the three other determined start points (Figure 4), and the assumed role of Box 2 as a -10 promoter element, the assignment of the transcription initiation point of the BoNT/C gene may require further investigation.

7.4.3
Transcription of other botulinum genetic loci

On the face of it, these results suggest that the transcription of botulinum toxin genes (at least in the case of gene loci types I, II, and V) may be controlled through the activity of a toxin complex specific promoter, positioned upstream of the encod-

STRAIN	GENE	BOX 1 (-35)		BOX 2 (-10)			RBS		START
NCTC 7272	NTNH/A	ATTTTAGGTTTACAAAA	aatagtgtggct	ATGTTATATATAA	atg	84	AGGAGG	10	ATG
A-NIH	NTNH/A	ATTTTAGGTTTACAAAA	aatagtgtggct	ATGTTATATATAA	atg	84	AGGAGG	10	ATG
NCTC 7273	NTNH/B	ATTTTAGGTTTACAAAA	aatagtgtggct	ATGTTATATATAA	atg	84	AGGAGG	10	ATG
NCTC 2916	NTNH/B	ATTTTAGGTTTACAAAA	aatagtgtggct	ATGTTATATATAA	at**G**	84	AGGAGA	10	ATG
667Ab	NTNH/B	ATTTTAGGTTTACAAAA	aatagtgtggct	ATGTTATATATAA	atg	84	AGGAGA	10	ATG
17B	NTNH/B	ATTTTAGGTTTACAAAA	aatagtgtggct	ATGTTATATATAA	atg	84	AGAGGG	10	ATG
Lamanna	NTNH/B	ATTTTAGGTTTACAAAA	aatagtgtggct	ATGTTATATATAA	atg	84	AGGAGG	10	ATG
ATCC 43757	NTNH/B	ATTTTAGGTTTACAAAA	aatagtgtggct	ATGTTATATATAA	atg	84	AGGAGG	10	ATG
C6814	NTNH/C	ATTTTAGGTTTACAAAA	aatgattgagaa	ATGTTATATATAA	gtg	82	AGGAGA	10	ATG
C468	NTNH/C	ATTTTAGGTTTACAAAA	gacgatgaagaa	**A**TGTTATATATAA	gtg	82	AGGAGA	10	ATG
C-st	NTNH/C	ATTTTAGGTTTACAAAA	gacgtagaagaa	ATGTTATATATAA	gtg	82	AGGAGA	10	ATG
D-CB16	NTNH/D	ATTTTAGGTTTACAAAA	gacgatgaagaa	ATGTTATATATAA	gtg	82	AGGAGA	10	ATG
D-1873	NTNH/D	ATTTTAGGTTTACAAAA	gacgatgaagaa	ATGTTATATATAA	gtg	82	AGGAGA	10	ATG
D-4947	NTNH/D	ATTTTAGGTTTACAAAA	aatgattgagaa	ATGTTATATATAA	gtg	82	AGGAGA	10	ATG
ATCC 27322	NTNH/G	gTTTTAGGTTTACAAAA	aataatgtagtc	ATGTTATATATAA	atg	97	AGGAGA	10	ATG
Tetanus	TeNT	ATTTTcaGTTTACAAAA	aataacctgatt	ATGTTATATgTTA	**T**tg	110	AGGAGA	8	ATG
NCTC 7272	HA33/A	ATTTTAGGTTTACAAAA	aataatttgatt	ATGTTATATgTtA	tat	21	AGGAGG	7	ATG
A-NIH	???/A	ATTTTAGGTTTACAAAA	aataatttgatt	ATGTTATATgTtA	ta	[not determined]			
NCTC 7273	HA33/B	ATTTTAGGTTTACAAAA	aataatttgatt	ATGTTATATgTtA	tat	21	AGGAGG	7	ATG
NCTC 2916	HA33/B	ATTTTAGGTTTACAAAA	aataacttgatt	ATGTTATATgTtA	t**A**t	21	AGGAGG	7	ATG
667Ab	HA33/B	ATTTTAGGTTTACAAAA	aataacttgatt	ATGTTATATgTtA	tat	21	AGGAGG	7	ATG
17B	HA33/B	ATTTTAGGTTTACAAAA	aataatttgatc	ATGTTATATATtA	tat	21	AGGAGG	7	ATG
Lamanna	HA33/B	ATTTTAGGTTTACAAAA	aataatttgatt	ATGTTATATgTtA	tat	18	TAAAGG	7	ATG
ATCC 4357	HA33/B	ATTTTAGGTTTACAAAA	aataatttgatt	ATGTTATATgTtA	tat	21	AGGAGG	7	ATG
C6814	HA33/C	ATTTTAGGTTTACAAAA	ttcggcttaaat	ATGTTATATgTAA	gtg	13	GGGAGG	7	ATG
C648	HA33/C	ATTTTAGGTTTACAAAA	tttggcttaaat	ATGTTATATgTAA	gtg	20	GGGAGG	7	ATG
C-St	HA33/C	ATTTTAGGTTTACAAAA	tttggcttaaat	ATGTTATATgTAA	gtg	20	GGGAGG	7	ATG
D-CB16	HA33/D	ATTTTAGGTTTACAAAA	tttggcttaaat	ATGTTATATgTAA	gtg	20	GGGAGG	7	ATG
D-1873	HA33/D	ATTTTAGGTTTACAAAA	tttggcttaaat	ATGTTATATgTAA	gtg	20	GGGAGG	7	ATG
D-4947	HA33/D	ATTTTAGGTTTACAAAA	tctagtttaaat	gTGTTATATATAA	gtg	20	GGGAGG	7	ATG
ATCC 27322	HA11/G	AATTTAGGTTTACAAAA	aataatataatt	ATGTTATATATAA	gtg	62	AAGAGG	8	ATG
F-Kyoto	???/A	ATTTTAGGTTTACAAAA	aataatatagag	ATGTTA-ATATAt	tat	[no obvious gene]			
Langeland	???/F	ATTTTAGGTTTACAAAA	aataatatagag	ATGTTA-ATATAt	tac	[no obvious gene]			
CONSENSUS		ATTTTAGGTTTACAAAA	- [12 nt] -	ATGTTATATATAA G					
Kyoto-F	P47/A	tTaaTAccTTTACAAAA	attaatttagat	AaGTTATATATtt	atg	13	AAGAGG	9	ATG
ATCC 43757	P47/B	tTaaTAaGTTTACAAAA	attgatttggat	AaGTTATATATtt	atg	14	AGGGGC	8	ATG
Langeland	P47/F	tTaaTAaGTTTACAAAA	attcatttagat	AAGTTATATATtt	atg	14	AGGGGT	8	ATG
E Iwanai	P47/E	aataTAaaTTTACAAGA	acaaaatcaaga	tgtaTATtTATAc	taa	5	AGGAGA	9	ATG
BL6340 but	P47/E	aataTAaaTTTACAAGA	acaaaatcaaga	tgtaTATtTATAc	taa	5	AGGAGA	9	ATG

Figure 4. Comparative alignment of the putative promoter regions of various botulinum toxin genes. The sequences have been arranged into three blocks, corresponding to the 5′ non-coding region preceding the NTNH gene (upper block), the region 5′ to the HA33 gene (middle block), and the sequence upstream of the P47 gene (lower block). In the case of the type A NIH strain, the illustrated promoter is assumed to precede the HA33 gene, but its presence in this strain has not been determined. In the case of strains F-Kyoto and Langeland the sequence shown is that which resides 3′ to their CntR genes. Where experimental data exists, the position at which transcription initiates is shown by "blocking" of the appropriate nucleotide base.

ing operons. What of the other genetic loci? Comparative analysis shows that the identified promoter is not present immediately upstream of NTNH in those organisms lacking HA-encoding genes. However, a similar sequence immediately precedes the upstream P47 gene (Figure 4). Homology is closest in those strains known to possess a *cntR* gene (i. e., strains with genetic loci corresponding to structure III), and most distant in those strains with a genetic loci corresponding to structure IV, which apparently lack a *cntR* gene (see below). The difference in promoter motif found in type III genetic loci is mirrored by a noticeable difference in CntR structure. Thus, the CntR factors of these strains form a discrete grouping differing significantly from the CntR of the other genetic loci, most notably in the absence of 19 amino acids from their COOH-termini (Figure 3.). These differences may have resulted in an alteration in DNA binding specificity, and as a consequence the observed alteration in promoter sequence motif.

The promoter motif upstream of P47 in the case of gene loci corresponding to structure IV exhibits the greatest divergence from the consensus CntR promoter motif. Strains carrying this locus also correspond to those in which the presence of a *cntR* gene could not be demonstrated by PCR [25]. This suggests that either *cntR* is absent, or it has significantly diverged from the sequence found in other clostridia. Either way, in these instances, it would seem likely that it would possess a different promoter motif.

If the promoter motif upstream of P47 does indeed represent the point at which CntR, a homologue in the case of gene loci structure IV, binds, then it would imply that P47 is transcribed on the same mRNA message as the NTNH and BoNT genes. If this is the case, then it would indicate that P47 plays some role in the formation of the botulinum complex. Indeed, it may even form a part of the complex itself, although as yet there is no experimental evidence available to suggest that M complexes contain anything other than NTNH and BoNT.

7.4.4
Transcription of the neurotoxin genes

During the analysis of the C468 botulinum gene locus, primer extension studies also demonstrated the presence of a transcript start point 100 bp upstream of the BoNT/C gene. In the case of NCTC 2916, a start point was also mapped to a position 20 bp lying 5′ to the BoNT/A gene ATG start codon [29] and to an equivalent position in front of the BoNT/B gene [63]. In addition, the analysis of NCTC 2916 mRNA by Northern blots using a BoNT/A-specific probe had detected a distinct 4 kb mRNA species, as well as the general smear of mRNA beginning at around 7.5 kb. As the distance between the NTNH/A and BoNT/A gene and the NTNH/B and BoNT/B gene is 42 bp and 25 bp, respectively, if promoter sequences are present, then they must overlap with the 3′ end of the preceding NTNH genes. Similarly, the start point of the BoNT/C gene was mapped entirely with the NTNH/C gene. Close proximity of NTNH and BoNT genes is a general feature of all botulinum loci. Thus, the distance between the stop codon of NTNH and the start codon of BoNT can be as little as 10-14 nucleotides in the case of type D and F strains, 24/25 nucleotides in strains producing type B and E toxin, and as great as 84 nucleotides in the case of BoNT/G. It follows that, with the possible exception of type G strains, any promoter element must be present within the NTNH coding sequence.

Examination of the different sequences suggests that no common sequence motif is present in the 3′ ends of the NTNH genes that could be acting as a promoter. This observation, taken with the requirement for positioning of the transcription initiation signal within the NTNH coding sequence may indicate that a promoter element *per se* does not exist upstream of the CNT genes. Rather, some form of specific RNA processing of the large 7.5 kb mRNA occurs. It is not clear why such a phenomenon would need to take place, but it may be necessary to ensure the stability of the mRNA encoding CNT. This might be needed, for instance, if ribosome stalling at the CNT ribosome site was required to allow a pause between CNT and NTNH, to facilitate effective binding of the NTNH to the CNT [63]. In this light, it is interesting to note that short polypurine tracts of mRNA, which resemble ribosome binding sites, are capable of attenuating mRNA degradation [64].

7.5
Gene transfer in neurotoxinogenic clostridia

The genetic tools for studying *Clostridium* species remain relatively under developed compared to their aerobic bacterial counterparts. To date, the main gene systems developed have been designed for use in saccharolytic, solvent producing industrial strains (e. g., *Clostridium acetobutylicum, Clostridium beijerinckii*) and in the pathogenic species *Clostridium perfringens* [65, 66]. However, in recent years progress has begun to be made in the development of genetic tools for neurotoxinogenic clostridia.

7.5.1
Conjugative transposons

The first reported example of gene transfer in a neurotoxic clostridial species involved the transfer of the conjugative transposon Tn*916* into *C. tetanus* from an *Enterococcus faecalis* donor [67]. Transfer was achieved using a filter mating procedure, and occurred at a frequency of approximately 10^{-4} per donor cell. Subsequent analysis of the transconjugants obtained by DNA/DNA hybridization techniques indicated that Tn*916* had integrated into the chromosome apparently at random. A similar frequency of transfer of this transposon was subsequently demonstrated to occur from *E. faecalis* to the Group I *C. botulinum* type A strains Hall A and 113B [68]. As with *C. tetani*, insertion of Tn*916* into the chromosome appeared to take place at different sites around the genome. Moreover, of the thirteen transconjugants examined, six had single insertions of the transposon, six had two copies and one strain was deduced to have three copies of Tn*916*.

In other clostridia, conjugative transposons have proven to be of limited utility, as they have been found to integrate at a single site within the genome, e. g., Tn*916* in *Clostridium difficile* and Tn*1545* in *Clostridium beijerinckii*. To further assess the behavior of Tn*916* in *C. botulinum*, 200 randomly selected transconjugants were screened for auxotrophic mutations. A total of eleven strains were shown to require casein hydrolysate for effective growth on minimal media. In four cases, the individual amino acid required was subsequently identified. Thus, the conjugative transposon Tn*916* shows promise as a mutational tool for the genetic analysis of this organism.

Its utility was subsequently demonstrated [53] through its application in the isolation of three mutants of the type A strain 62A which were no longer able to produce botulinum toxin (so-called *tox-* strains). All three strains were deficient in the production of both the neurotoxin (BoNT) and the associated nontoxic non-hemagglutinin (NTNH) protein and furthermore their isolated genomic DNA failed to hybridize to PCR amplified regions derived from either gene. More detailed analysis, showed that the transposon of two of the mutants had integrated into the same site in the clostridial genome in a region which was not obviously encoding. One of the two transposons of the third strain had also integrated at a similar site. Subsequent analysis of all three strains by PFGE and parallel attempts to specifically amplify regions form the BoNT- and NTNH-encoding and flanking regions indicated that a large deletion had occurred in these strains, amounting to some approximately 12 kb. This apparent loss of the entire region encoding botulinum toxin components may indicate that they exist on a mobile element, although the link between insertion of Tn*916* and loss of this element remains unclear.

7.5.2
Introduction of autonomous cloning vehicles

As in many other bacterial species the successful introduction of plasmids into strains of *C. botulinum* has been almost exclusively reliant on transformation by electroporation. Zhou and Johnson were the first to describe a successful transformation system [45], using a type A strain (Hall A) and a vector, pGK12, based on the broad host range plasmid pWV01. The highest transformation frequencies (approximately 10^3 transformants per μg of DNA) were obtained using mid-log cells, an electroporation buffer composed of 10 % PEG 8000, and a field strength of 2.5 kV/cm. Similar frequencies for the transfer of plasmid pGK12 and plasmid pAT19, were subsequently reported for strains NCTC 2916 and 62A, but in this instance an alternative electroporation buffer (composed of 7 mM Na_2PO_4, pH 7.4, and containing 1 mM $MgCl_2$ and 270 mM sucrose) was employed. A similar method [63] has also been used to transform the type A strain ATCC 3502 (I. Henderson and N. P. Minton, unpublished data) with plasmids based on the *C. butyricum* plasmid pCB102 [69].

In the case of the Group II *C. botulinum* strain 17B (ATCC 25765) initial attempts to obtain transformation proved relatively unproductive [70]. Thus, while the occasional antibiotic resistant colonies were obtained using pGK12, no transformants were obtained when clostridial vectors based either on the broad host range plasmid pAMβ1 (pMTL500E) or the *C. butyricum* plasmid pCB102 (pMTL540E) were employed. Through a combination of classical and molecular approaches it proved possible to demonstrate that strain ATCC 25765 possessed a restriction endonuclease (*Cbo*I) and a methylase activity (M.*Cbo*I) which has the same specificity as *Msp*I and M.*MspI*, respectively. *Cbo*I cleaves the palindrome 5′-CCGG-3′ to generate a 3′-GC overhang, while M.*Cbo*I specifically methylates the external C residue. To circumvent this apparent restriction barrier, an *E. coli* host was generated which expressed a *Bacillus subtilis* methylase enzyme (M.*Bsu*F1) with equivalent specificity to M.*Cbo*I. DNA of plasmids pMTL540E and pMTL500E prepared in this strain were subsequently shown to be capable of transforming ATCC 25765. The highest frequencies (0.8 X 10^4 transformants per μg of DNA) were obtained when cells were cultivated in media supplemented with 1 % (w/v) glycine, and when the electroporation was undertaken at 10 kV/cm, 25 μF, and at 400 Ω [70].

An alternative approach to circumvention of the restriction barrier is to introduce recombinant shuttle vectors by conjugal transfer. Such a strategy was successfully adopted by Bradshaw et al. [71] to introduce shuttle vector previously devised for use in *C. perfringens*. Based on the pIP404 replicon, this plasmid (pJIR1457) carries the origin of transfer region (*oriT*) of the plasmid RP4. Such a plasmid may be mobilized from *E.coli* donors to both Gram-negative and Gram-positive bacteria provided the donor strain is endowed with the necessary plasmid encoded transfer conjugation functions. This plasmid was shown to be transferable to both strain Hall A and 62A at frequencies of between 10^{-3} and 10^{-4} per donor cell.

7.6 CNT structure and function

7.6.1 Protein architecture

The CNT proteins are each synthesized as a single polypeptide chain without a leader peptide and are released from the bacterial cell following lysis. The polypeptides are post-translationally modified by protease hydrolysis to form covalently linked di-chain polypeptides, each with three discrete structural and functional domains. This common architecture (Figure 5) comprises a light chain (LC, 50 kDa)

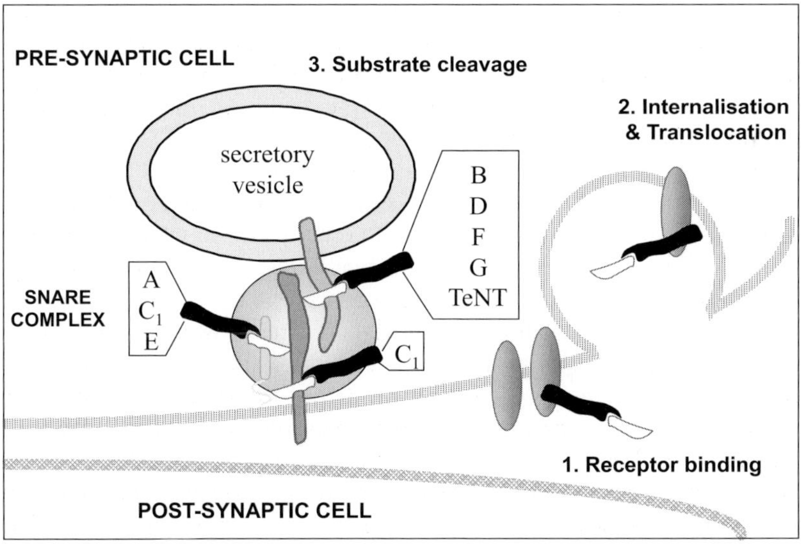

Figure 5. Diagrammatic representation of the three-step intoxication process of CNTs. The CNT binds to a high affinity receptor on the cell surface by virtue of an interaction between its H_C domain and protein/ganglioside components on the target cell surface. Receptor internalisation introduces the CNT into early endosomes – an event which, following acidification, promotes a conformational change in the H_N and subsequent translocation of the LC into the cytosol. The CNTs interact with and hydrolyse the uncomplexed forms of their respective substrates. The hydrolysis events destabilise the SNARE complex such that a failure of the vesicle docking and fusion mechanism ensues

with metalloprotease activity, linked via a disulphide bridge to a receptor binding and translocating heavy chain (HC, 100 kDa). The HC is further comprised of two regions; the amino-terminal 50 kDa (H_N) is required for the translocation of the toxin into the neuronal cell while the carboxyl-terminal 50 kDa region (H_C or fragment-C) facilitates the pivotal interaction with high affinity receptors on the target cell. The H_C domain is itself comprised of two similarly sized sub-domains – an N-terminal H_{CN} and a C-terminal H_{CC} [72]. Whereas the H_{CC} domain has been clearly associated with ganglioside binding, the role of the "lectin-like" H_{CN} domain has yet to be elucidated [73, 74]. The three CNT domains are thus independently responsible for neuronal target cell binding, translocation and intracellular activity and all three components are necessary for the toxic activity of the CNT proteins (Figure 5).

7.6.2
Associated complex proteins

Of the CNT family only TeNT is released from the parent cell as a single protein species. The seven BoNTs are released as so-called "progenitor toxins" with a variety of other proteins as part of a variable size, high molecular weight, non-covalent complex. In these complexes the BoNTs are co-operatively protected from the rigors of the gastrointestinal (GI) tract [75]. Some members of the protein complex have haemagglutination properties. There are some reports that the non-neurotoxin proteins of the complex may also assist in BoNT uptake from the GI and transcytosis to the bloodstream [76]. This structural difference between the CNTs may be attributed to the different etiology of tetanus, a wound infection with direct access to the blood stream, compared with the GI intoxication of botulism.

Three forms of "progenitor toxins" are recognized: M toxin (300 kDa), L toxin (500 kDa) and LL toxin (900 kDa). The smaller M toxin is composed of a BoNT (150 kDa) in association with a similarly sized nontoxic protein (NTNH, 150 kDa). The larger L and LL progenitor toxins additionally contain a variable number of proteins which possess hemagglutinin (HA) activity. The type of progenitor toxin produced varies between the different toxinogenic strains. More than one form may be produced by a single strain. Thus, all three forms have been found in type A *C. botulinum* cultures, while only the L and M forms are produced by type B, C, and D *C. botulinum* strains. *C. botulinum* type G strains produce the L toxin. The toxin of type E and F strains is exclusively composed of M progenitor toxin. Progenitor toxins exhibit greater stability to pH extremes and high temperature than purified BoNT. Botulism, unlike tetanus which is not produced in association with nontoxic proteins, is principally caused by the ingestion of food in which contaminating clostridial cells have produced BoNT. It has therefore been argued that the major role of the nontoxic proteins of the botulinum complex may therefore be to aid the neurotoxin in withstanding the acidic and proteolytic environment of the stomach [19, 20]. Consistent with this hypothesis is the observation that the larger the progenitor toxin complex the greater its oral toxicity [77].

7.6.3 Sequence homologies

The deduced primary amino acid sequences of the encoded genes shows that in general they are composed of highly conserved amino acid domains interspersed with amino acid tracts exhibiting little overall similarity [24]. In specific instances, little sequence divergence is evident. Thus, the sequenced pairs of TeNT, *C. butyricum* BoNT/E, and the *C. botulinum* BoNT/G genes are identical, while the two BoNT/C genes and the BoNT/A genes of *C. botulinum* NCTC 2916 and 62A, diverge by only 1 and 2 nucleotides, respectively. In other instances, quite substantial differences are apparent. This is particularly the case when, although encoding the same CNT serotype, the genes are carried by a different member of the 6 neurotoxic clostridial Groups. This is most evident among the 3 BoNT/F genes, where the encoded neurotoxins of the Group II and Group VI strains differ from the BoNT/F of the Group I strain Langeland by 151 and 341 aa, respectively. Similarly, BoNT/B produced by the Group II strain Eklund and the Group I strain Danish diverge by 93 aa. The lowest level of divergence is seen between the BoNT/E of the Group II *C. botulinum* strains and that of Group V *C. butyricum* strains. Thus, there are only 34 aa differences between BoNT/E of ATCC 43181 and ATCC 11219.

With notable exceptions, the degree of identity among the different serotypes generally does not exceed 40 %. The exceptions include, the identity observed between BoNT/C1 and BoNT/D (47 % and 56 % between the L and H chains, respectively), that shared by the L chain of BoNT/B with those of BoNT/G and TeNT (61 % and 50 %, respectively), and the identity between the L and H chains of BoNT/E and all type F BoNTs (e. g., 57 % and 68 % identity to the BoNT/F of strain Langeland). Distance matrix analysis has shown a different genealogical relationship among serotypes, dependent on which portion of the molecule is compared. Thus, for instance, whereas comparisons of L chains places BoNT/A as the most divergent CNT, with TeNT being relatively closely related to many BoNTs, a complete reversal in relative relatedness is seen between H chains. BoNT/C1 and BoNT/D, on the other hand, remain distantly related to the other CNTs regardless of which di-chain component is compared. The greatest variation between the CNTs is found in the receptor binding H_C regions [72]. Within these regions the sequence homology variation is significantly more pronounced in the H_{CC} sub-domains as compared to the H_{NN} sub-domains [78].

The interchain disulphide bond is an attribute common to all of the CNTs. De Paiva et al. (1993) examined the role of this feature in the toxicity of BoNT/A in neurons of the giant sea slug *Aplysia californica* [79]. Unlike the free cysteine residues which occur at various positions in the BoNT/A sequence, those involved in the interchain linkage were demonstrated to be essential for translocation of the LC into the neuronal cytosol and hence neurotoxicity.

7.6.4
Crystal structures

Three-dimensional structures have now been resolved for TeNT H_C (or "fragment-C") [80], BoNT/A [81], recombinant BoNT/A LC (*rec*LCA) [82] and BoNT/B [83]. Additionally, BoNT/B-LC has been co-crystallized with its substrate synaptobrevin-II (VAMP-2) [84]. These studies have revealed that BoNT/A and BoNT/B both adopt a linear arrangement of the three functional domains with no contact between the LC catalytic and the H_C binding domains. The three domains are spatially distinct with the exception of an unusual loop of the H_N domain which wraps around the outside of the LC [81].

7.6.5
L chains

The L chains of the CNTs are known to be proteolytic enzymes and are proposed to constitute a new family of zinc-dependent endopeptidases [85-87]. The CNT endopeptidase active sites are negatively charged, deeply buried and characterized by containing a common Zn^{2+}-binding helical HEXXH motif. The catalytic domains share little homology outside of the CNT family. Thermolysin and leishmanolysin are closest relations in this respect [81] but the similarities are weak and appear to be limited to primary sequence alignment. The Zn^{2+} is considered to contribute both to the catalytic function of the L chain and also to its tertiary conformation and thermostability [88]. Removal of the zinc completely abolishes the enzyme activity and compromises the tertiary structure such that replenishment with zinc can only partially restore enzymatic function. Close to the metal ion-binding motif, BoNT/A and BoNT/B have highly similar structures. However, significant differences between BoNT/A and BoNT/B are evident at only 15 from the zinc-binding motif and it has been proposed that these may reflect differences in enzyme substrate specificity [72].

In both BoNT/A and BoNT/B the LC is wrapped within a long span of the H_N domain. The rationale for this wrapping is unknown. At the primary sequence level this H_N "belt" is longer in BoNT/A than in BoNT/B and analysis of the crystal structures indicates that the BoNT/A "belt" occludes the active site such that reduction of the LC-H_N interchain disulphide bond is required for activation of the toxin. In BoNT/B the belt region does not completely occlude the active site which may therefore be accessible to small molecule inhibitors. As with BoNT/A, however, separation of the LC and H_N introduces changes in the folding of the polypeptides which are critical to catalytic activity.

Recombinant techniques have facilitated the dissection of these functional domains. Substitution mutations in the TeNT LC [89, 90] have confirmed for the TeNT zinc binding motif, the role of H_{232}, $H_{237,}$ and E_{234} in the functional activity of this molecule. Kurazono et al. have determined the minimal essential functional domain of the BoNT/A and TeNT LCs by *in vitro* micro-injection of mRNA truncates into neurons of *Aplysia californica* [91]. In these studies it was shown that

BoNT/A was still functional following an 8 aa truncation at the N-terminus and a 32 aa deletion at the C-terminus; TeNT could tolerate a C-terminal truncation of 65 aa residues. *In vitro* peptide cleavage studies using *E. coli* expressed recombinant LC proteins (*rec*LCs), however, indicate that in this system TeNT can only tolerate a C-terminal deletion of 16 aa. The reason for this discrepancy is unclear but it is most likely related to the different assay systems used to characterise the mutants.

The corresponding recombinant minimum functional domain of BoNT/A and a 30 aa C-terminal deleted TeNT *rec*LC have recently been expressed from *E. coli* and crystallized [82, 86]. Interestingly, the truncated *rec*LCs from both toxins exhibited greater stability than the isolated parental protein domains and the TeNT *rec*LC had a slightly enhanced *in vitro* catalytic activity. The BoNT/A *rec*LC had a lower K_{cat} than the native toxin and reducing agents did not enhance *rec*LC enzymatic activity in a synthetic peptide substrate assay. The observation that BoNT/A *rec*LC crystallizes as a dimer was proposed as a possible explanation for this reduction in catalytic activity [92].

7.6.6 Binding domains

Much of the current CNT research is focused on understanding the nature of the cellular receptors. While the existence and nature of a protein component of the neurotoxin receptors is still under investigation, the involvement of target cell gangliosides and the location of intramolecular CNT ganglioside binding sites is better understood [73].

Shone et al. demonstrated [93], using limited trypsin digestion, that the 30 C-terminal residues of BoNT/A were involved in receptor binding and Swaminathan and Eswaramoorthy have provided evidence that the interaction of BoNT/B with gangliosides also occurs at the extreme H_C terminal domain – the region designated H_{CC} [83]. Photoaffinity labeling studies have demonstrated that the homologous region of TeNT is also capable of ganglioside binding [94]. Both TeNT and BoNT/B have been shown to bind disialo- and trisialogangliosides (*e.g.*, GD1a, GT1b and GD1b) [95-98]. The binding of TeNT to ganglioside GT1b has also been reported to be inhibited by monoclonal antibodies raised against different epitopes in the H_C domain [99].

Ganglioside interactions, however, are not considered to be solely responsible for neurospecific binding of the CNTs and, consequently, Montecucco has proposed the double-receptor model of CNT activity suggesting that a protein receptor component must also be involved [95]. In this scenario the CNTs bind with high affinity and specificity to the target cell membranes as a consequence of multiple interactions with both carbohydrate and protein components. Supporting evidence comes from Nishiki et al. [96] who have reported the high affinity binding of BoNT/B to synaptotagmin-II in the presence of polysialogangliosides, an observation that has now been extended to BoNT/A and BoNT/E [100]. Furthermore, Herreros et al. have now reported that the H_{CC} domain of TeNT contains all of the structural information necessary for binding to neuronal cells *in vitro* and have identified a candidate protein receptor for this CNT [74].

In their crystal structures the H_C domain structures of BoNT/A and BoNT/B, unlike their respective LC and H_N domains, do not superimpose very well in terms of orientation to the entire neurotoxin molecule [83]. The BoNT/A, BoNT/B and the TeNT H_C sub-domains all present as two distinct β-strand regions of roughly equal size joined by a single prominent α-helix (Figure 6). The H_{CN} sub-domains have similarities to several lectin proteins and the majority of the highly conserved residues in the H_{CN} sub-domain are postulated to preserve the hydrophobic core of a β-barrel structure. Although the major spatial differences between the binding domains of these three CNTs are found within the H_{CC} region, all of their C-terminal sub-domains adopt a β-trefoil conformation [73, 80, 101]. This has also been predicted to be the case for the other CNT family members [101].

Halpern and Loftus demonstrated for TeNT that mutations in the H_{CC} affect ganglioside binding *in vitro* [102]. This is supported by the observations of Shapiro et al. who reported that the C-terminal 34 aa residues of TeNT were capable of binding gangliosides and that ganglioside photoaffinity labeling was primarily targeted at residue H_{1293} which lies 23 residues from the TeNT C-terminus [94]. Compelling evidence implicating gangliosides as at least part of the CNT receptor components was provided who demonstrated that ganglioside binding to BoNT/A quenched tryptophan fluorescence and inhibited the activity of the toxin [103]. It is of note that tryptophan residues are highly conserved in this region of the toxins and this residue position (W_{1265} in BoNT/A, W_{1289} in TeNT) is considered to define the ganglioside binding site (Fig 6). The major differences in the H_{CC} regions appear in the loops of the β-trefoil motif with TeNT having a slightly longer primary sequence and consequently longer loop lengths [81]. More recently, Swaminathan and Eswaramoorthy have identified E_{1688}, E_{1189} H_{1240} and Y_{1262} as the contact residues between the H_{CC} domain of BoNT/B and sialyllactose and have described this interaction in detail [83].

Figure 6. Schematic representation of BoNT/A illustrating the three-domain protein architecture and the key structural features with amino acid residue positions. The BoNT/A neurotoxin comprises two polypeptide chains which are covalently linked by a disulphide bond. Full activation of the toxin results from reduction of this disulphide link in the intracellular compartment. The protein architecture comprises a light chain (LC, 50kDa) which contains the metalloprotease activity, and a translocating heavy chain (HC, 100kDa). The HC is further comprised of two regions; the amino-terminal 50kDa (H_N) is required for the translocation of the toxin into the neuronal cell while the carboxyl-terminal 50kDa region (H_C or fragment-C) facilitates the pivotal interaction with high affinity receptors on the target cell. The H_C domain is itself comprised of two similarly sized sub-domains – an N-terminal H_{CN} and a C-terminal H_{CC}, both of which appear to play discrete but inter-related receptor recognition functions. The three CNT domains are thus independently responsible for neuronal target cell binding, translocation and intracellular activity and all three components are necessary for the toxic activity of the CNT proteins.

Regions of identity with other CNTs

Regions of high homology with other CNTs

"Belt" region

Key α-helical features

Other key residues and structural features are indicated.

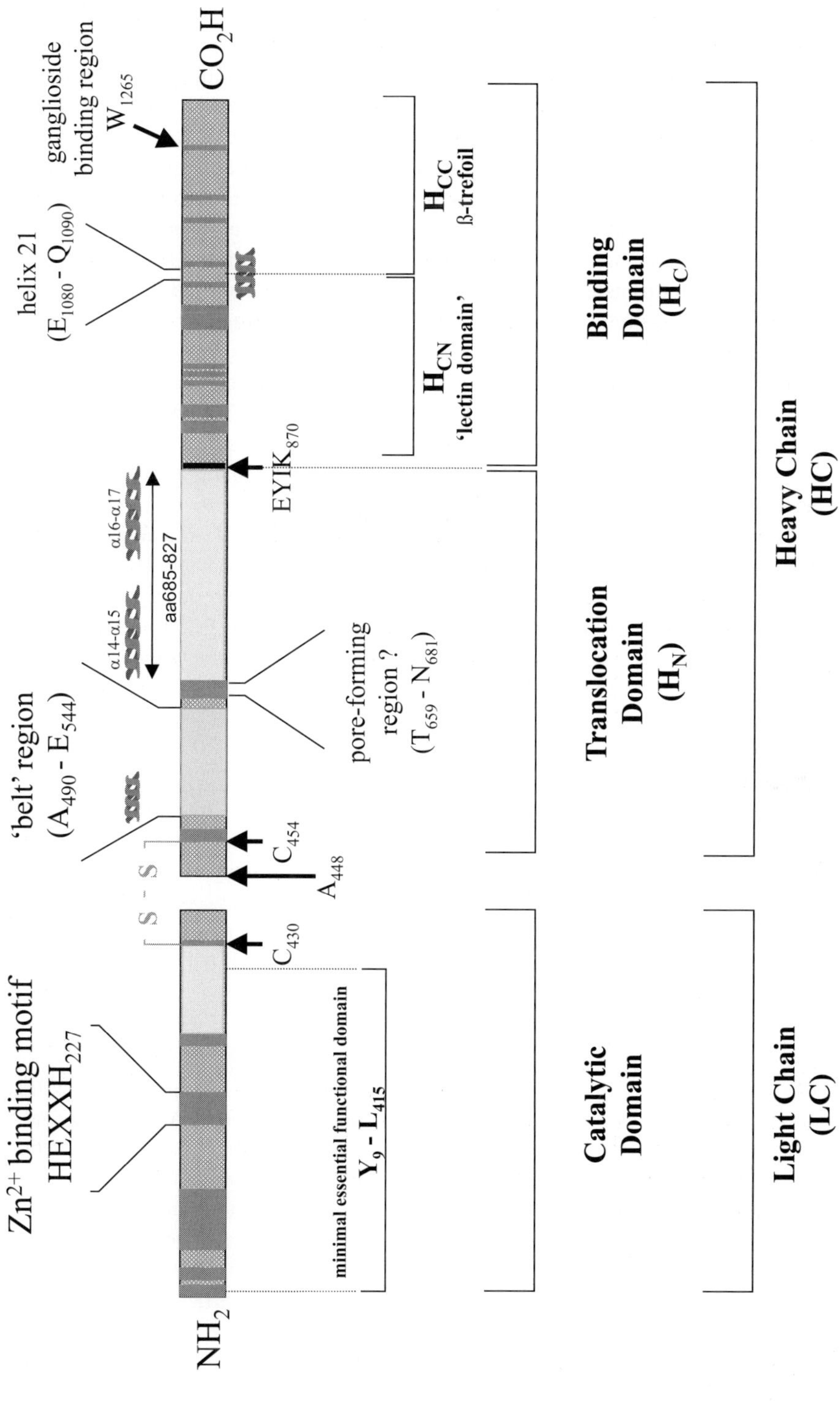
NH_2
CO_2H
Zn^{2+} binding motif
$HEXXH_{227}$
minimal essential functional domain
Y_9 - L_{415}
C_{430}
S - S
A_{448}
C_{454}
'belt' region
(A_{490} - E_{544})
pore-forming
region ?
(T_{659} - N_{681})
α14-α15
α16-α17
aa685-827
$EYIK_{870}$
helix 21
(E_{1080} - Q_{1090})
ganglioside
binding region
W_{1265}
H_{CN}
'lectin domain'
H_{CC}
β-trefoil
Catalytic
Domain
Translocation
Domain
(H_N)
Binding
Domain
(H_C)
Light Chain
(LC)
Heavy Chain
(HC)

Sinha et al. have used a mutagenesis strategy to confirm the essential role of the TeNT C-terminal domain (H_{CC}) for ganglioside binding [104]. Deletion of 10- aa from the H_{CC} β-trefoil C-terminus reduced ganglioside binding by 20 %, but did not appear to affect binding to primary spinal cord neurons. Deletions of 4-aa had little effect on ganglioside binding whereas H_{1293}S and H_{1293} A substitutions at the residue implicated by ganglioside photoaffinity labeling suppressed ganglioside binding by almost 60 % and 90 %, respectively. However, non-competitive analyses of these mutants indicated that their cell surface interaction and retrograde transport functions did not appear to be compromised. It is interesting to speculate whether this indicates a potential uncoupling of ganglioside binding from retrograde transport or that the residual ganglioside binding was sufficient to facilitate proper function at the cell surface.

Ginalski et al. have recently used a structure-based β-trefoil alignment approach to further analyse and compare the H_{CC} domains of the CNTs and to address some of the inconsistencies arising from linear sequence comparisons [101]. Using this model Ginalski et al. proposed that the key residues for ganglioside binding are located in a shallow surface pocket of the β-trefoil, a region distinct from the deep, positively charged cleft between the H_{CN} and H_{CC} sub-domains [101]. These authors additionally suggest that the deep cleft is more likely involved in interaction with the protein component of the neurotoxin receptors. This proposal is in keeping with Herreros et al. who demonstrate that, in PC12 pheochromocytoma cells, the H_{CC} sub-domain of TeNT is sufficient for interaction with a 15 kDa putative receptor protein [74].

7.6.7
H_N domains

The LH_N domain has been demonstrated to be capable of forming channels in liposomes at low pH [105], in artificial bilayers [106-108] and in cell membranes [109]. With the exception of the "belt" region (Figure 6) which extends and wraps itself around the LC domain, primary sequence homology between the H_N domains of different CNT is high. Crystallographic analyses indicate that the H_N domains of both BoNT/A [81] and BoNT/B [83] each contain an unusual pair of long alpha-helices (residues 685-827). These anti-parallel amphipathic helices twist around each other, as in coiled-coils, to form the cylindrical main body of the H_N domain and have triple helix bundles at each end [81]. The long helices may each be further sub-divided into two smaller helices (α14-α15 and α16-α17 respectively). Within this helical region the conserved residues of the CNTs are regularly spaced and it is proposed that these preserve the order and spatial arrangements of the helical structures [78]. The core regions of the long helices show some limited homology to the translocation domains of other proteins. However, it is suggested that the fold of this translocation domain is indicative of a mechanism of pore formation different from that of other bacterial protein toxins and it is more in keeping with coiled-coil viral proteins having membrane disruption properties [81].

It is considered that low pH within the endosomal compartment of the target cell triggers a conformational change in H_N which then forms the LC "transporter" [72,

78]. Using synthetic peptides, which have been shown to form channels in lipid bilayers [110, 111], and computer algorithm searches for amphipathic membrane-spanning sequences, the pore forming regions of the BoNT/A H_N domain has been tentatively identified to reside – at least partially – within the sequence spanning residues 659-681. These regions are distinct from the BoNT/A H_N long helical regions identified from the crystal structure.

For BoNT/A, the H_C domain may be selectively removed by extended trypsin digestion [93] to generate a di-chain LH_N molecule of much reduced toxicity. Chaddock et al. demonstrated that LH_N of BoNT/A, when coupled to wheat-germ agglutinin lectin, could translocate *in vitro* to effect inhibition of neurotransmitter release [112]. This is the first demonstration that full HC, specifically H_C, is not required for translocation and creates opportunities for therapeutic derivatives of neurotoxin that have had the C-terminal and perhaps the entire HC domains deleted or replaced.

7.7 CNT mechanism of action

7.7.1 The "SNARE" proteins

Release of neurotransmitters from secretory vesicles in presynaptic neurons is dependent on a series of membrane fusion events that promote and allow fusion of the secretory vesicle with the cell outer membrane [113]. These fusion events are dependent on the presence, assembly into a complex and activity of a conserved set of membrane proteins referred to as "SNARE" proteins [soluble NSF (N-ethyl maleimide sensitive fusion protein) attachment protein receptor proteins] [114, 115].

These proteins and their isotypes can be grouped into several small protein families. Increasing numbers of isotypes within these families are being discovered and they are all considered to be pivotal components of a "universal" mechanism for vesicle fusion and secretory processes in neurotransmitting and other secretory cells [116]. In nerve terminals the SNARE set comprises vesicle associated membrane protein (VAMP, also-called synaptobrevin), syntaxin and a 25 kDa synaptosomal protein (SNAP-25).

The mechanism by which the CNTs inhibit neurotransmitter secretion is the Zn-dependent hydrolysis of one or more isotypes of VAMP, syntaxin, or SNAP-25. Tetanus toxin and botulinum neurotoxin types B, D, F, and G cleave VAMP, whereas BoNT types A and E cleave SNAP-25. BoNT type C1 appears to be unique within the CNT family in that it cleaves two of the three substrate proteins SNAP-25 and syntaxin. The specificity of the toxins for their substrates extends to the residue position of substrate hydrolysis (Table 2), with only TeNT and BoNT/B having the same peptide bond specificity although their clinical manifestations are very different. Even with this exception, the exquisitely potent neurotoxins are thus also highly specific in terms of both their target cell interactions and their substrate cleavage requirements.

Table 2. CNT substrate specificity and peptide cleavage sites

Toxin type	*Intracellular target*	*Peptide bond cleaved* P_4-P_3-P_2-P_1-//-P_1'-P_2'-P_3'-P_4'
TeNT	VAMP	GLSQ//FETS
BoNT/A	SNAP-25	EANQ//RATK
BoNT/B	VAMP	GLSQ//FETS
BoNT/C_1	Syntaxin	DTKK//AVKF
BoNT/C_1	SNAP-25	ANQR//ATKM
BoNT/D	VAMP	RDQK//LSEL
BoNT/E	SNAP-25	QIDR//IMEK
BoNT/F	VAMP	ERDQ//KLSE
BoNT/G	VAMP	ETSA//AKLK

TeNT, tetanus toxin; BoNT, botulinum toxin serotype (from [4]).

7.7.2
The "SNARE motif"

A common structural element within the SNARE structure has been identified as a recognition motif for the CNTs. The "SNARE motif" is considered to be a 10-residue consensus XHAAXHAXHP where "X" is any amino acid, "H" is a hydrophobic residue, "A" is an acidic residue and "P" is a polar residue [117]. The motif adopts an α-helical conformation with a negatively charged surface flanked by a hydrophobic face [118-120]. Multiple copies of this CNT recognition site are present in SNAP-25, VAMP, and syntaxin [2] and synthetic peptides must contain at least one copy of the motif before they are recognized as CNT substrates [121, 122]. The SNARE motif is thus considered to be a major determinant of the specificity of CNTs for their intracellular substrates.

7.7.3
Substrate interaction

Using BoNT/B as a model for the other CNT family members, the toxin-substrate interaction has lower entropy than the helical conformation of synaptobrevin-II (or VAMP II) that is required for formation of the SNARE complex. Consequently, binding of synaptobrevin to BoNT/B occurs preferentially to SNARE formation. Interaction of BoNT/B with its substrate synaptobrevin-II is followed by conformational changes in the toxin LC, which expose the substrate binding residues and effect active site positioning. Each catalytic domain has two substrate recognition sites – both of which are required for proteolysis to occur [84].

Once cleaved, the synaptobrevin is no longer available for the exocytotic process and, reciprocally, synaptobrevin that has already formed the SNARE complex is no longer susceptible to cleavage by the toxin. This indicates a possible conformation dependency or stearic accessibility requirement for BoNT/B-substrate interaction [84] and as a generalization it applies to all of the CNT endopeptidases. The hydrolysis of SNARE proteins does not, however, appear to prevent complex formation [123] rather it has a destabilizing effect such that a failure of the vesicle docking and fusion mechanism ensues [124].

Vaidyanathan et al. studied the interaction of BoNTs A, C and E with SNAP-25 and also with two non-neuronal SNAP isoforms (murine SNAP-23 (syndet) and human SNAP-23) [125]. They determined that the C-terminal region of SNAP-25 was critical to interaction with BoNT/A, that the minimal SNAP-25 region required for binding of BoNT/A incorporated M_{146}-Q_{197} and that an inactive form of BoNT/A *rec*LC could still bind its substrate with high affinity. Murine SNAP-23 was cleaved by BoNT/E, but was only a poor substrate for BoNT/A. The human SNAP-23 isoform was resistant to both enzymes but, interestingly, it required only single amino acid mutations to convert it to susceptibility [125].

With the exception of TeNT, for which variably a signal transduction modulatory activity [126] and a transglutaminase stimulatory activity have been reported [127-130] and debated [131], this proteolysis is the sole reported intracellular activity of the CNTs.

7.8
Therapeutic development

7.8.1
"The molecular scalpel"

As described above, the clostridial neurotoxins and in particular their endopeptidase (L chain) domains are exquisitely potent and precise tools for the molecular dissection of neuroendocrine secretory mechanisms. Analogous to a surgical instrument in the hands of a skilled practitioner, they may be regarded as "molecular scalpels".

Despite their reputations as the most poisonous biological substances known [132], the clostridial toxins are not actually cytotoxic and when used in therapeutic doses, unlike a surgical blade, they do not inflict permanent damage. Neuronal intoxication with BoNTs causes a transient and reversible atrophy [133, 134] but CNT lethality is generally a consequence of a whole-organism effect of neuronal modulation, usually respiratory failure, rather than cell killing. Even following prolonged exposure in cellular systems the CNTs do not promote cell death [135] and victims of botulism may recover if supported by mechanical ventilation [136]. Injections with BoNTs appear to be well tolerated with only limited spread to muscle beds near to the site of the primary injection [132]. Incidents of "flu"-like reactions have been reported, but these appear to be rare. Treatments may be repeated several times without major side effects and should a neutralizing immune response develop, a different neurotoxin serotype can be used to allow continuation of effective treatment.

7.8.2
Clinical indications

The characteristic potencies, specificities, duration of action and reversal of intoxication all contribute to the attraction of using CNTs as biopharmaceuticals. Since the pioneering work of Scott in the late 1960s [136-138] native botulinum neurotoxin type A (BoNT/A) has been exploited and subsequently licensed for treating the aberrant muscle spasms of blepharospasm, hemifacial spasm, strabismus, and torticollis. The effectiveness of BoNT in these conditions led to its further use in the treatment of limb dystonias, vocal disorders, gastrointestinal disorders and tremors [139].

The range of possible new indications is continually growing and now includes conditions such as cosmetic removal of wrinkles, alleviation of the spastic paralysis of cerebral palsy in children, tension-type headache and autonomic conditions such as hyperhydrosis [140]. Furthermore, BoNT/A therapy provides significant and prolonged relief from the pain associated with muscle dystonias. Whereas this effect was initially considered to be a consequence of reduced muscle tone, attention is now focussing on the possible direct analgesic pharmacological action of the toxin. Thus, there exists the potential for use of BoNTs in the treatment of various pain syndromes such as myofascial pain, migraine [136] and severe chronic pain [140].

7.8.3
Engineered toxin fragments

The development of "second generation" therapeutics from CNT endopeptidases has also been advanced by the observations of Chaddock et al. [112, 141]. In these studies it was demonstrated that the LH_N domain of BoNT/A, when chemically coupled to wheat germ agglutinin lectin (WGA) or nerve growth factor (NGF), could be effectively retargeted *in vitro* to cells that are normally recalcitrant to BoNT/A intoxication. The effects were potent, endopeptidase dependent and ligand mediated. Thus the potential has now been demonstrated to exploit clostridial endopeptidases in the modulation of cell secretory disorders for which they were previously unsuited. This is a particularly valuable observation in view of the retrograde transport capabilities of both WGA and NGF [142].

All of the CNTs appear to be highly selective for their target cells and are capable of high-affinity binding to target cell receptors. TeNT, however, appears to be unique among the CNTs in that it is capable of entering two different types of neurons – motorneurons and inhibitory interneurons – while intoxicating only the latter. When used in *in vitro* cell culture the CNTs appear to lose some of this selectivity. Nonetheless, the potential applications of the CNT heavy chains, particularly TeNT H_C, for delivery of therapeutic proteins and DNA molecules to the central nervous system have been explored [143-146]. Furthermore, Schneider et al. have reported that the tropism of adenovirus vectors may be engineered using a TeNT H_C conjugate chemically coupled to an adenovirus-neutralizing antibody [147]. *In vivo* intramuscular injection of this conjugate in mice resulted in selective gene transfer to the neurons of the hypoglossal nucleus and this was taken as an indication of the potential to use such molecules for gene therapy. The motorneuron to interneuron retrograde transport characteristic of TeNT is particularly attractive in the context of a therapeutic vehicle. Although the active tetanus vaccination policies of many countries may present a possible limitation of this technology, this may not necessarily preclude their use, for example, as carriers of other epitopes not compromised by anti-TeNT antibodies or by introduction of therapeutic constructs into immuno-privileged sites.

7.9 Future prospects

For many years the driving force behind scientific research on the neurotoxinogenic clostridia has been to develop more effective measures for the prevention of botulism from foodstuffs. During 2001, the sequencing of the genome of at least one strain of *C. botulinum* will initiate at the Sanger Centre in the UK. The data obtained will facilitate a greater understanding of the organism's physiology, particularly as it relates to growth and survival in food. However, in recent years the major focus of attention with regard to *C. botulinum* is the therapeutic potential of its neurotoxins. The continued developing use of clostridial neurotoxins for therapy and as molecular tools for the dissection of intracellular secretory processes provides an exciting prospect in both research and clinical medicine. The development of inhibitors [148] should also broaden the scope for toxin application because there will be a concomitant reduction in the risk of toxin overdose and additionally the need to wait for the effects of overdose to wear off will no longer apply. With the recent expression and purification of stable, soluble, reduced domain, recombinant BoNT and TeNT LCs of comparable enzymatic activity to the native toxins [82, 92], the development of high affinity *in vitro* substrate binding assays and the demonstration that single amino acid substitutions can convert insensitive SNAP isoforms into CNT substrates [125], an exciting prospect for development of novel CNT therapeutics is emerging.

As we learn increasingly more about these toxins and their mode of action so we also learn about the intracellular activities and downstream events for which they have evolved so specifically to interact with. This in turn increases our understanding of cellular mechanisms of secretion and protein trafficking. It is surely ironic that the most toxic of toxins is increasingly becoming the most potent of medicines.

References

[1] Schiavo, G., Montecucco, C., The structure and mode of action of botulinum and tetanus toxins, in: *The Clostridia: Molecular Biology and Pathogenesis*, Academic Press, London, **1997**, pp. 295-322.

[2] Herreros, J., Lalli, G., Montecucco, C., Schiavo, G., Pathophysiological properties of clostridial neurotoxins, in: *The Comprehensive Sourcebook of Bacterial Protein Toxins* (Alouf, J. E., Freer, J. H., Eds.), Academic Press, London, **1999**, pp. 202-228.

[3] Popoff, M. R., Marvaud, J.-C., Structural and genomic features of clostridial neurotoxins, in: *The Comprehensive Sourcebook of Bacterial Protein Toxins* (Alouf, J. E., Freer, J. H., Eds.), Academic Press, London, **1999**, pp. 174-201.

[4] Schiavo, G., Matteoli, M., Montecucco, C., Neurotoxins affecting neuroexocytosis, *Physiol. Rev.* **2000**, 80, 717-766.

[5] Lund, B. M., Peck, M. W., *Clostridium botulinum*, in: *The Microbiological Safety and Quality of Food* (Lund, B. M., Baird-Parker, A. C., Gould, G. W., Eds.), Aspen, Gaithersburg, USA, **2000**, pp. 1057-1109.

[6] Werner, S. B., Passaro, D., McGee. J., Schechter, R., Vugia, D. J. Wound botulism in california, 1951-1998: recent epidemic in heroin injectors, *Clin. Infect. Dis.* **2000**, 31, 1018-24.

[7] Arnon, S. S., Botulism as an intestinal toxaemi in *Infections of gastrointestinal Tract* (Blaser, M. J., Smith, P. D., Ravdin, J. I., Greenberg, H. B., Guerrant, R. L., Eds.), Raven Press, New York, **1995**, pp. 257-271.

[8] Minton, N. P., Molecular genetics of clostridial neurotoxins, *Curr. Topics Microbiol. Immunol.* **1995**, 195, 161-194.

[9] Hatheway, C. L., *Clostridium botulinum* and other clostridia that produce botulinum neurotoxin, in: *Clostridium botulinum: Ecology and Control in Food* (Hauschild, A. H. W., Dodds, K. L., Eds.), Marcel Dekker, New York, **1993**, pp. 3-20.

[10] Aureli, P., Fenica, L., Pasolini, B., Gianfranceschi, M., McCroskey, L. M., Hatheway C. L., Two cases of type E infant botulism caused by neurotoxigenic *Clostridium butyricum* in Italy, *J. Infect. Dis.* **1986**, 154, 207-211.

[11] Hall, J. D., McCroskey, L. M., Pincomb, B. J., Hatheway, C. L., Isolation of an organism resembling *C. barati* which produces type F Botulinal toxin from an infant with botulism, *J. Clin. Microbiol.* **1985**, 21, 105-115.

[12] McCroskey, L. M., Hatheway, C. L., Fencia, L., Pasolini, B., Aureli, P., Characterization of an organism that produces type E botulinal toxin but which resembles *Clostridium butyricum* from the faeces of an infant with type E botulism, *J. Clin. Microbiol.* **1986**, 23, 201-202.

[13] McCroskey, L. M., Hatheway, C. L., Woodruff, B. A., Greenberg, J. A., Jurgenson, P., Type F botulism due to neurotoxigenic *Clostridium barati* from an unknown source in an adult, *J. Clin. Microbiol.* **1991**, 29, 2618-2620.

[14] Hutson, R. A., Thompson, D. E., Lawson, P. A., Schocken-Itturino, R. P., Bottger, E. C., Collins, M. D., Genetic interelationships of proteolytic *Clostridium botulinum* types A, B and F and other members of the *Clostridium botulinum* complex as revealed by small subunit rRNA gene sequences, *Antonie van Leewenhoek* **1993**, 64, 278-283.

[15] Hutson, R. A., Thompson, D. E., Collins, M. D., Genetic interrelationships of saccharolytic *Clostridium botulinum* types B, E and F and related clostridia as revealed by small-subunit rRNA gene sequences, *FEMS Microbiol. Lett.* **1993**, 108, 103-110.

[16] Lin, W-J., Johnson, E. A., Genome analysis of *Clostridium botulinum* type A by pulsed-field gel electrophoresis, *Appl. Environ. Microbiol* .**1995**, 61, 4441-4447.

[17] Hielm, S., Bjorkroth, J., Hyytia, E., Korkeala, H., Genomic analysis of *Clostridium botulinum* group II by pulsed-field gel electrophoresis, *Appl. Environ. Microbiol.* **1998**, 64, 703-708.

[18] Hielm, S., Bjorkroth, J., Hyytia, E., Korkeala, H., Prevalence of *Clostridium botulinum* in Finnish trout farms: pulsed-field gel electrophoresis typing reveals extensive genetic diversity among type E isolates, *Appl. Environ. Microbiol.* **1998**, 64, 4161-4167.

[19] Ohishi, I., Sughii, S., Sakaguchi. G., Oral toxicities of *Clostridium botulinum* toxins in response to molecular size, *Infect. Immun.* **1977**, 16, 107-109.

[20] Ohishi, I., Sakaguchi, G., Oral toxicities of *Clostridium botulinum* type C and D toxins of different molecular size, *Infect. Immun.* **1980**, 28, 303-309.

[21] Eisel, U., Jarausch, W., Goretzki, K., Henschen, A., Engels, J., Weller, U., Hudel, M., Habermann, E., Niemann, H., Tetanus toxin: primary structure, expression in *Escherichia coli*, and homology with botulinum toxins, *EMBO J.* **1986**, 5, 2495-2502.

[22] Fairweather, N. F., Lyness, V. A., The complete nucleotide sequence of tetanus toxin. *Nucleic Acids Res.* **1986**, 14, 7809-7812.

[23] Thompson, D. E., Brehm, J. K., Oultram, J. D., Swinfield, T. J., Shone, C. C., Atkinson, T., Melling, J., Minton, N. P., The complete amino acid sequence of the *Clostridium botulinum* type A neurotoxin, deduced by nucleotide sequence analysis of the encoding gene, *Eur. J. Biochem.* **1990**, 189, 73-81.

[24] Henderson, I., Davis, T., Elmore, M., Minton, N. P., The genetic basis of toxin production, in: *Clostridium botulinum* and

Clostridium tetani. The Clostridia: Molecular Biology and Pathogenesis, Chapter 17, London, Academic Press, **1997**.

[25] Kubota, T., Yonekura, N., Hariya, Y., Isogai, E., Isogai, H., Amano, K-I., Fujii, N., Gene arrangement in the upstream region of *Clostridium botulinum* type E and *Clostridium butyricum* BL6340 progenitor toxin genes is different from that of other types, *FEMS Microbiol. Lett.* **1998**, 158, 215-221.

[26] Franciosa, G., Ferreira, J. L., Hatheway, C. L., Detection of type A, B, and E botulism neurotoxin genes in *Clostridium botulinum* and other *Clostridium* species by PCR: evidence of unexpressed type B toxin genes in type A toxigenic organisms, *J. Clin. Microbiol.* **1994**, 32, 1911-1917.

[27] Hutson, R. A., Zhou, Y., Collins, M. D., Johnson, E. A., Hatheway, C. L., Sugiyama, H. Genetic characterization of *Clostridium botulinum* type A containing silent type B neurotoxin gene sequences, *J. Biol. Chem.* **1996**, 271, 10786-10792.

[28] Franciosa, G., Hatheway, C. L., Aureli, P., The detection of a deletion in the type B neurotoxin gene of *Clostridium botulinum* A(B) strains by a two-step PCR, *Lett. Appl. Microbiol.* **1998**, 26, 442-446.

[29] Henderson, I., Whelan, S. M., Davis, T. O., Minton, N. P., Genetic characterization of the botulinum toxin complex of *Clostridium botulinum* strain NCTC 2916, *FEMS Microbiol. Lett.* **1996**, 140, 151-158.

[30] Moriishi, K., Koura, M., Abe, N., Fujii, N., Fujinaga, Y., Inoue, K., Ogumad, K., Mosaic structures of neurotoxins produced from *Clostridium botulinum* types C and D organisms, *BBA* **1996**, 1307, 123-126.

[31] Moriishi, K., Koura, M., Fujii, N., Fujinaga, Y., Inoue, K., Syuto, B., Oguma, K., Molecular cloning of the gene encoding the mosaic neurotoxin, composed of parts of botulinum neurotoxin types C1 and D, and PCR detection of this gene from *Clostridium botulinum* type C organisms, *Appl. Environ. Microbiol.* **1996**, 62, 662-667.

[32] Scott, V. N., Duncan, C. L., Cryptic plasmids in *Clostridium botulinum* and *Clostridium botulinum*-like organisms, *FEMS Microbiol. Lett.* **1978**, 4, 55-58.

[33] Strom, M. S., Eklund, M. W., Poysky, F. T., Plasmids in *Clostridium botulinum* and related *Clostridium* species, *Appl. Environ. Microbiol.* **1984**, 48, 956-963.

[34] Weickert, M. J., Chambliss, G. H., Sugiyama, H., Production of toxin *by Clostridium botulinum* type A strains cured by plasmids, *Appl. Environ. Microbiol.* **1986**, 51, 52-56.

[35] Zhou, Y., Sugiyama, H., Nakano, H., Johnson, E. A., The genes for *Clostridium botulinum* type G toxin complex are on a plasmid, *Infect. Immunol.* **1995**, 63(5), 2087-2091.

[36] Finn, C. W. Jr., Silver, R. P., Habig, W. H., Hardegree, M. C., The structural gene for tetanus neurotoxin is on a plasmid, *Science* **1984**, 224, 881-884.

[37] Hauser, D., Gibert, M., Boquet, P., Popoff, M. R., Plasmid localisation of a type E botulinal neurotoxin gene homologue in toxigenic *Clostridium butyricum* strains, and absence of this gene in non-toxigenic *Clostridium butyricum strains, FEMS Microbiol. Lett.* **1992**, 99, 251-256.

[38] Wang, X., Maegawa, T., Karasawa, T., Kozaki, S., Tsukamoto, K., Gyobu, Y., Yamakawa, K., Oguma, K., Sakaguchi, Y., Nakamura, S., Genetic analysis of type E botulinum toxin-producing *Clostridium butyricum* strains, *Appl. Environ. Microbiol.* **2000**, 66, 4992-4997.

[39] Inoue, K., Iida, H., Bacteriophages of *Clostridium botulinum, J. Virol.* **1968**, 2, 537-540.

[40] Eklund, M. W., Poysky, F. T., Boatman, E. S., Bacteriophages of *Clostridium botulinum* types A, B, E, and F and nontoxigenic strains resembling type E, *J. Virol.* **1969**, 3, 270-274.

[41] Dolman, C. E., Chang, E., Bacteriophages of *Clostridium botulinum, Can. J. Microbiol.* **1972**, 18, 67-76.

[42] Eklund, M. W., Poysky, F. T., Habig, W. H., Bacteriophages and plasmids in *Clostridium botulinum* and *Clostridium tetani* and their relationships to production of toxins in *Botulinum neurotoxin and tetanus toxin* (Simpson, L. L., Ed.), Academic Press, San Diego, CA, **1989**, pp. 26-52.

[43] Binz, T., Kurazono, H., Popoff, M. R., Eklund, M. W., Sakaguchi, G., Kozaki, S., Krieglstein, K., Henschen, A., Gill, D. M., Niemann H., Nucleotide sequence of the gene encoding *Clostridium* botulinum neurotoxin type D, *Nucleic Acids Res.* **1990**, 18, 5556.

[44] Hauser, D., Eklund, M. W., Kurazono, H., Binz, T., Niemann, H., Gill, D. M., Boquet,

P., Popoff, M. R., Nucleotide sequence of *Clostridium botulinum* C1 neurotoxin, *Nucleic Acids Res.* **1990**, 18, 4924.

[45] Zhou, Y., Johnson, E. A., Genetic Transformation of *Clostridium botulinum* Hall A by electroporation, *Biotechnol. Lett.* **1993**, 15, 121-126.

[46] Croux, C., Garcia, J. L., Sequence of the *lyc* gene encoding the autolytic lysozyme of *Clostridium acetobutylicum* ATCC 824: comparison with other lytic enzymes, *Gene* **1991**, 104, 25-31.

[47] Lichenstein, H. S., Hastings, A. E., Langley, K. E., Mendiaz, E. A., Rohde, M. F., Elmore, R., Zukowski, M. M., Cloning and nucleotide sequence of the N-acetylmuramidase MI-encoding gene from *Streptomyces globisporus*, *Gene* **1990**, 88, 81-86.

[48] Boizet,B., Lahbib-Mansais ,Y., Dupont, L., Ritzenthaler, P., Mata, M., Cloning, expression and sequence analysis of an endolysin-encoding gene of *Lactobacillus bulgaricus* bacteriophage mv1, *Gene* **1990**, 94, 61-67.

[49] Garcia, E., Garcia, J. L., Garcia, P., Arraras, A., Sanchez-Puelles, J. M., Lopez, R., Molecular evolution of lytic enzymes of *Streptococcus pneumoniae* and its bacteriophages, *Proc. Natl. Acad. Sci. USA* **1988**, 85, 914-918.

[50] Whelan, S. M., Garcia, J. L., Elmore, M. J., Minton, N. P., The botulinum neurotoxin gene of the type a *Clostridium botulinum* strain NCTC 2916 is followed by a gene *(lycA)* encoding a lysozyme, in: *Bacterial Protein Toxins* (Freer et al., Eds.) **1994**, Suppl 24, 162-163.

[51] Bonventre, P. F., Kempe, L.L, Physiology of toxin production of Clos*tridium botulinum* types A and B. I. Growth, autolysis and toxin production, *J. Bacteriol.* **1960**, 79, 18-23.

[52] Hauser, D., Gibert, M., Eklund, M. W., Boquet, P., Popoff, M. R., Comparative analysis of C3 and botulinal neurotoxin genes and their environment in *Clostridium botulinum* types C and D, *J. Bacteriol.* **1993**, 175, 7260-7268.

[53] Johnson, E. A., Lin, W.-J., Zhou, Y.-T., Bradshaw, M., Characterization of neurotoxin mutants in *Clostridium botulinum* type A, *Clin. Infect.* **1997**, 25(Suppl. 2), S168-70.

[54] Bowers, L. E., Williams, O. B., Effect of arginine on the growth and lysis of *Clostridium botulinum*, *J. Bacteriol.* **1963**, 85, 1175-1176.

[55] Patterson-Curtis, S. I., Johnson, E. A., Regulation of neurotoxin and protease formation in *Clostridium botulinum* Okra B and Hall A by arginine, *Appl. Environ. Microbiol.* **1989**, 55, 1544-1548.

[56] Schantz, E. J., Johnson, E. A., Properties and use of botulinum toxin and other microbial neurotoxins in medicine, *Microbiol. Rev.* **1992**, 56, 80-99.

[57] Garnier T., Cole, S. T., Studies of UV-inducible promoters from *Clostridium perfringens in vivo* and *in vitro, Mol. Microbiol.* **1988**, 2, 607-614.

[58] Moncrief, J. S., Barroso, L. A., Wilkins, T. D., Positive regulation of *Clostridium difficile* toxins, *Infect. Immun.* **1997**, 65, 1105-1108.

[59] Russell, R. R. B., Dduse-Opoku, J., Sutclife, I. C., Tao, L., Ferretti, J. J., A binding protein-dependent transport system in *Streptococcus mutans* responsible for multiple sugar metabolism, *J. Biol. Chem.* **1992**, 267, 4631-4637.

[60] Marvaud, J. C., Gibert, M., Inoue, K., Fujinaga, Y., Oguma, K., Popoff, M. R., *botR/A* is a positive regulator of botulinum neurotoxin and associated non-toxin protein genes in *Clostridium botulinum, Mol. Microbiol.* **1998**, 29, 1099-1018.

[61] Marvaud, J-C., Eisel, U., Binz, T., Niemann, H., Popoff, M. R., TetR is a positive regulator of the tetanus toxin gene in *Clostridium tetani* and is homologous to BotR, *Infect. Immunol.* **1998**, 66, 5698-5702.

[62] Hauser, D., Eklund, M. W., Boquet, P., Popoff, M. R., Organization of the botulinum neurotoxin C1 gene and its associated non-toxic protein genes in *Clostridium botulinum* C 468, *Mol. Gen. Genet.* **1994**, 243, 631-640.

[63] Davis, T. O., Regulation of Botulinum Toxin Complex Formation in *Clostridium botulinum, Thesis,* Open University, UK, **1998**.

[64] Agaisse, H., Lereclus, D. STAB-SD: a Shine-Dalgarno sequence in the 5′ untranslated region is a determinant of mRNA stability, *Mol. Microbiol.* **1996**, 20, 633-643.

[65] Rood, J. I., Cole, S. T., Molecular genetics and pathogenesis of *Clostridium perfringens, Microbiol. Rev.* **1991**, 55, 621-648.

[66] Mauchline, M. L., Davis, T. O., Minton, N. P., Clostridia, in: *Manual of Industrial Microbiology and Biotechnology,* (Demain, A. L., Davies, J. E., Eds.), ASM Press, Washington, DC, **1999**, pp. 475-490.

[67] Volk, W. A., Bizzini, B., Jones, K. R., Macrina, F. L., Inter- and intrageneric transfer to Tn*916* between *Streptococcus faecalis* and *Clostridium tetani*, *Plasmid* **1988**, 19, 255-259.

[68] Lin, W-J., Johnson, E. A., Transposon Tn*916* mutagenesis in *Clostridium botulinum*, *Appl. Environ. Microbiol.* **1991**, 57, 2946-2950.

[69] Minton, N. P., Morris, J. G., Isolation and partial characterisation of three cryptic plasmids from strains of *Clostridium butyricum*, *J. Gen. Microbiol.* **1981**, 127, 325-331.

[70] Davis, T. O., Henderson, I., Brehm, J. K., Minton, N. P., Development of transformation and gene reporter system for Group II, non-proteolytic *Clostridium botulinum* type B strains, *J. Mol. Microbiol.* **2000**, 2, 59-69.

[71] Bradshaw, M, Goodnough, M. C., Johnson, E. A., Conjugative transfer of *Escherichia coli*-*Clostridium perfringens* shuttle vector pJIR1457 to *Clostridium botulinum* type A strains, *Plasmid* **1998**, 40, 233-237.

[72] Singh, B. R., Intimate details of the most poisonous poison, *Nature Struct. Biol.* **2000**, 7, 617-619.

[73] Emsley, P., Fotinou, C., Black, I. Fairweather, N., Charles, I. G., Watts, C., Hewitt, E., Isaacs, N. W., The structures of the H_C fragment of tetanus toxin with carbohydrate subunit complexes provide insight into ganglioside binding, *J. Biol. Chem.* **2000**, 275, 8889-8894.

[74] Herreros, J., Lalli, G., Shiavo, G., C-terminal half of tetanus toxin fragment C is sufficient for neuronal binding and interaction with a putative protein receptor, *Biochem. J.* **2000**, 347, 199-204.

[75] Chen, F., Kuziemko, G. M., Stevens, R. C., Biophysical characterization of the stabiity of the 150-kilodalton botulinum toxin, the nontoxic component, and the 900-kilodalton botulinum toxin complex species, *Infect. Immun.* **1998**, 66, 2420-2425.

[76] Fuginaga,Y., Inoue, K., Watanabe, S., Yokota, K., Hirai, Y., Nagamachi, E., Oguma, K., The hemagglutinin of *Clostridium botulinum* type C progenitor toxin plays an essential role in binding of toxin to the epithelial cells of guinea pig small intestine, leading to the efficient absorption of the toxin, *Microbiology* **1997**, 143, 3841-3847.

[77] Sugii, S., Sakaguchi, J. Botulinogenic properties of vegetables with special reference to the molecular size of the toxin in them, *J. Food Safety* **1977**, 1, 53-65.

[78] Borden Lacy, D., Stevens, R. C., Sequence homology and structural analysis of the clostridial neurotoxins, *J. Mol. Biol.* **1999**, 291, 1091-1104.

[79] de Paiva, A., Poulain, B., Lawrence, G., Shone, C. C., Tauc, L., Dolly, J. O., A role for the interchain disulfide or its participating thiols in the internalization of botulinum neurotoxin A revealed by a toxin derivative that binds to ecto-acceptors and inhibits transmitter release intracellularly, *J. Biol. Chem.* **1993**, 268, 20838-20844.

[80] Umland, T. C., Wingert, L. M., Swaminathan, S., Furey, W. F., Schmidt, J. J., Sax, M., Structure of the receptor binding fragment H_C of tetanus toxin, *Nature Struct. Biol.* **1997**, 4, 788-792.

[81] Borden Lacy, D., Tepp, W., Cohen, A. C., DasGupta, B. R., Stevens, R. C., Crystal structure of botulinum neurotoxin type A and implications for toxicity, *Nature Struct. Biol.* **1998**, 5, 898-902.

[82] Kadkhodayan, S. Knapp, M. S., Schmidt, J. J., Fabes, S. E., Rupp, B., Balhorn, R., Cloning, expression and one-step purification of the minimal essential domain of the light chain of botulinum neurotoxin type A, *Protein Expr. Purif.* **2000**, 19, 125-130.

[83] Swaminathan, S., Eswaramoorthy, S., Structural analysis of the catalytic and binding sites of Clostridium botulinum neurotoxin B, *Nature Struct. Biol.* **2000**, 7, 693-699.

[84] Hanson, M. A., Stevens, R. C., Cocrystal structure of symaptobrevin-II bound to botulinum neurotoxin type B at 2.0 Å resolution, *Nature Struct. Biol.* **2000**, 7, 687-692.

[85] Montecucco, C., Schiavo, G., Tetanus and botulism neurotoxins: a new group of zinc proteases, *Trends Biochem. Sci.* **1993**, 18, 324-327.

[86] Tonello, F., Morante, S., Rossetto, O., Schiavo, G., Montecucco, C., Tetanus and botulism neurotoxins: a novel group of zinc-endopeptidases, *Adv. Exp. Med. Biol.* **1996**, 389, 251-260.

[87] Pellizzari, R., Rossetto, O., Schiavo, G., Montecucco, C., Tetanus and botulinum neurotoxins: mechanism of action and therapeutic uses, *Philos. Trans. R. Soc. Lond. B Biol. Sci.* **1999**, 354, 259-68.

[88] Li, L., Singh, B. R., Role of zinc binding in type A botulinum neurotoxin light chain's toxic structure, *Biochemistry* **2000**, 39, 10581-10586.

[89] Yamasaki, S., Baumeister, A., Binz, T., Blasi, J., Link, E., Cornille, F., Roques, B., Fykse, E. M., Sudhof, T. C., Jahn, R., Niemann, H., Cleavage of members of the synaptobrevin/VAMP family by types D and F botulinal neurotoxins and tetanus toxin, *J. Biol. Chem.* **1994**, 269, 12764-12772.
[90] Li, Y., Foran, P., Fairweather, N., de Paiva, A., Weller, U. Dougan, G. Dolly, J. O., A single mutation in the recombinant light chain of tetanus toxin abolishes proteolytic acitivty and removes the toxicity seen after reconstitution with native heavy chain, *Biochemistry* **1994**, 33, 7014-7020.
[91] Kurazono, H., Mochida, S., Binz, T., Eisel, U., Quanz, M., Grebenstein, O., Wernars, K., Poulain, B., Tauc, L., Niemann, H., Minimum essential domains specifying toxicity of the light chains of tetanus toxin and botulinum neurotoxin type A, *J. Biol. Chem.* **1992**, 267, 14721-14729.
[92] Tonello, F., Pellizzari, R., Pasqualato, S., Grandi, G., Peggion, E., Montecucco, C. Recombinant and truncated tetanus neurotoxin light chain: cloning, expression, purification and proteolytic activity, *Protein Expr. Purif.* **1999**, 15, 221-227.
[93] Shone, C. C., Hambleton, P., Melling, J., Inactivation of Clostridium botulinum type A neurotoxin by trypsin and purification of two tryptic fragments. Proteolytic action near the COOH-terminus of the heavy subunit destroys toxin-binding activity, *Eur. J. Biochem.* **1985**, 151, 75-82.
[94] Shapiro, R. E., Specht, C. D., Collins, B. E., Woods, A. S., Cotter, R. J., and Schnaar, R. L., Identification of a ganglioside recognition domain of tetanus toxin using a novel ganglioside photoaffinity ligand, *J. Biol. Chem.* **1997**, 272, 30380-30386.
[95] Montecucco, C., How do tetanus and botulinum toxins bind to neuronal membranes? *Trends Biochem. Sci.* **1986**, 11, 315-317.
[96] Nishiki, T., Tokuyama, Y., Kamata, Y., Nemoto, Y., Yoshida, A., Sato, K., Sekiguchi, M., Takahashi, M., Kozaki, S., The high affinity binding of *Clostridium botulinum* type B neurotoxin to synaptotagmin II associated with gangliosides GT1b/GD1a, *FEBS Lett.* **1996**, 378, 253-257.
[97] Kozaki, S., Kamata, Y. Watarai, S., Nishiki, T., Mochida, S., Ganglioside GT1b as a complementary receptor component for *Clostridium botulinum* neurotoxins, *Microbiol. Pathog.* **1998**, 25, 91-99.
[98] Kitamura, M., Takamiya, K., Aizawa, S., Furukawa, K., Furukawa, K., Gangliosides are the binding substances in neural cells for tetanus and botulinum toxins in mice, *Biochim. Biophys. Acta* **1999**, 1441, 1-3.
[99] Fitzsimmons, S. P., Clark, K. C., Wilkerson, R., Shapiro, M. A., Inhibition of tetanus toxin fragment C binding to ganglioside G(T1b) by monoclonal antibodies recognizing different epitopes, *Vaccine* **2000**, 19, 114-21.
[100] Li, L., Singh, B. R., Isolation of synaptogamin as a receptor for type A and type E botulinum neurotoxin and analysis of their comparative binding using a new microtitre plate assay, *J. Nat. Toxins* **1998**, 7, 215-226.
[101] Ginalski, K., Venclovas, C., Lesyng, B., Fidelis, K., Structure-based sequence alignment for the β-trefoil subdomain of the clostridial neurotoxin family provides residue level information about the putative ganglioside binding site, *FEBS Lett.* **2000**, 482, 119-124.
[102] Halpern, J. L., Loftus, A., Characterization of the receptor binding domain of tetanus toxin, *J. Biol. Chem.* **1993**, 268, 11188-11192.
[103] Kamata, Y., Yoshimoto, M. Kozaki, S., Interaction between botulinum neurotoxin type A and ganglioside: ganglioside inactivates the neurotoxin and quenches its tryptophan fluorescence, *Toxicon* **1997**, 35, 1337-40.
[104] Sinha, K., Box, M., Lalli, G., Schiavo, G., Schneider, H., Groves, M., Siligardi, G., Fairweather, N., Analysis of mutants of tetanus toxin H_C fragment: ganglioside binding and retrograde axonal transport properties, *Mol. Microbiol.* **2000**, 37, 1041-1051.
[105] Shone, C. C., Hambleton, P., Melling, J., A 50-kDa fragment from the NH_2 terminus of the heavy subunit for *Clostridium botulinum* type A neurotoxin forms channels in lipid membranes, *Eur. J. Biochem.* **1987**, 167, 175-180.
[106] Hoch, D. H., Romero-Mira, M., Ehrlich, B. E., Finkelstein, A., Dasgupta, B. R., Simpson, L. L., Channels formed by botulinum, tetanus, and diphtheria toxins in planar lipid bilayers: relevance to translocation of proteins across membranes, *Proc. Nat. Acad. Sci. USA* **1985**, 82, 1692-1696.

[107] Donovan, J. J., Middlebrook, J. L., Ion-conducting channels produced by botulinum toxin in planar lipid membranes, *Biochemistry* **1986**, 25, 2872-2876.
[108] Blaustein, R. O., Germann, W. J., Finkelstein, A., Dasgupta, B. R., The N-terminal half of the heavy chain of botulinum type A neurotoxin forms channels in planar phospholipid bilayers, *FEBS Lett.* **1987**, 226, 115-120.
[109] Sheridan, R. E., Gating and permeability of ion channels produced by botulinum toxin types A and E in PC12 cell membranes, *Toxicon* **2000**, 36, 703-17.
[110] Lebeda, F. J., Olsen, M. A., Structural predictions for the channel forming region of botulinum neurotoxin heavy chain, *Toxicon* **1995**, 33, 559-567.
[111] Oblatt-Montal, M., Yamazki, M., Nelson, R., Montal, M., Formation of ion channels in lipid bilayers by a peptide with the predicted transmembrane sequence of botulinum neurotoxin A, *Protein Sci.* **1995**, 4, 1490-1497.
[112] Chaddock, J. A., Purkiss, J., Friis, L., Broadbridge, J., Duggan, M., Shone, C., Quinn, C. P., Foster, K. A., Inhibition of neurotransmitter release by a retargeted endopeptidase derivative of *C. botulinum* neurotoxin type A, *Infect. Immun.* **2000**, 68, 2587-2593.
[113] Rizo, J., Südhof, T. C., Mechanics of membrane fusion, *Nature Struct. Biol.* **1998**, 5, 839-842.
[114] Fasshauer, D., Brünger, A. T., Jahn, R., Conserved structural features of the synaptic fusion complex: SNARE proteins reclassified as Q- and R-SNAREs, *Proc. Nat. Acad. Sci. USA* **1998**, 95, 15781-15786.
[115] Sutton R. B., Fasshauer, D., Jahn, R., Brünger, A. T., Crystal structure of a SNARE complex involved in symaptic exocytosis at 2.4 Å resolution, *Nature* **1998**, 395, 347-353.
[116] Linial, M., SNARE proteins – why so many, why so few? *J. Neurochem.* **1997**, 69, 1781-1792.
[117] Rossetto, O., Schiavo, G., Montecucco, C., Poulain, B., Deloye, F., Lozzi, L., Shone, C. C., SNARE motif and neurotoxins, *Nature* **1994**, 372, 415-416.
[118] Pellizzari, R., Rossetto, O., Lozzi, L., Giovedi, S., Johnson, E., Shone, C. C., Montecucco, C., Structural determinants of the specificity for synaptic vesicle associated membrane protein/synaptobrevin of tetanus and botulinum type B and G neurotoxins, *J. Biol. Chem.* **1996**, 271, 20353-20358.
[119] Pellizzari, R., Mason, S., Shone, C. C., Montecucco, C., The interaction of synaptic vesicle-associated membrane protein/synaptobrevin with botulinum neurotoxins D and F, *FEBS Lett.* **1997**, 408, 339-342.
[120] Washbourne, P., Pellizzari, R., Baldini, G., Wilson, C., Montecucco, C., Botulinum neurotoxin type A and type E require the SNARE motif in SNAP-25 for proteolysis, *FEBS Lett.* **1997**, 418, 1-5.
[121] Cornille, F., Goudreau, N., Ficheux, D., Niemann, H., Roques, B. P., Solid-phase synthesis, conformational analysis and *in vitro* cleavage of synthetic human synaptobrevin II 1-93 by tetanus toxin L chain, *Eur. J. Biochem.* **1994**, 222, 173-81.
[122] Foran P, Shone C. C. and Dolly, J. O., Differences in the protease activities of tetanus and botulinum B toxins revealed by the cleavage of vesicle-associated membrane protein and various sized fragments, *Biochemistry* **1994**, 33, 15365-74.
[123] Hayashi, T., Yamasaki, S., Nauenburg, S., Binz, T., Niemann, H. , Disassembly of the reconstituted synaptic vesicle membrane fusion complex *in vitro*, *EMBO J.* **1994**, 14, 2317-2325.
[124] Pellegrini, L. L., O'Connor, V., Betz, H., Clostridial neurotoxins compromise the stability of a low energy SNARE complex mediating NSF activation of synaptic vesicle fusion, *EMBO J.* **1995**, 14, 4705-4713.
[125] Vaidyanathan, V. V., Yoshino, K., Jahnz, M., Dörries, C., Bade, S., Nauenburg, S., Niemann, H., Binz, T., Proteolyis of SNAP-25 isoforms by botulinum neurotoxin types A, C, and E: domains and amino acid residues controlling the formation of enzyme substrate complexes and cleavage, *J. Neurochem.* **1999**, 72, 327-337.
[126] Gil, C., Chaib-Oukadour, I., Pelliccioni, P., Aguilera, J., Activation of signal transduction pathways involving trkA, PLCgamm-1, PKC isoforms and ERK-1/2 by tetanus toxin, *FEBS Lett.* **2000**, 481, 177-182.
[127] Facchiano, F., Luini, A., Tetanus potently stimulates tissue transglutaminase: a possible mechanism of neurotoxicity, *J. Biol. Chem.* **1992**, 267, 13267-13271.
[128] Facchiano, F., Valtorta, F., Benfenati, F., Luini, A, The transglutaminase hypothesis

for the action of tetanus toxin, *Trends Biochem. Sci.* **1993**, 18, 327-329.

[129] Facchiano, F., Valtorta, F., Benfenati, F., Luini, A., Covalent modification of synapsin I by a tetanus toxin-activated transglutaminase, *J. Biol. Chem.* **1993**, 268, 4588-4591.

[130] Ashton, A. C., Li, Y., Doussau, F., Weller, U., Dougan, G., Poulain, B., Dolly, J. O., Tetanus toxin inhibits neuroexocytosis even when its Zn^{2+}-dependent protease activity is removed, *J. Biol. Chem.* **1995**, 270, 31386-31390.

[131] Coffield, J. A., Considine, R. V., Jeyapaul, J., Maksymowych, A. B., Zhang, R. D., The role of transglutaminase in the mechanism of action of tetanus toxin, *J. Biol. Chem.* **1994**, 269, 24454-24458.

[132] Münchau, A., Bhatia, K. P., Uses of botulinum toxin injecion in medicine today, *BMJ* **2000**, 320, 161-165.

[133] Duchen, L. W., Tonge, D. A., The effects of tetanus toxin on neuromuscular transmission and on the morphology of motor endplates in slow and fast skeletal muscle of the mouse, *J. Physiol.* **1973**, 228, 157-172.

[134] Bambrick, L., Gordon, T., Neurotoxins in the study of neural regulation of membrane proteins in skeletal muscle, *J. Pharmacol. Toxicol. Methods* **1994**, 32, 129-138.

[135] Shone, C. C., Melling, J., Inhibition of calcium-dependent release of noradrenaline from PC12 cells by botulinum type-A neurotoxin. Long-term effects of the neurotoxin on intact cells, *Eur. J. Biochem.* **1992**, 207, 1009-1016.

[136] Schantz, E. J., Johnson, E. A., Botuliinum toxin: the story of its development for the treatment of human disease, *Persp. Biol. Med.* **1997**, 40, 317-327.

[137] Scott, A. B., Rosenbaum, A., Collins, C. C., Pharmacologic weakening of extraocular muscles, *Invest. Opthamol.* **1973**, 12, 924-927.

[138] Scott, A. B., Clostridial neurotoxins as therapeutic agents, in: *Botulinum Neurotoxin and Tetanus Toxin* (Simpson, L. L., Ed.), Academic Press, San Diego, CA, **1989**.

[139] Jankovic, J., Hallet, M., *Therapy with botulinum toxin,* Marcel Dekker, New York, **1994**.

[140] Guyer, B. M., Mechanism of botulinum toxin in the relief of chronic pain, *Curr. Rev. Pain* **1999**, 3, 427-431.

[141] Chaddock, J. A., Purkiss, J. R., Duggan, M. J., Quinn, C. P., Shone, C. C., Foster, K. A., A conjugate composed of nerve growth factor coupled to a nontoxic derivative of *Clostridium botulinum* neurotoxin type A can inhibit neurotransmitter release *in vitro, Growth Factors* **2000**, 18, 147-55.

[142] Stoeckel, K. Schwab, M., Thoenen, H., Role of gangliosides in the uptake and retrograde axonal transport of cholera and tetanus toxin as compared to nerve growth factor and wheat germ agglutinin, *Brain Res.* **1977**, 132, 273-285.

[143] Knight, A., Carvajal, J., Schneider, H., Coutelle, C., Chamgerlalin, S., Fairweather, N., Non-viral neuronal gene delivery mediated by the HC fragment of tetanus toxin, *Eur. J. Biochem.* **1999**, 259, 762-769.

[144] Figueiredo, D. M., Hallewell, R. A., Chen, L. L., Fairweather, N. F., Dougan, G., Savitt, J. M., Parks, D. A., Fishman, P. S., Delivery of recombinant tetanus-superoxide dismutase proteins to central nervous system neurons by retrograde axonal transport, *Exp. Neurol.* **1997**, 145, 546-554.

[145] Francis, J. W., Hosler, B. A., Brown, R. H. Jr, Fishman, P. S., CuZn superoxide dismutase (SOD-1): tetanus toxin fragment C hybrid protein for targeted delivery of SOD-1 to neuronal cells, *J. Biol. Chem.* **1995**, 270,15434-42.5

[146] Francis, J. W., Brown, R. H. Jr, Figueiredo, D., Remington, M.P, Castillo, O., Schwarzschild, M. A., Fishman, P. S., Murphy, J. R., vanderSpeck, J. C., Enhancement of diphtheria toxin potency by replacement of the receptor binding domain with tetanus toxin C-fragment: a potential vector for delivering heterologous proteins to neurons, *J. Neurochem.* **2000**, 74, 2528-36.

[147] Schneider, H., Groves, M., Muhle, C., Reynolds, P. N., Knight, A., Themis, M., Carvajal, J., Scaravilli, F., Curiel, D. T., Fairweather, N., Coutelle, C., Retargeting of adenoviral vectors to neurons using the H_C fragment of tetanus toxin, *Gene Ther.* **2000**, 7, 1584-1592.

[148] Schmidt, J. J., Stafford, R.G, Bostian, K. A., Type A botulinum neurotoxin proteolytic activity: development of competetitive inhibitors and implications for substrated specificity at the S1′ binding site, *FEBS Lett.* **1998**, 435, 61-64.

8
Clostridia in Cancer Therapy

Nigel P. Minton, J. Martin Brown, Philippe Lambin and Jozef Anné

8.1
Introduction

It is widely recognized that if currently available anti-cancer drugs could be administered at an appropriate therapeutic dose then they would be able to entirely eliminate tumor cells. However, the concentrations that would need to be applied would simultaneously cause devastation to normal cells throughout the body. In recognition of this fact attention has focused on attempting to deliver high doses of the therapeutic agent specifically to the tumor vicinity, while maintaining relatively innocuous levels of drug in normal, healthy tissues. A number of strategies and delivery vehicles have been investigated including tumor-specific antibodies and liposomes. In those instances where the targeted drug is of a proteinacous nature (e. g., an enzyme) attention has more recently focused on the delivery of the encoding gene. Much of this latter work has involved the use of viral vectors. However, the use of such vehicles has encountered significant difficulties, particularly with regard to target specificity.

Many of the current difficulties could be overcome through the use of a delivery system based on clostridia. This is because intravenously injected clostridial spores exclusively germinate in the hypoxic region of solid tumors. Here we discuss how this novel property may be exploited in cancer therapy.

8.2
Clostridial oncolysis – a historical perspective

Members of the genus *Clostridium* are obligate anaerobes. Thus, while clostridial spores are widespread in the environment, they are unable to germinate unless they encounter the requisite anaerobic conditions. As a consequence, they cannot grow in normal healthy tissue. Their ability to germinate in necrotic tissue has, however, long been recognized. Thus, the diseases tetanus and gas gangrene are caused by the direct colonization of necrotic tissues by *Clostridium tetani* and *Clostridium perfringens*, respectively. This apparent selectivity by clostridia for tumors received significant attention over the last 50 years.

8.2.1
Clostridial oncolysis in animals

The first attempt to use clostridia in the treatment of cancer was reported in 1935 where clinical improvements were reported in cases of advanced cancer following treatment with sterile filtrates of *Clostridium histolyticum* [1]. These beneficial effects were attributed to the production of enzymes which preferentially proteolyzed neoplastic tissue without effecting normal tissue. This led to experiments which sought to produce such proteins *in situ* by deliberately infecting the tumor tissue. Thus, Parker et al. injected spore suspensions of *Clostridium histolyticum* directly into the transplanted sarcomas of mice [2], where upon significant lysis of the tumor resulted from the subsequent vegetative growth of the organism. Furthermore, while the growth of the organism in the tumor accelerated death, co-treatment with penicillin and anti-toxin extended the survival of some animals compared to tumor-bearing mice which were not injected with clostridial spores. These early studies provided the first indication that clostridial-induced lysis of tumors could have a therapeutically beneficial effect.

Thereafter, Malmgren and Flanigan went on to show that direct injection of spores into the tumor was unnecessary, and that spores could be administered intravenously [3]. In their studies, both tumor-bearing and healthy mice were injected with spores, and their survival monitored. Those mice with tumors succumbed to tetanus poisoning within 48 h. In contrast, the healthy controls remained unaffected, and demonstrated no signs of the disease over the period monitored (40 days). Furthermore, vegetative organisms could neither be detected in, following microscopic examination, nor cultured from, the healthy tissue of mice carrying colonized tumors. Aside from demonstrating that direct injection into the tumor mass was not required for colonization, the study of Malmgren and Flanigan also served to highlight the exquisite selectivity of the process. Thus, the benign nature of the spores in normal healthy mice clearly demonstrated that the bacterium was incapable of establishing itself in the animal in the absence of appropriate anaerobic conditions.

Möse and Möse were the first investigators to conceive the notion of effecting the colonization of tumors with a clostridial species as a means of treating cancer through clostridial-induced lysis [4, 5]. For the therapy to be effective, they proposed the use of a clostridial species that was non-pathogenic, was able to effectively lyse tumor tissue, and allowed the preparation of high spore titers. Accordingly, they isolated a strain from soil which they originally designated *Clostridium butyricum* M-55, but later re-named *Clostridium oncolyticum* [6], and used it in studies with mice bearing Ehrlich carcinomas [4, 5]. Following intravenous injection of M-55 spores, the carcinomas were seen to soften, become fluctuant and eventually to break open to release a brown liquid containing large numbers of vegetative bacteria. Animals rarely survived past this stage of oncolysis, and of those that did 65 % had a recurrence of the tumor at the same site. Thus, while liquefaction of the tumor led to the destruction of large parts of the tumor, a viable outer rim frequently remained from which tumor regrowth occurred. Similar experiments undertaken with other animal models confirmed these results [7-10].

8.2.2
Enhancement of oncolysis

The inability of clostridial oncolysis to completely eliminate a tumor led to the suggestion that the co-application of other noxae could enhance the observed therapeutic effect [11]. This notion was investigated by Thiele et al. through the co-administration with spores of a number of drugs to mice bearing sarcoma 180 [9]. Most of the drugs tested (e. g., 6-mercaptopurine, thioguanine, 4-aminopyrazole-3(3,4-*d*)pyrimidine, 5-fluorouracil, azaserine, actinomycin D, various quinones and peroxides, and nitrogen mustard bifunctional alkylating agents) did not enhance the effects of spores alone. However, alkylating agents of the ethyleneimino type in all instances produced positive results. Thus, the use of tetramin, E-39, trenimon, and mitomycin C in combination with spore treatment all caused a pronounced suppression of tumor growth. Moreover, while these reductions were accompanied by increased toxicity, these side effects could be significantly reduced through the adjustment of the dose and timing of drug/spore administration [9].

In other studies, using the Jensen-Sarcoma of rat and the Fortner-Melanoma Amel-3 of Syrian hamsters, the use of M-55 spores in combination with cyclophosphamide, significant improvements on survival time were also demonstrated [12, 13]. A similar beneficial effect was not observed in an equivalent study undertaken by Schlechte et al. [14]. However, neither the animal model employed (mouse UVT 15264), nor the clostridia strain used (*Clostridium butyricum* H8), were the same.

Other workers have sought to increase the degree of hypoxia in tumors through various routes. One method employed was to raise the temperature of the tumor through the application of microwaves. Using this treatment (high-frequency hyperthermia, H-FH) it proved possible to increase both the frequency of occurrence and the extent of lysis with a number of different mouse tumors [15], although the effects on survival rates were not significant. However, the additional irradiation of the tumor, combined with H-FH, led to a 20 % survival rate of the animals treated. Through repeated cycles of these three treatments (spores/microwaves/irradiation) it subsequently proved possible to cure tumors in 60 % of the animals [16]. In a different approach, Möse investigated the effect of reducing the oxygen content of the respiratory air supplied to tumor bearing rats to 11-12 % over a one-week period. In both of the tumor models tested (Ehrlich and Hardy Passey Melanoma) the extent of lysis was significantly improved (by over 60 %), and, more remarkably, effected complete tumor eradication in 30 % of the tumors treated [17].

8.2.3
Clostridial oncolysis in humans

The early animal studies were rapidly followed by experiments involving human subjects. Initially Möse and Möse established the benign nature of spores of *C. butyricum* M-55 by administration to themselves, and then treated some thirty-six patients with spores (10^6 to 10^9) injected either intratumorally or intravenously (cited in [18]). These injections were reported to be well tolerated. Administration was followed some six hours later by a moderate to high grade fever, and thereafter was characterized by a low grade fever for one to three days. Lysis of the tumor, when it occurred, took place some five to eight days later. The results of these studies led Carey et al. to evaluate the potential of this form of treatment in the US [18]. Treatment was undertaken with five patients with advanced neoplastic disease in which all established forms of then available conventional therapies had been exhausted. Oncolysis occurred with three of the subjects in their largest tumors, but not in their smaller tumors, and in the one instance a transient clinical benefit was attributed to the treatment.

Of particular focus has been the possibility of treating inoperable malignant brain tumors. In the study of Kretschmer, the effects of the administration of various doses of spores to 5 patients with glioblastomas were evaluated [19]. In one of these patients, an eight-year-old boy with a posterior fossa tumor, an apparent cure was effected, with the patient showing sign of the tumor some nine months post treatment. In a second case, a considerable improvement in the patients' condition was evident following treatment, although the individual concerned succumbed as a result of a ventricle rupture. This later consequence led Kretschmer to conclude that the danger of such a rupture would necessarily confine this form of treatment to a limited number of cases. During this same period (1964-1973), a total of 49 patients with brain tumors were similarly treated with *C. butyricum* M-55 spores by Heppner and Möse [20]. In all cases where oncolysis occurred the body temperature of the patient rose some three to four days following administration of the spores and was accompanied by leucocytosis. By the week the patients became drowsy and an increase of focal signs and mounting intracranial pressure made the surgical removal of the tumor imperative. The rate of tumor recurrence, however, remained unaffected, and 16 of the patients died during the treatment. In a later study, Heppner et al. combined the treatment of individuals with brain tumors with CT scanning as a means of more effectively assessing tumor lysis [21]. This allowed the timing of surgical removal of the tumor to be delayed, rather than the routine removal of the glioma one week after their administration. However, while CT scanning proved a highly effective means of monitoring oncolysis (in 14 of 17 cases), no data on any beneficial effect on treatment was reported.

8.3
Clostridia in cancer diagnosis

An intriguing second use has been proposed for clostridial spores within the arena of diagnosis as opposed to therapy. This is based on the premise that, following the administration of clostridial spores, growth of clostridia in tumors results in the production of antibodies directed against vegetative cells. These are distinct from those that might be raised against spores. Their detection in sera is therefore indicative of actively growing clostridial cells, and as spores may only germinate in hypoxic regions, provides strong evidence for the presence of vegetative rods and therefore of the existence of a tumor. The utility of this system has been extensively tested in both animals and humans. The organism used was a "non-oncolysing"

Clostridium butyricum strain CNRZ 528, which has now been re-classified as *Clostridium beijerinckii.*

The initial studies undertaken were with tumor-bearing hamsters [22], mice [23], and horses before moving on to dogs [24], and eventually cattle and humans [25-27]. In early work with dogs [28], a total of 189 dogs were examined (28 healthy, 32 benign tumors, 118 with malignant tumors, and 11 dogs with non-tumorous diseases). A dose of 2×10^8 spores of CNRZ 528 per kg body weight were administered, and serum samples taken before and 14 and 21 days after administration. Antibodies to vegetative cells were detected using both a complement fixation and hemagglutination test. These were shown to be 67 % and 49 % effective, respectively, at predicting the presence of a tumor. A combination of the two tests showed a marginal improvement in efficiency (70 % effective). In the case of the cattle tests, the system was tested with 6 cattle infected with bovine leucosis virus (BLV) [25]. Of these 6 animals, 3 were shown to be positive for vegetative antibodies. These 3 positive examples corresponded to those animals which were both hematologically and serologically leucosis-positive and which had enlarged lymph nodes. The 3 negative animals, while they were also hematologically and serologically leucosis-positive, were all younger and lacked any sign of tumorogenic lymph nodes.

The test was later applied to human patients [26]. The test was applied to a total of 77 patients in a phase I and II clinical study. The effectiveness of the test was compared to diagnosis of tumors based on the detection of tumor antigens, principally carcinoma embryonic antigen (CEA). These studies showed that the test was over 80 % effective, compared to approximately 60 % for the detection of specific tumor markers, such as CEA.

8.4
Overcoming the limitations of oncolysis

The combined outcome of both the animal and human experiments was that clostridial spore treatment was remarkably well tolerated and that growth of the organism frequently led to the destruction of large parts of the tumor. Invariably, however, an outer viable rim remained from which tumor regrowth frequently occurred. From these observations it may be concluded, that spore treatment alone is not sufficient to effect complete tumor regression. The advent of recombinant technology, however, provides a possible solution to this deficiency. This is because it opens up the possibility of using genetic techniques to endow the clostridial strains used with the ability to produce anti-tumor agents. The production of such therapeutic molecules, in combination with oncolysis, has the potential to effect a complete cure. This strategy was first realized by Schlechte and Elbe [29], who attempted to produce recombinant strains of *C. butyricum* M-55 able to produce Colicin E3, an *E. coli*-derived bacteriocin which was reported to have canceriostatic properties. However, the evidence presented to support the creation of the desired recombinant clostridial strains was not convincing, and the generation of true recombinant strains has only proven possible a decade later with the considerable advances made in the ability to genetically manipulate clostridial species.

While the strain (M-55) favored in the early studies of tumor spore treatment was originally classified as *Clostridium butyricum*, and later as *C. oncolyticum*, taxonomic studies have now clearly established that it is a strain of *Clostridium sporogenes* (ATCC 13732). This is a proteolytic species, and goes someway to explaining why this strain is particularly effective at bringing about the liquefaction of colonized tumors. Although this species is not a pathogen *per se*, certain strains may be found in association with disease, *e.g.*, in cases of gangrene. From a safety perspective, therefore, it may be more acceptable to utilize a host which has no known association with disease. It may also be advantageous to use a host which was less aggressive with regard to destruction of the tumor mass, as extensive lysis all too often results in toxemia. In more recent years, therefore, the ability of other strains which may be unequivocally considered as non-pathogenic have been examined as potential delivery systems for therapeutic agents.

One group of strains that represents particularly attractive candidates in this regard are those saccharolytic strains of clostridia able to produce the industrial solvents acetone and butanol, such as *Clostridium acetobutylicum* and *Clostridium beijerinckii* [30]. These organisms have no known association with human disease and furthermore have an extended period of safe use on an industrial scale in the so-called acetone-butanol-ethanol (ABE) fermentation process [31]. In addition, numerous vector systems have been developed for the introduction of heterologous DNA into a number of strains [32, 33].

8.5 Colonization of tumors by solvent-producing *Clostridium* spp.

Prior to the adoption of *C. oncolyticum* M-55 as the principle strain used in tumor studies, the ability of a number of other clostridial species to colonize tumors was investigated, including *C. acetobutylicum* and *C. beijerinckii* [7, 9]. Their effectiveness has, however, been recently re-assessed [34, 35]. In one study the ability of four different saccharolytic clostridial strains, *C. beijerinckii* ATCC 17778, *C. limosum* DSM 1400, *C. acetobutylicum* strains ATCC 824, and NI-4082 (now re-classified as *C. saccharoperbutylacetonicum*), to colonize tumors was compared to that of *C. oncolyticum* [34]. In general, it was found that sufficient infiltration into tumors (syngenic rhabdomyosarcomas) of WAG/Rij rats required the systemic administration of at least 10^7 spores. The efficiency with which *C. acetobutylicum* colonized tumors was found to be significantly greater than the other three saccharolytic strains, and achieved comparable populations to *C. oncolyticum*. Analysis of tumor tissues some four to five days after injection showed that colony forming units (cfu) of up to 10^9 per gram of tissue were present compared to only 10^4 to 10^6 in normal tissues (Figure 1). The cfu present in tumor tissue was reduced some two orders of magnitude by heating (73 °C for 20 min). A similar treatment had no effect on normal tissue samples, indicating that whereas significant number of vegetative cells are present in the tumor, the cfus in normal tissues are due to the presence of spores. Consistent with this conclusion was the finding that no vegetative cells could be microscopically detected in healthy tissues following

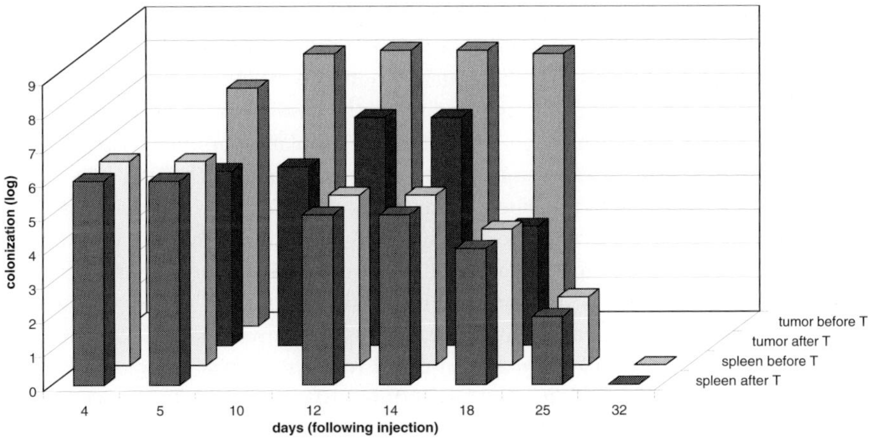

Figure 1. Quantification of spores and vegetative cells in tumors and spleen as a function of time following systemic injection of *C. acetobutylicum*. Heat treatment at 73 °C (T) kills vegetative cells but not spores. The difference in cfu before and after heating gives thus an indication of the number of vegetative cells (adapted from [34]).

Gram-staining. Histochemical staining confirmed that vegetative cells were confined to the hypoxic and acellular necrotic areas of the tumor, but could not be detected in the oxygenated regions of the tumor.

Within healthy tissues, then cfu was seen to progressively decline with time (up to 25 days) while the number in tumors remained at the high levels seen four days after administration of spores. This is consistent with the clearance of the spores from the rest of the body and the maintenance of high numbers of actively growing vegetative cells in the tumor.

In a separate study the ability of *C. beijerinckii* NCIMB 8052 (formerly *C. acetobutylicum*) to colonize tumors was also examined [35]. In this instance EMT6 tumor-bearing mice were employed. Animals were injected intravenously with 10^8 spores (0.1 ml of a saline suspension containing 10^9 spores per ml), killed 24 h later, and tissues examined by Gram-staining for the presence of vegetative rods. High numbers of bacterial rods were evident in the necrotic regions of tumors and the surrounding region. Isolated rods were also occasionally found in the more oxygenated regions of the tumor, but no evidence for the presence of bacterial cells was obtained in other tissues, such as heart, kidney, liver, lung, and spleen.

Very recently, treatment with angiogenesis inhibitors or vascular targeting treatment was suggested to increase the degree of hypoxia in tumors in order to improve the tumor colonization. This would also allow small tumors to become colonized by *Clostridium* after intravenous injection. Angiogenesis is a complex process leading to the growth of new blood vessels, an important natural process occurring in the body both in health and in disease. The healthy body controls angiogenesis through a series of "on" and "off" switches. When angiogenic growth factors are produced in excess of angiogenesis inhibitors, the balance is tipped in favor of blood vessel growth. When inhibitors are present in excess of stimulators, angiogenesis is stopped. The normal, healthy body maintains a perfect balance of angiogenesis modulators. Excessive angiogenesis occurs in diseases such as cancer, diabetic blindness, age-related macular degeneration and others. In these conditions, new blood vessels feed diseased tissues, and in the case of cancer, the new vessels allow tumor cells to grow and to escape into the circulation and lodge in other organs (tumor metastases). Antiangiogenic therapies, aimed at halting new blood vessel growth, are being developed. Combretastatin (OXiGENE, Lund, Sweden) is the first in a new class of tumor vascular targeting drugs that are intended to selectively attack and destroy tumor-specific blood vessels formed by angiogenesis [36]. This results in a massive, rapid and irreversible shutdown of these blood vessels while leaving normal vasculature unharmed. It was investigated if treatment of tumor bearing rats with Combretastatin could result in increased colonization of tumors, especially of small tumors. Therefore, 10^8 spores of *C. acetobutylicum* NI4082 were administered to WAG/Rij rats with rhabdomyosarcomas of different sizes that ranged between 0.2 and 10 cm^3, followed 4 hours later by an injection of a single dose of Combretastatin A-4 phosphate (25 mg/kg). It was shown that treatment with Combretastatin resulted in a significant improvement of colonization both of small (1-3 cm^3) and very small (< 1 cm^3) tumors. Whereas in Combretastatin-untreated animals, tumors smaller than 4 cm^3 were only poorly colonized ($10^{0.5}$-10^3 cfu per g tissue), a dramatic increase in the number of cfu was noticed with a colonization of 10^7-10^8 cfu per g tissue for Combretastatin-treated animals. This was obviously a consequence of the massive shutdown of tumor blood vessels in the treated rats, which successfully induced hypoxia within tumors of varying size and allowed anaerobic bacteria to proliferate within the tumors [37]. The Combretastatin-induced hypoxia will be used to further test this model and to potentially develop the anaerobic vector for use in targeted tumor treatment.

8.6 Delivery of therapeutic drugs

In considering the types of therapeutic molecules that could be delivered to tumors, two classes of agents are evident. There are those therapeutic agents that bring about more or less direct effects on the tumor cell, such as toxins or cytokines, and there are those proteins which bring about the conversion of a secondary agent into a tumorogenic killing agent, such as prodrug-activating enzymes.

8.6.1 Clostridial-directed enzyme prodrug therapy (CDEPT)

The concept of directed-enzyme prodrug therapy (DEPT) seeks to bring about high levels of therapeutic drug specifically at the tumor site. In this approach, the anticancer drug is rendered harmless, through its chemical conversion into a "prodrug", prior to its introduction into the bloodstream. Its transformation into the active drug is achieved through the action of a protein "enzyme". This enzyme component is specifically targeted ("directed") to tumor cells, and is administered prior to injection of the prodrug. The generation of active drug, therefore, occurs only within the vicinity of the tumor. This allows high therapeutic doses of the drug to be achieved within the tumor, while maintaining harmless levels of drugs elsewhere in the body. Furthermore, because a single enzyme molecule is able to catalyze the generation of large quantities of a therapeutic drug, there is no necessity to target every tumor cell.

In the original DEPT strategy [38], targeting of the enzyme was achieved through the fusion of the enzyme component to a monoclonal antibody directed against a tumor specific/enriched antigen, ie., antibody-directed enzyme prodrug therapy (ADEPT). For its application, antibody-enzyme conjugate is administered and sufficient time then allowed to pass for the conjugate to become localized to tumors and for clearance of "free" conjugate from the circulatory system and other tissues. A relatively innocuous nontoxic prodrug is then administered which is converted to a highly cytotoxic drug through the action of the targeted enzyme localized at the tumor site [39, 40]. More recently, attention has switched to strategies that are reliant on the delivery of the gene encoding the therapeutic enzyme using viral vectors, in so-called gene-DEPT (GDEPT) approach [41].

Both the ADEPT and GDEPT approach have a number of limitations. GDEPT suffers from a lack of tumor specificity, poor levels of transgene expression, and the inefficient distribution of the vector throughout the tumor mass. ADEPT is disadvantaged by antigen heterogeneity between tumors and the absence of suitable antigen targets in certain tumor types. The deficiencies of ADEPT and GDEPT with regard to specificity would not apply to a delivery system based on clostridia (clostridial-directed enzyme prodrug therapy, CDEPT). This is because the germination of spores is merely reliant on the presence of hypoxic/necrotic regions within the tumor (Figure 2).

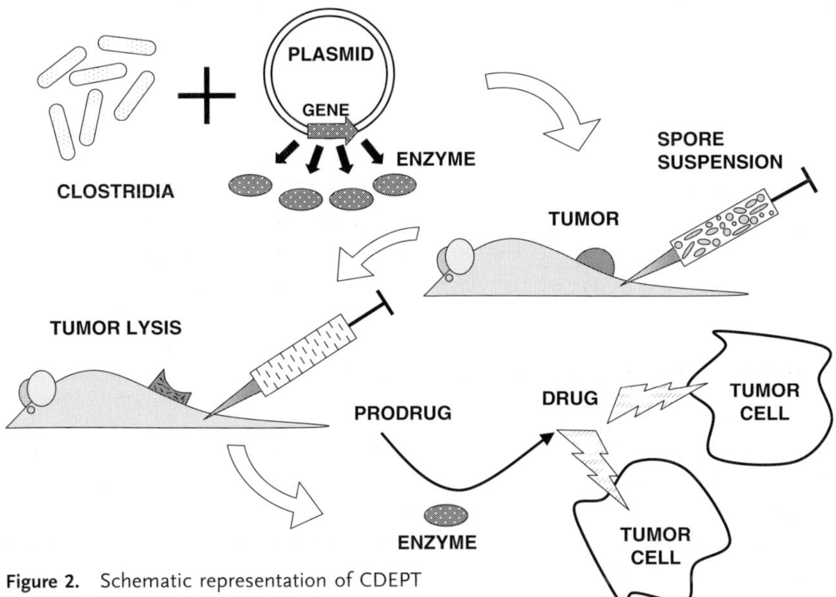

Figure 2. Schematic representation of CDEPT (clostridial-directed enzyme prodrug therapy).

8.6.2
Enzymes useful in DEPT strategies

A number of enzymes and prodrug/drug combinations have been proposed for use in DEPT strategies. In the case of ADEPT, one particularly promising reagent has been made by fusing a bacterial enzyme carboxypeptidase G_2 (CPG_2), isolated from a *Pseudomonas* sp. (now re-classified as *Variovorax paradoxus*) to a monoclonal antibody (A5B7) directed against carcinoma embryonic antigen [42, 43]. CPG_2 cleaves the glutamate moiety from several inactivated nitrogen mustards, bringing about a 100-fold increase in drug toxicity. First generation systems, involving the chemical coupling of enzyme to monoclonal antibodies, have been shown to bring significant reduction in tumor growth [44] and even complete eradication of tumor xenografts in certain animal models [45]. Second generation systems, in which the CPG_2 gene has been genetically fused to a sFv raised against CEA, have also now been constructed [46], and the recombinant fusion protein obtained shown to be even more effective in tumor localization studies [47].

In GDEPT approaches, genes encoding enzymes such as nitroreductase and cytosine deaminase have been favored. Enzymes of the former type are able to convert the monofunctional alkylating agent CB1954 (5-(aziridin-1-yl)-2,4-dinitrobenzamide) to a cytotoxic DNA interstrand crosslinking agent by reduction of its 4-nitro group to the corresponding hydroxylamino species [48, 49]. On a dose by dose basis, the 4-hydroxylamine species is 10^4- to 10^5-fold more cytotoxic than the CB1954 progenitor. Though this activity was first attributed to a rat DT diaphorase, one of the most effective nitroreductase enzymes was subsequently isolated, from *Escherichia coli* B. This enzyme catalyzed the reduction of CB1954 at a rate 60 times greater than that seen with the rat enzyme [50].

The enzyme cytosine deaminase converts 5-fluorocytosine (5-FC) into 5-fluorouracil (5-FU). The latter is toxic because it is further metabolized into 5-fluorouridine-5′-triphosphate and 5-fluoro-2′-deoxyuridine 5′-monophophate, which inhibit DNA and RNA synthesis [51]. This particular enzyme prodrug system has a number of advantages. As with nitroreductase and CB1954, the enzyme is not found in human tissue, and the differential toxicity between prodrug (5-FC) and drug (5-FU) is large (10^4). An additional attraction of this system is, however, that both 5-FU and 5-FC are currently approved for clinical applications, the former in the treatment of breast and gastrointestinal cancers.

8.6.3
Clostridial strains can produce prodrug-converting enzymes

To establish that clostridia may be modified such that they produce anti-cancer enzymes, use has been made of the clostridial expression vectors pMTL500F and pMTL540FT (Figure 3), which are respectively based on the broad-host range streptococcal plasmid pAMβ1 [52] or the higher copy *Clostridium butyricum* plasmid pCB102 [53]. Both plasmids carry multiple cloning sites within a *lacZ'* gene into which heterologous genes may be cloned. To facilitate the expression of cloned

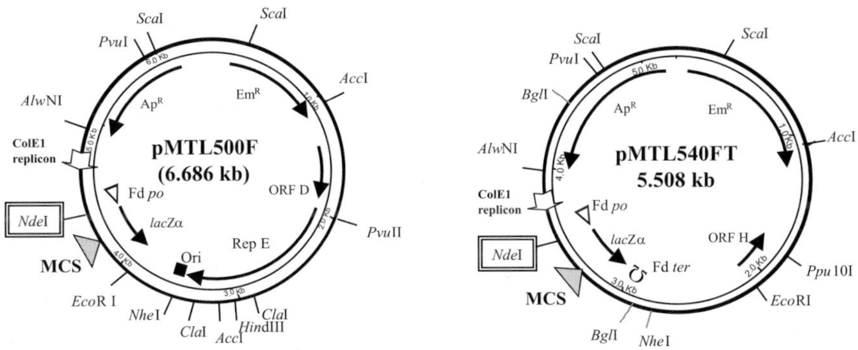

Figure 3. Clostridial expression vectors used to manipulate *C. beijerinckii* NCIMB 8052. Plasmid pMTL500F is based on the replicon of the enterococcal plasmid pAMβ1, while pMTL540FT carries the replicon of the *C. butyricum* plasmid pCB102 [33]. For optimum results the heterologous gene to be expressed is derivatized such that an *Nde*I (CAT**ATG**) is created over its translational start codon (**ATG**) and then cloned between the *Nde*I site of the vector (boxed) and an appropriate site with the multiple cloning site (MCS) with the *lacZα* gene. This places the gene at an optimal distance from the ribosome binding site of the upstream promoter region of the clostridial ferredoxin gene.

genes, the promoter and RBS of the *lacZ'* have been replaced in pMTL500F by the equivalent signals of the ferredoxin (Fd) gene of *Clostridium pasteurianum* [54]. The provision of an *Nde*I recognition sequence (CATATG), whereby the ATG of this palindrome represents the AUG start codon of *lacZ'*, allows the heterologous gene containing equivalent sites to be precisely positioned in pMTL500F relative to the RBS of Fd. This ensures that both transcription and translation is optimized for the clostridial host. This is particularly important when the heterologous genes are of an *E. coli* origin, as the transcription and translation signals of a Gram-negative bacterium are extremely inefficiently utilized by Gram-positive bacteria.

Capitalizing on the specialized features of these vectors, recombinant plasmids capable of directing the expression of both the *codA* and *nfnB* genes (encoding cytosine deaminase and nitroreductase, respectively), in *C. beijerinckii* were generated [35, 55]. In essence, an *Nde*I site was created "over" the translational start codon of the two genes, allowing their subsequent insertion into pMTL500F and pMTL540FT immediately adjacent to the clostridial Fd promoter and RBS. In all instances, lysates of cells carrying the recombinant plasmids encoding the heterologous genes were shown to contain the expected enzymatic activity. Furthermore, this activity was shown to be able to turnover the appropriate prodrug into the corresponding cytotoxic species active against tumor cells (Figures 4 and 5). In the case of cells carrying pNTR500F (pMTL500F + *nfnB*), the abundant lysate protein identified as nitroreductase using anti-NfnB antibody was estimated by densitometric scanning of electrophoretograms to equate to 8 % of the cells' soluble protein [56].

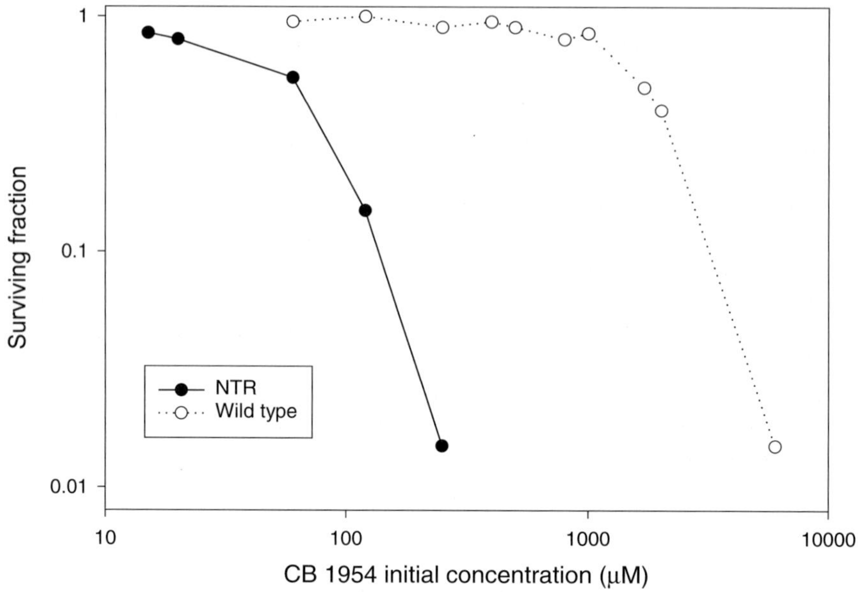

Figure 4. Activation of the prodrug CB1954 by clostridial lysates derived from recombinant strains carrying the *E. coli nfnB* (nitroreductase) gene. EMT6 tumor cells were exposed to varying concentrations of the prodrug CB1954 in the presence of clostridial lysates derived from cells carrying pMTL500F plasmid only (wild-type) or carrying a recombinant plasmid encoding the *E. coli nfnB* gene (NTR). EMT6 cell survival was measured after a 2 hour exposure.

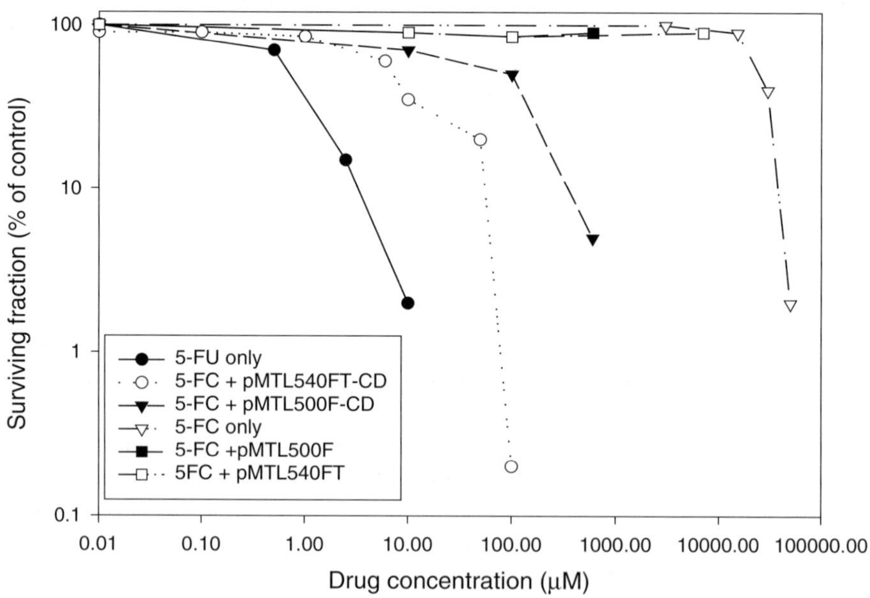

Figure 5. Conversion of the prodrug 5-FC by clostridial lysates expressing the *E. coli codA* (cytosine deaminase) gene. Survival of EMT6 cells was determined following exposure to 5-FC in the presence of the indicated clostridial lysates.

8.6.4
Clostridial strains can be used to deliver enzymes to tumors

Having shown that clostridial cells may be engineered to produce an anti-cancer enzyme, it was interest to determine whether such modified bacteria could be employed to deliver the enzyme to tumor cells *in vivo*. Spore suspensions prepared from the recombinant strain generated were therefore administered to tumor bearing mice, and tumor homogenates prepared some four days later and subjected to Western blot analysis using anti-NfnB antibodies [35]. These antibodies clearly demonstrated the presence of a protein of the expected size in EMT6 tumor homogenates of animals which had received recombinant NCIMB 8052 spores carrying the plasmid pNTR500F. No such protein was evident in tumor homogenates injected with plasmid-free NCIMB 8052 spores.

Also *C. acetobutylicum* strains NI4082 and DSM792 engineered to produce cytosine deaminase are able to express and secrete this enzyme at the tumor site [57]. To obtain this, the *E. coli codA* gene, preceded by the clostripain regulatory and signal sequences [58], was ligated into the *E. coli/C. acetobutylicum* shuttle vector pKNT19 [59] and introduced into *C. acetobutylicum* by electroporation. *In vitro* tests showed that cytosine deaminase could be secreted by the recombinant *Clostridium* into the culture medium as a functional enzyme converting more than 700 pmol 5-FC to 5-FU per minute per ml. Moreover, functional cytosine deaminase enzyme was detected in the tumor of rhabdomyosarcoma bearing WAG/Rij rats that were injected with the recombinant *C. acetobutylicum,* but not in control animals. Besides its killing effects by interference with RNA and DNA synthesis, 5-FU has also a radiosensitizer effect, thus enhancing tumor effects in radiotherapy [60]. As calculated from a theoretical model using parameters available in the literature [61], the enzymatic activity detected at the tumor site would be high enough to obtain satisfactory 5-FC to 5-FU conversion efficiencies to achieve clinically significant radiosensitization of radiotherapy. The *in vivo* effect of this cytosine deaminase producing *Clostridium* on tumor control is still to be investigated.

8.6.5
In vivo effects of delivered enzymes

Despite the successful delivery of engineered *C. beijerinckii* NCIMB 8052 cells producing NfnB, no turnover of prodrug could be detected. The reasons for this were not clear, but some loss of the recombinant plasmid by vegetative cells growing in the tumor was evident. It was also apparent that this particular strain attained significantly lower cell populations in the tumor compared to other clostridial species. Thus, in a comparative experiment it was shown that the cfu obtained from tumors colonized by *C. beijerinckii* NCIMB 8052 was two orders of magnitude lower than comparable tumors colonized by *C. oncolyticum* (Lemmon and Brown, unpublished data). Attempts were therefore made to capitalize on the higher cell number achieved by the latter strain through the introduction of the recombinant plasmids carrying the *nfnB* gene. These plasmids could not, however, be electroporated into

C. oncolyticum. Indeed, this strain could not be transformed with any of the plasmids commonly used to introduce genes into clostridial species.

In contrast, it has recently proven possible to transform an alternative *C. sporogenes* strain, NCIMB 10696, with pMTL500F and pMTL540FT plasmids carrying the *codA* gene [62]. Lysates prepared from such transformants increased the sensitivity of murine tumor cells to FC by more than 1000-fold, indicating that FU was being generated by cytosine deaminase present in the clostridial lysate. Preliminary therapy experiments have been undertaken with this recombinant strain in a rat tumor model. These have shown that recombinant enzyme may readily be detected in colonized tumor by Western blotting of macerated tissue against cytosine deaminase antibody. Furthermore, a significant reduction in tumor growth rate was evident following the administration of FC to animals carrying colonized tumors. Equivalent reductions were not seen if animals were treated with either spores or FC alone.

8.6.6 Engineered clostridia can produce human cytokines

An alternative strategy to the use of prodrug converting enzymes would be to deliver a therapeutic molecule that is anti-tumorogenic in its own right. One particularly attractive candidate is TNFα (tumor necrosis factor α), which is cytotoxic to tumor cells through the induction of apoptosis [63]. TNFα additionally has a selective effect on the neovasculature of tumors and stimulates T-cell mediated immunity. Its inhibitory effect on the proliferation of tumors has been demonstrated *in vivo* [64]. However, its clinical use is complicated by systemic toxicity when administered at an appropriate therapeutic dose. These side effects would be minimized if TNFα production could be confined to the tumor through its localized production by an engineered clostridial cell population.

In order to test this possibility it has first been necessary to establish whether clostridia could be engineered to produce biologically active TNFα. Accordingly, the gene encoding mouse TNFα was cloned and derivatized such that the mature coding sequence was fused to the signal peptide sequence of the *eglA* gene (encoding endo-β1,4-glucanase) of *C. acetobutylicum* P262 [65], together with the *eglA* promoter region. The derivatized TNFα gene was then cloned into pIM13-derived vectors pIMP1 and pKNT19 [59, 66] and the resulting constructs introduced into *C. acetobutylicum* DSM792 [67]. The transformants obtained where found to retain the introduced plasmid vectors in the absence of antibiotic selection, and their lysates and supernatants shown by immunoblots to contain TNFα. Lysates contained both the mature (17 kDa) and preprotein (21 kDa), whereas the supernatants contained only the mature protein.

The recombinant TNFα produced was shown to be biologically active through its cytotoxicity towards WEH164 clone 13 cell lines. Time course experiments demonstrated that the activity of the TNFα present in the supernatant declined over time [67]. A similar decrease in activity was not observed in the TNFα present in the lysate. Buffering of the culture media with MOPS prevented the decline indicating that it was the reduction in pH as a result of fermentative metabolism, as opposed to protease production, which was causing loss of TNFα activity.

8.7 Phospholipase C enhancement of liposome entrapped drug delivery

While not directly reliant on clostridia *per se*, a further potential application of a clostridial product in cancer therapy is worthy of note, and that is "Lipoburst" technology [68] based on the phospholipase C (PLC) of *Clostridium perfringens* [69]. Liposomes have found extensive use as carriers of specific drugs to tumor tissues as a means of reducing systemic levels [70, 71] or as a means of increasing drug half life and improving stability [72, 73]. The specific delivery of liposomes to tumors has, however, proven difficult to achieve. Lipoburst technology seeks to circumvent this deficiency through the fusion of phospholipase C (PLC) to an antibody raised against a tumor-specific antigen, such as carcinoembryonic antigen (CEA) and the localization of the fusion protein to the tumor surface. A circulatory liposomal drug is then administered whereupon the tumor-associated PLC cause local degradation of the liposome to bring about release of the encapsulated drug.

The potential of the system was demonstrated through the use of a chemical conjugate between PLC and either an anti-CEA or anti-EGFR (epidermal growth factor receptor) monoclonal antibody and liposomes containing daunorubicin. The effect of these reagents was initially tested on HeLa cells, where it was shown that the prior localization of conjugate to cells increased the subsequent anti-proliferation effects of liposome entrapped daunorubicin by a factor of 3 and 10 for anti-CEA-PLC and anti-EGFR-PLC, respectively. This enhancement was not seen if unconjugated PLC was employed. To facilitate reagent production, a recombinant fusion protein was also constructed through the fusion of a gene encoding a sFv directed against EGFR to the PLC gene. This recombinant fusion protein, however, proved less effective than the chemical conjugate, causing only a 3-fold inhibition of cell proliferation.

The therapeutic potential of anti-EGFR-PLC conjugate was subsequently tested in athymic nu/nu mice. Injections of immunoconjugate were given by the i. p. route on days 1-3, followed by i. v. administration of daunorubicin liposomes on day 5. Examination of animals after 50 days showed that, while no eradication of tumors had occurred, growth had been arrested in 5 out of 6 animals.

8.8 Concluding remarks

Cancer is a disease with a high incidence in the western world. High lethality rates and the poor prognosis for many cancer patients prescribed conventional drug or radiation treatment has led to recent interest in the investigation of clinical protocols based on gene therapy. However, strategies devised to date are ineffective, most notably due to deficiencies in the delivery systems employed. These are devoid of tumor specificity and lack the high efficiency of transduction required to avoid tumor relapse. Alternative strategies are clearly required. As a consequence, in recent years there has been a renewed interest in the use of bacteria such as *Clostridium* and *Salmonella* as novel delivery systems.

Although not immediately apparent, an association between bacteria and tumors date back more than 100 years (see [74]) when William Coley found that certain patients who contracted bacterial infections recovered remarkably well from certain cancers. Currently, *Clostridium spp., Bifidobacterium longum* [76] as well as an attenuated *Salmonella typhimurium* auxotrophs [75, 76], are being investigated as systems to deliver antitumor compounds specifically to the tumor site. The latter strain grows under aerobic and anaerobic conditions. Its selectivity for tumors is reportedly a consequence of its auxotrophic nature. The specificity of clostridia for tumors resides in its obligate requirement for anaerobic conditions. This gives *Clostridium* a clear advantage over *Salmonella*. Thus, intravenously injected clostridial spores localize to, and germinate in, the hypoxic regions of solid tumors. This process is exquisitely selective, as the spores are demonstrably incapable of germinating in normal healthy tissues. While growth alone in the tumor is not sufficient for complete cure, the possibility now exists to engineer *Clostridium spp.* to produce a variety of heterologous proteins with anti-cancer properties. Clostridia can thus be used as a highly selective *in situ* cell factory able to produce and secrete antitumor therapeutics specifically at the tumor site. This possibility has given *Clostridium*-mediated tumor therapy new impetus.

It seems clear that a delivery system based on *Clostridium spp.* has several advantages compared to the classical approach using viruses or liposomes. It is highly tumor-specific, since in mammals hypoxia is a feature of solid tumors, not of normal tissues. The heterologous gene is not transduced into the genome of the tumor cell, since the anti-cancer gene will be expressed and secreted from the bacteria. Targeted gene expression can be stopped at any time by administration of suitable antibiotics. This is obviously not the case when a gene is inserted within the genome of mammalian cells, as with traditional gene therapeutic approaches. As a consequence, *Clostridium* in cancer therapy holds considerable promise in the fight against cancer. The only apparent obstacle for the further development of clinical applications would appear to be concerns over the concept of deliberately "infecting" patients with a live clostridial species. It is to be hoped that rational arguments will allay these fears, and the full potential of the system is realized through appropriate clinical evaluation.

References

[1] Connell, H. C., The study and treatment of cancer by proteolytic enzymes. A preliminary report, *Can. Med. Ass. J.* **1935**, 33, 364-370.

[2] Parker, R. C., Plumber, H. C., Siebenmann, Chapman, M. G., Effect of histolyticus infection and toxin on transplantable mouse tumors, *Proc. Soc. Exp. Biol. Med.* **1947**, 66, 461-465.

[3] Malmgren, R. A., Flanigan, C. C., Localization of the vegetative form of *Clostridium tetani* in mouse tumors following intravenous spore administration, *Cancer Res.* **1955**, 15, 473-478.

[4] Möse, J. R. Zur Beeinflussbarkeit verschiedener Tiertumoren durch einen apathogenen Clostridienstamm, *Z. Krebsforsch.* **1960**, 63, 447-455.

[5] Möse, J. R., Möse, G., Onkolyseversuche mit apathogen, anaerogen Sporenbildern am Ehrlich-Tumor der Maus, *Z. Krebsforsch.* **1959**, 63, 63-74.

[6] Schlechte, H., Elbe, B., Recombinant plasmid DNA variation of *Clostridium oncolyticum* – model experiments of cancerostatic gene transfer, *Zbl. Bakt. Hyg. A.*, **1988**, 268, 347-356.

[7] Möse, J. R., Möse G., Oncolysis by clostridia. I. Activity of *Clostridium butyricum* (M-55) and other nonpathogenic clostridia against the Ehrlich carcinoma, *Cancer Res.* **1964**, 24, 212-216.

[8] Gericke, D., Engelbart, K., Oncolysis by clostridia. II. Experiments of a tumor spectrum with a variety of clostridia in combination with heavy metal, *Cancer Res.* **1964**, 24, 217-221.

[9] Thiele, E. H., Arison, R. N., Boxer, G. E., Oncolysis by clostridia. III. Effects of clostridia and chemotherapeutic agents on rodent tumors, *Cancer Res.* **1964**, 24, 222-233.

[10] Engelbart, K., Gericke, D., Oncolysis by clostridia: V. Transplanted tumors of the hamster, *Cancer Res.* **1964**, 24,239-243.

[11] Warburg, O., Über die fakultative Anaerobiose der Krebszellen und ihre Anwendung auf die Chemotherapie, in: *New Methods of Cell Physiology*, Georg Thieme Verlag, Stuttgart, **1962**, pp. 627-630.

[12] Kretschmer, H., Ludewig, R., Hambsch, K., Therapieversuche mit *Clostridium butyricum-Sporen* und zyklophosphamid am Jensen-Sarkom der Ratte, *Arch. Geschwulstfrsch.* **1975**, 45, 16-18.

[13] Kretschmer, H., Wohlrab, W., Lübbe, D., Peker, J., Ludewig, R., Hambsch, K., Möglichkeiten mit clostridialen Sporen in der Behandlung des malignen Melanoms, *Dermatol. Wochenschr.* **1975**, 161, 584-587.

[14] Schlechte, H., Schwabe, K., Mehnert, W. H., Schulze, B., Bräuniger, H. Chemotherapy for tumors using clostridial oncolysis, antibiotics and cyclophosphamide: Model trial on the UVT 15264 tumor, *Arch. Geschwulstforsch.* **1982**, 52, 41–48.

[15] Dietzel, F., Gericke, D., Schumacher, L., Linhart, G., Combination of radiotherapy, microwave-hyperthermia and clostridial oncolysis on experimental mouse tumors, in: *Cancer Therapy by Hyperthermia and Radiation* (C. Streffer, Ed.), Urban & Schwarzenberg, Baltimore, Munich, Essen, **1978**, pp. 689-694.

[16] Gericke, D., Dietzel, F., König, W., Rüster, L., Schumacher, L., Further progress with oncolysis due to apathogenic clostridia, *Zbl. Bakt. 1. Abt. Orig. A,* **1979**, 243, 102-112.

[17] Möse, J. R., Versuche zur Verbesserung der Onkolyse mit dem Clostridienstamm, *Zbl. Bakt. Hyg., I. Abt. Orig. A* **1979**, 244, 541-545.

[18] Carey, R. W., Holland, J. F., Whang, H. Y., Neter, E., Bryant, B., Clostridial oncolysis in man, *Eur. J. Cancer* **1967**, 3, 37-46.

[19] Kretschmer H., Treatment of malignant brain tumors by *Clostridium butyricum* M55. *Proceedings of the VIIth International Congress of Chemotherapy*, **1972**, 721-723.

[20] Heppner, F., Möse, J. R., The liquefaction (Oncolysis) of malignant gliomas by a nonpathogenic *Clostridium, Acta. Neurochir.* **1978**, 42, 123-125.

[21] Heppner, F. Möse, J. R., Ascher, P. W., Walter, G., Oncolysis of malignant gliomas of the brain, *Proceedings of the 13th International Congress of Chemotherapy* **1983**, 38-44.

[22] Schau, H.-P., Fabricius, E. M., Schneeweiss, U., Benedix, A., Untersuchungen zur Charakterisierung eines Clostridienstammes für die Krebsdiagnostik, *Zbl. Bakt. Hyg.* **1980**, *I. Abt. Orig.* 246, 80-97.

[23] Schlechte, H., Mehnert, W.-H., Schulze, B., Bräuniger, H., Schwade, Measurement of specific multiplication of an oncolytic and a nononcolytic strain of *Clostridium butyricum* inside the tumor UVT 15264 of the mouse, *Zbl. Bakt. Hyg.* **1981**, *I. Abt. Orig.* 249, 512-519.

[24] Fabricius, E.-M., Schneeweiss, U., Dietz, O., Tudyka, G., Wildner, G. P., Schmidt, W., Benedix, A., Schubert, E., Der Einsatz eines serologischen Klostridientestes in der Tumordiagnostik, *Z. Erkr. Atm. Org.* **1980**, 155, 292-304.

[25] Wittman, W., Fabricius, E.-M., Schneeweiss, U., Schaepe, C. H. R., Benedix, A., Weissbrich, C. H. R., Schwanbeck, U., Application of microbiological cancer test to cattle infected with bovine leucosis virus, *Arch. Exp. Vet. Med.* **1990**, Leipzig 44, 205-212.

[26] Fabricius, E.-M., Schneeweiss, Benedix, A., Weisbrich, C., U., Dietz, O., Gutz, H-J., Jacobasch, K.-H., Seifart, W., Lübbe, D., Wildner, G.-P., Tumor diagnosis with *Clostridium* tumor test, in: *Tumor Associated Antigens, Oncogenes, Receptors, Cytokines in Tumor Diagnosis and Therapy at the Beginning of the Nineties. Cancer of the Breast – State and Trends in Diagnosis and Therapy* (Klapdor, R., Ed.), W. Zuckschwerdt Verlag, München, **1992**.

[27] Fabricius, E.-M., Schneeweiss, U., Schau, H. P., Schmidt, W., Benedix, A., Quantitative investigations into the elimination of *in vitro*-obtained spores of the non-pathogenic *Clostridium butyricum* strain CNRZ 528, and their persistence in organs of different species following intravenous spore administration, *Institut Pasteur* **1993**, 144, 741-753.

[28] Fabricius, E.-M., Schneeweiss, U., Schmidt, U., Methodological aspects of a serodiagnostic *Clostridium* tumor test – experience with spontaneous canine tumors, *Zbl. Bakt. Hyg. A.* **1987**, 265, 99-112.

[29] Schlechte, H., Elbe, B., Recombinant plasmid DNA variation of *Clostridium oncolyticum* – model experiments of cancerostatic gene transfer, *Zbl. Bakt. Hyg A.* **1988**, 268, 347-356.

[30] Minton, N.P, Clarke, D. J., *Clostridia Biotechnology Handbooks* Vol. 3, Plenum Press, New York, **1989**.

[31] Woods, D. R., *The Clostridia and Biotechnology*, Butterworths, Stoneham, MA, **1993**.

[32] Minton, N. P., Brehm, J. K., Swinfield, T.-J., Whelan, SM., Mauchline, ML., Bodsworth, N., Oultram, J. D., Clostridial cloning vectors, in: *The Clostridia and Biotechnology* (Woods, D. R., Ed.), Butterworth-Heinemann Publishing, Stoneham, MA **1993**, pp. 119-150.

[33] Mauchline, M. L., Davis, T. O., Minton, N. P., Genetics of Clostridia, in: *Manual of Industrial Microbiology and Biotechnology*, 2nd Edn. (Demain, A. L., Davies, J. E., Eds.), ASM Press, Washington, DC, **1999**, pp. 475-490.

[34] Lambin, P., Theys, J., Landuyt, W., Rijken, P., van der Kogel, A., van der Schueren, E., Hodgkiss, R., Fowler, J., Nuyts, S., de Bruijn, E., van Mellaert, L., Anné, J., Colonization of the *Clostridium* in the body is restricted to hypoxic and necrotic areas of tumors, *Anaerobe* **1998**, 4, 183-188.

[35] Lemmon, M. J., van Zijl, P., Fox, M. E., Mauchline, M. L., Giaccia, A. J., Minton, N. P., Brown, J. M., Anaerobic bacteria as a gene delivery system that is controlled by the tumor microenvironment, *Gene Ther.* **1997**, 4, 791-796.

[36] Horsman M. R., Murata R., Breidahl T., Nielsen F. U., Maxwell R. J.O, Stodkiled-Jorgensen H., Overgaard J., Combretastatins novel vascular targeting drugs for improving anti-cancer therapy. Combretastatins and conventional therapy, *Adv. Exp. Med. Biol.* **2000**, 476, 311-323.

[37] Theys, J., Landuyt, W., Nuyts, S., Van Mellaert, L., Bosmans, E, Rijnders, A., Van den Bogaert, W., van Oosterom, A., Anné, J., Lambin, P., Improvement of *Clostridium* tumor targeting vectors evaluated in rat rhabdomyosarcoma, *FEMS Immunol. Med. Microbiol.* **2000**, 1284, 1-5.

[38] Bagshawe, K. D., Antibody directed enzymes revive anti-cancer prodrug concept, *Br. J. Cancer* **1987**, 56, 531-532.

[39] Bagshawe, K. D., Springer, C. J., Searle, F., Antoniw, P., Sharma, S. K., Melton, R. G., Sherwood, R. F., A cytotoxic agent can be generated selectively at cancer sites, *Br. J. Cancer* **1988**, 58, 700-703.

[40] Springer, C. J., Bagshawe, K. D., Sharma, S. K., Searle, F., Boden, J. A., Antoniw, P., Burke, P. J., Rodgers, G. T., Sherwood, R. F. and Melton, R. G., Ablation of human choriocarcinoma xenografts in nude mice by antibody-directed enzyme prodrug therapy (ADEPT) with three novel compounds, *Eur. J. Cancer* **1991**, 27, 1361-1366.

[41] McNeish, I. A., Searle, P. F., Young, L. S., Kerr, D. J., Gene directed enzyme prodrug therapy for cancer, *Adv. Drug Delivery Rev.* **1997**, 26, 173-184.

[42] Minton, N. P., Atkinson, T., Bruton, C. J., Sherwood, R. F., Molecular cloning of the *Pseudomonas* carboxypeptidase G_2 gene and its expression in *Escherichia coli* and *Pseudomonas putida, Gene* **1984**, 31, 31-38.

[43] Harwood, P. J., Britton, D. W., Southall, P. J., Boxer, G. M., Rawlins, G., Rogers, G. T., Mapping epitope characteristics on carcinoembryonic antigen, *Br. J. Cancer* **1986**, 54, 75-82.

[44] Blakey, D. C., Burke, P. J., Davies, D. H., Dowell, R. I., East, S. J., Eckersley, K. P., Fitton, J. E., McDaid, J., Melton, R. G., Niculescu-Duvaz, I. A., Pinder, P. E., Sharma, S. K., Wright, A. F., Springer, C. J., ZD2767, an improved system for antibody-directed enzyme prodrug therapy that results in tumor regressions in colorectal tumor xenografts, *Cancer Res.* **1996**, 56, 3287-3292.

[45] Eccles, S. A., Court, W. J., Box, G. A., Dean, C. J., Melton, R. G., Springer, R. J., Regression of established breast carcinoma xenografts with antibody-directed enzyme prodrug therapy against c-*erb* B2 p185, *Cancer Res.* **1994**, 54, 5171-5177.

[46] Michael, N. P., Chester, K. A., Melton, R. G., Robson, L., Nicholas, W., Boden, J. A., Pedley, R. B., Begent, R. H. J., Sherwood, R. J., Minton, N. P., *In vitro* and *in vivo* characterisation of a recombinant carboxypeptidase G_2-anti-CEA scFv fusion protein, *Immunotechnology* **1996**, 2, 47-57.

[47] Bhatia, J., Sharma, S. K., Chester, K. A., Pedley, R. B., Bodem, R. W., Read, D. A., Boxer, G. M., Michael, N. P., Begent, R. H. J., Catalytic activity of an *in vivo* tumor targeted anti-CEA scFv::carboxypeptidase G2 fusion protein, *Int. J. Cancer* **2000**, 85, 571-577.

[48] Knox, R. J., Friedlos, F., Jarman, M., Roberts, J. J., A new cytotoxic DNA interstrand crosslinking agent, 5-(aziridin-1-yl)-4-hydroxylamino-2-nitrobenzamide, is formed from 5-(aziridin-1-yl)-2,4-dinitrobenzamide (CB 1954) by a nitroreductase enzyme in Walker carcinoma cells, *Biochem. Pharmacol.* **1988**, 37, 4661-4669.

[49] Knox, R. J., Boland, M. P., Friedlos, F., Coles, B., Southan, C., Roberts, J. J., The nitroreductase in Walker cells that activates 5-(aziridin-1-y1)-4-hydroxylamino-2-nitrobenzamide is a form of NAD(P)H dehydrogenase (quinone) (EC 1.6.99.2), *Biochem. Pharmacol.* **1988**, 37, 4671-4677.

[50] Anlezark, G. M., Melton, R. G., Sherwood, R. F., Coles, B., Friedlos, F, Knox, R. J., The bioactivation of 5-(aziridin-1-y1)-2,4-dinitrobenzamide (CB1954). I. Purification and properties of a nitroreductase enzyme from *Escherichia coli*: a potential enzyme for antibody-directed enzyme prodrug therapy (ADEPT), *Biochem. Pharmacol.* **1992**, 44, 2289-2295.

[51] Polak, A, Eschenhof, A., Fernex, M, Scholer, H. J., Metabolic studies with 5-fluorocytosine-6-^{14}C in mouse, rat, rabbit, dog and man, *Chemotherapy* **1976**, 22, 137-153.

[52] Swinfield, T.-J., Oultram, J. D., Thompson, D. E., Brehm, J. K., Minton, N. P., Physical characterisation of the replication region of the *Streptococcus faecalis* plasmid pAMβ1, *Gene* **1990**, 87, 79-90.

[53] Minton, N. P., Morris, J. G., Isolation and partial characterisation of three cryptic plasmids from strains of *Clostridium butyricum, J. Gen. Microbiol.* **1981**, 127, 325-331

[54] Graves M. C., Rabinowitz J. C., *In vivo* and *in vitro* transcription of the *Clostridium pasteurianum* ferrodoxin gene, *J. Biol. Chem.* **1986**, 261, 11409-11415.

[55] Fox, M. E., Lemmon, M. J., Mauchline, M. L., Davis, T. O., Giaccia, A. J., Minton, N. P., Brown, J. M., Anaerobic bacteria as a delivery system for cancer gene therapy: *in vitro* activation of 5-fluorocytosine by genetically engineered clostridia, *Gene Ther.* **1996**, 3, 173-178.

[56] Minton, N. P., Mauchline, M. L., Lemmon, M. J., Brehm, J. K., Fox, M., Michael, N. P., Giaccia, A. J., Brown, J. M., Chemotherapeutic tumor targeting using clostridial spores, *FEMS Microbiol. Rev.* **1995**, 17, 357-364.

[57] Theys J., Landuyt W., Nuyts S., Van Mellaert L., de Bruijn E., Lambin P., Anné J., Specific targeting of cytosine deaminase to solid tumors by engineered *Clostridium acetobutylicum, Cancer Gene Ther.* **2001**, 8, (4) in press.

[58] Dargatz H, Diefenthal T, Witte V, Reipen G, von Wettstein D., The heterodimeric protease clostripain from *Clostridium histolyticum* is encoded by a single gene, *Mol. Gen. Genet.* **1993**, 240, 140-145.

[59] Azzedoug, H. J., Hubert, J., Reysset, G., Stable inheritance of shuttle vectors based on

plasmid p1M13 in a mutant strain of *Clostridium acetobutylicum*, *J. Gen Microbiol.* **1992**, 138, 1371-1378.

[60] O'Connell, M. J., Martenson, J. A., Wieand, H. S., Krook, J. E., Macdonald, J. S., Haller, D. G., Mayer, R. J., Gunderson, L. L., Rich, TA., Improving adjuvant therapy for rectal cancer by combining protracted-infusion fluorouracil with radiation therapy after curative surgery, *N. Engl. J. Med.* **1994**, 331, 502-507.

[61] Lambin, P., Nuyts, S., Landuyt, W., Theys, J., De Bruijn, E., Anné, J., Van Mellaert, L., Fowler, J., The potential therapeutic gain of radiation-associated gene therapy with the suicide gene cytosine deaminase, *Int. J. Radiat. Biol.* **2000**, 285-93.

[62] Liu, S. C., Shibata, T., Giaccia, A. J., Minton, N. P., Brown, N. P., Genetically engineered clostridia expressing a prodrug activating enzyme for cancer gene therapy. *Abstract of the 3rd International Meeting on the Molecular Genetics and Pathogenesis of Clostridia* p. 62, Chiba, Japan, June **2000**.

[63] Sidhu, R. S., Bollon, A, P., Related Tumor necrosis factor activities and cancer therapy – a perspective, *Pharmacol. Ther.* **1993**, 57, 79-128.

[64] Sersa, G., Willingham, V., Milas, L., Antitumor effects of tumor necrosis factor alone or combined with radiotherapy, *Int. J. Cancer* **1988**, 42, 129-134.

[65] Zappe, H., Jones, W. A., Jones, D. T., Woods, D. R., Structure of an endo-β-1, 4-glucanase gene from *Clostridium acetobutylicum* P262 showing homology with endoglucanase genes from *Bacillus* spp., *Appl. Environ. Microbiol.* **1988**, 54, 1289-1292.

[66] Mermelstein, L. D., Welker, H. E., Bennett, G. N., Papotsakis, E. T., Expression of cloned homologous fermentative genes in *Clostridium acetobutylicum* ATCC 824, *Bio/Technology* **1992**, 10, 190-195.

[67] Theys, J., Nuyts, S., Landuyt, W., van Mellaert, L., Dillen, C., Böhringer, M., Dürre, P., Lambin, P., Anné, J., Stable *Escherichia coli-Clostridium acetobutylicum* shuttle vector for secretion of murine tumor necrosis factor alpha, *Appl. Environ. Microbiol.* **1999**, 65, 4295-4300.

[68] Carter, G., White, P., Fernie, M., King, S., McLean, G, Titball, R. W., Carr, F. J., Enhanced antitumor effect of liposomal daunorubicin using antibody-phospholipase C conjugates or fusion protein, *Int. J. Oncol.* **1998**, 13, 819-825.

[69] Titball, R. W., Hunter, S. E. C., Martin, K. L., Morris, B. C., Shuttleworth, A. D., Rubidge, T., Anderson, D. W., Kelly, D. C., Molecular cloning and nucleotide sequence of the alpha-toxin (phospholipase C) of *Clostridium perfringens*, *Infect. Immun.* **1989**, 57, 367-376.

[70] Ranade, V. V., Drug delivery systems l. Site-specific drug delivery using liposomes as carriers, *J. Cancer Pharmacol.* **1989**, 29, 685-694.

[71] Forssen, E. A., Coulter, D. M., Proffit, R. T., Selective *in vivo* localization of daunorubicin small unilamellar vesicles in solid tumors, *Cancer Res.* **1992**, 52, 3255-3261.

[72] Forssen, E. A., Ross, M. E., Daunoxome® treatment of solid tumors: preclinical and clinical investigations, *J. Liposome Res.* **1994**, 4, 481-512.

[73] Rutman, R. J., Ritter, C. A., Avadhani, N. G., Hansel, J., Liposomal potentiation of the antitumor activity of alkylating drugs, *Cancer Ther. Rep.* **1976**, 60, 617-618.

[74] Fox J. L., Harnessing *Salmonella*'s positive powers against tumors, *ASM News* **2000**, 66, 332-334.

[75] Pawelek J. M., Low K. B., Bermudes D., Tumor-targeted *Salmonella* as a novel anticancer vector, *Cancer Res.* **1997**, 57, 4537-4544.

[76] Low K. B., Ittensohn M., Le T., Platt J., Sodi S., Amoss M., Ash O., Carmichael E., Chakraborty A., Fischer J., Lin SLO, Luo X., Miller SI., Zheng L., King I., Pawelek JM., Bermudes D., Lipid A mutant *Salmonella* with suppressed virulence and TNFalpha induction retain tumor-targeting *in vivo*, *Nature Biotechnol.* **1999**, 17, 37-41

[77] Yazawa, K., Fujimori, M., Amano, J., Kano, Y., Taniguchi, S., *Bifidobacterium longum* as a delivery system for cancer gene therapy: selective localization and growth in hypoxic tumors, *Cancer Gene Ther.* **2000**, 7, 269-274.

Stichwortverzeichnis